Charles Taylor

## An Introduction to the Ancient and Modern Geometry of Conics

Being a Geometrical Treatise on the Conic Sections with a Collection of...

Charles Taylor

**An Introduction to the Ancient and Modern Geometry of Conics**
*Being a Geometrical Treatise on the Conic Sections with a Collection of...*

ISBN/EAN: 9783337014810

Printed in Europe, USA, Canada, Australia, Japan

Cover: Foto ©berggeist007 / pixelio.de

More available books at **www.hansebooks.com**

# AN INTRODUCTION

## TO THE

## ANCIENT AND MODERN

# GEOMETRY OF CONICS

BEING A GEOMETRICAL TREATISE ON THE CONIC SECTIONS WITH
A COLLECTION OF PROBLEMS AND HISTORICAL NOTES
AND PROLEGOMENA

By CHARLES TAYLOR M.A.

FELLOW OF ST JOHN'S COLLEGE CAMBRIDGE

CAMBRIDGE
DEIGHTON BELL AND CO
LONDON GEORGE BELL AND SONS
1881

# PREFACE.

Geometry, at once ancient and modern, is the science of Euclid, Archimedes and Apollonius, of Kepler, Desargues, Newton and Poncelet. Geometrical processes have indeed been simplified and their applications greatly extended in recent times, but the modern methods may be traced to the ancient as their germ and source, and thus it remains in a sense still true that there is but one road for all, ἐν τῇ γεωμετρίᾳ πᾶσίν ἐστιν ὁδὸς μία. The modern infinitesimal calculus is an adaptation of the ancient method of exhaustions, the method of Descartes differs only in the manner of its application from that of Apollonius, the idea of perspective was already formulated by Serenus, and the principle of anharmonic section with the leading properties of transversals are found in the lemmas of Pappus to the lost three books of Porisms of Euclid. There is not however in the works of the Greek geometers any distinct foreshadowing of Kepler's doctrine of the infinite, of his principles of analogy and continuity, or of the theory of ideal chords and points, at length completed by Poncelet's discovery of the so-called circular points at infinity in any plane.

In the present as in a former work (1863) I have commenced with an elementary treatment of the general conic *in plano*, following out a suggestion made by Professor Adams in a course of lectures on the Lunar Theory delivered in 1861. This department of the subject has now been made more complete with the help of the Eccentric Circle, the characteristic feature of a masterly though neglected work of Boscovich. In the chapter on the Cone the focal spheres are more fully discussed, and the angle-properties of the sections as well as their metric properties are deduced. The chapter on Orthogonal Projection contains proofs of Lambert's theorem* in elliptic motion. To the chapter on Conical Projection is appended some account of the homographic method of Reversion, which springs out of the above mentioned construction of Boscovich.

---

* It has been remarked that, so far as relates to the parabola, Lambert's theorem is implicitly contained in Newton's *Principia* lib. III. lemma 10. See Lagrange *Mécanique Analytique* tome II. p. 28, ed. 3 (1853—5); Brougham and Routh *An analytical view of Sir Isaac Newton's Principia* p. 436 (Lond. 1855).

PREFACE. vii

Abundant references will be found to the works of authors to whom I am indebted. Suffice it here to add that my warmest thanks are due to the Reverend Professor Richard Townsend, F.R.S., Fellow of Trinity College, Dublin, who has been at all times ready, in the midst of pressing engagements, to aid me with his criticism and advice, and has from first to last shewn as great an interest in the work now brought to a close as if it had been his own.

C. TAYLOR.

*St. John's College,*
    *December* 31, 1880.

### ERRATA.

| | for | read |
|---|---|---|
| Page 42, Ex. 73 | is | varies as. |
| Page 82, line 25 | 1657 | 1710. |
| Page 136, note* | Le Sueur | Le Seur. |
| Page 194, Scholium | Erastosthenes | Eratosthenes. |
| Page 206, Ex. 545 | two or more | two. |
| Page 257, line 29 | *Dynamics* | *Dynamic.* |

# CONTENTS.

## PROLEGOMENA.

| | PAGE |
|---|---|
| Section I. Geometry before Euclid | xvii |
| Section II. From Euclid to Serenus | xxxiv |
| Section III. Kepler. Desargues. Newton | lvi |
| Section IV. Modern Geometry | lxxiii |

## DEFINITIONS.

| | |
|---|---|
| A conic defined by focus, directrix and eccentricity | 1 |
| The eccentric circle | 3 |

## CHAPTER I.
### DESCRIPTION OF THE CURVE.

| | |
|---|---|
| The three conics traced | 5 |
| The circle a limit of the ellipse | 7 |
| A conic is a curve of the second order | 8 |
| To describe a conic by the eccentric circle | 9 |
| The intersections of any line with a conic determined | 10 |
| A conic is a curve of the second class | 11 |
| Every conic is concave to its axis | 12 |
| Examples 1—10. | ,, |

## CHAPTER II.
### THE GENERAL CONIC.

| | |
|---|---|
| Tangent subtends right angle at focus | 14 |
| The directrix the polar of the focus | 15 |
| Geometrical equation to a tangent | ,, |
| The tangent as defined by Euclid | 16 |
| Construction for tangents from given point | ,, |
| Two tangents from any point subtend equal angles at focus | 17 |

|   | PAGE |
|---|---|
| Harmonic division of any chord by focal distance of its pole and directrix | 18 |
| Geometrical equation to any chord | 19 |
| Length of intercept by normal on the axis | " |
| Projection of normal on focal distance constant | 20 |
| Properties of angles subtended at focus | " |
| Angle-properties of the circle deduced | 22 |
| Parallel chords bisected by a right line | 23 |
| Ordinates to any diameter parallel to tangents at its extremities | 24 |
| The diameter of any chord passes through its pole | 25 |
| The second focus and directrix, and the centre | " |
| The centre of the parabola is at infinity | 26 |
| Condition for conjugate diameters | " |
| Sum of reciprocals of segments of focal chord constant | 27 |
| Ratio of products of segments of intersecting chords | 29 |
| Intersection of a circle with a conic | " |
| Properties of polars | 30 |
| The straight line at infinity | 32 |
| Envelope of polar of any point on a conic with respect to a conic having the same focus and directrix | " |
| Polar of external point meets the conic in real points | 34 |
| Polar equations of tangent, normal, chord and polar | " |
| Examples 11—80 | 35 |

## CHAPTER III.

### THE PARABOLA.

|   |   |
|---|---|
| The parabola regarded as a central conic | 44 |
| Square of principal ordinate varies as abscissa | " |
| Construction for two mean proportionals | 45 |
| Every diameter is parallel to the axis | 46 |
| Length of any focal chord determined | 47 |
| Square of any ordinate varies as abscissa | 48 |
| Two sides of any triangle on base parallel to axis are as the parallel tangents | 49 |
| Ratio of segments of any chord by a diameter | " |
| Products of segments of intersecting chords are as the parallel focal chords | 50 |
| The tangent equally inclined to axis and focal distance | 51 |
| Tangents at right angles meet on directrix | 52 |
| The principal subtangent double of the abscissa | " |
| Normal bisects angle between focal distance and diameter | " |
| The subnormal constant and equal to half latus rectum | 53 |
| The focal perpendicular on the tangent | " |
| The angle between two tangents, and their inclinations to the axis | 54 |
| Two tangents with focus determining similar triangles | 55 |
| Envelope of line cut in constant ratios by three fixed lines | " |
| Lambert's circle through intersections of three tangents | " |
| Steiner's property of directrix of parabola inscribed in triangle | 56 |
| The parabola touching four given lines determined | " |
| Proportionality of segments of two tangents by a third | 57 |
| Steiner's theorem in relation to Pascal's | 58 |
| The subtangent on any diameter double of the abscissa | " |

|                                                                                      | PAGE |
| ------------------------------------------------------------------------------------ | ---- |
| Tangent and diameter at any point cut any chord proportionally                       | 59   |
| Area of triangle circumscribed to a parabola determined                              | ,,   |
| Theorem of Archimedes on quadrature of parabola                                      | ,,   |
| Examples 81—200                                                                      | 61   |

## CHAPTER IV.

### CENTRAL CONICS.

|                                                                                      |      |
| ------------------------------------------------------------------------------------ | ---- |
| The principal ordinate                                                               | 75   |
| Length of minor axis of hyperbola defined                                            | 76   |
| Projective property of central conics                                                | 77   |
| The axis or its complement a degenerate form of conic                                | ,,   |
| Square of any ordinate varies as product of abscissae                                | ,,   |
| Length of imaginary diameter defined                                                 | 78   |
| Sum or difference of focal distances constant                                        | ,,   |
| Relations between segments of the axis                                               | 80   |
| Lengths of latus rectum and any focal chord                                          | 81   |
| The names parabola, ellipse, hyperbola explained                                     | 82   |
| Tangent bisects angle between focal distances                                        | 83   |
| Confocal conics intersect at right angles                                            | 84   |
| The focal perpendiculars upon the tangent                                            | ,,   |
| Constant intercept on focal distance by diameter parallel to tangent                 | 86   |
| Any diameter cut harmonically by tangent and ordinate at any point                   | 87   |
| The director circle or orthocycle                                    88, 165, 280    |      |
| Locus of vertex of right angle circumscribed to two confocals                        | 89   |
| Property of polars                                                                   | 90   |
| Normal bisects angle between focal distances                                         | 92   |
| Length of intercept on normal by either axis                                         | 93   |
| The subnormal varies as the abscissa                                                 | 94   |
| Supplemental chords parallel to conjugate diameters                                  | 95   |
| Conjugate diameters correspond to diameters at right angles in auxiliary circle      | 96   |
| One of every two conjugate diameters meets hyperbola                                 | ,,   |
| Construction for conjugate diameters containing a given angle                        | ,,   |
| Sum or difference of squares of conjugate diameters constant                         | ,,   |
| Conjugate diameter expressed in terms of normal, and of focal distances              | 98   |
| Area of conjugate parallelogram constant                                             | 99   |
| Product of segments of fixed tangent by conjugate diameters constant                 | 100  |
| Product of segments of parallel tangents by any third                                | 101  |
| The hyperbola identified with the ellipse                                            | ,,   |
| The bifocal definition                                                               | 102  |
| Magnitude of angle between two tangents to a conic                                   | 105  |
| Two tangents equally inclined to focal distances of their point of concourse         | 106  |
| Property of the triangle formed by lines equal to the axis and to the focal distances of an external point | 107  |
| Projection of normal on focal radius constant                                        | 108  |
| Property of directrix deduced from bifocal definition                                | 110  |
| Construction for normal from point on minor axis                                     | 111  |
| Apollonius on foci                                                                   | ,,   |
| Examples 201—390                                                                     | 112  |

## CONTENTS

### CHAPTER V.

#### THE ASYMPTOTES.

| | PAGE |
|---|---|
| The asymptotes are self-conjugate diameters and tangents at infinity | 142 |
| Conjugate hyperbolas defined | ,, |
| The asymptotes a degenerate similar conic | 143 |
| Construction for normal at given point on hyperbola | 144 |
| Product of segments of chord parallel to axis by the asymptotes constant | 145 |
| Product of distances of point on curve from the asymptotes constant | 146 |
| The curve and its asymptotes make equal intercepts on any chord or tangent | ,, |
| Product of intercepts by the curve on any chord of the asymptotes constant | 147 |
| Any tangent contains a constant area with the asymptotes | 148 |
| Difference of squares on conjugate diameters constant | 149 |
| Product of segments of any chord by a parallel to an asymptote | ,, |
| The hyperbola as defined with reference to its asymptotes | ,, |
| Any tangent and its normal meet asymptotes and axes on circle through centre | 150 |
| The conjugate parallelogram has its diagonals on the asymptotes | 151 |
| Construction for hyperbola from given pair of conjugate diameters | 152 |
| The two branches of a hyperbola meet at infinity | 153 |
| Examples 391—460 | 154 |

### CHAPTER VI.

#### THE EQUILATERAL HYPERBOLA.

| | |
|---|---|
| The equilateral as a limit of the general hyperbola | 167 |
| Every diameter equal to its conjugate | ,, |
| The principal ordinate a mean proportional to the abscisses | 168 |
| Conjugate diameters equally inclined to either asymptote | 169 |
| Diameters at right angles equal | 170 |
| The normal equal to the semi-diameter or its conjugate | ,, |
| Any two diameters and their conjugates contain equal angles | ,, |
| Any chord subtends equal angles at the extremities of a diameter | 171 |
| Locus of centres of equilateral hyperbolas through three points | ,, |
| Circle round self-polar triangle passes through centre | ,, |
| Conjugate parallelogram equal to square on axis | 172 |
| Locus of vertex of triangle the difference of whose base angles is constant | ,, |
| Any ordinate a mean proportional to the abscisses | 173 |
| Product of segments of chord through any point | 174 |
| Equilateral hyperbola through three points passes through their orthocentre | 175 |
| Four given points determine an equilateral hyperbola | 176 |
| Construction for tangent and normal at any point | ,, |
| Product of intercepts on any diameter by a tangent and ordinate, or on any tangent by conjugate diameters | ,, |
| Mechanical description of rectangular hyperbola | 177 |
| Examples 461—540 | 178 |

### CHAPTER VII.

#### THE CONE.

| | |
|---|---|
| Definitions | 192 |
| The ordinate | 193 |

# CONTENTS.

|  | PAGE |
|---|---|
| Minor axis of section a mean proportional to diameters of circular sections through its vertices | 194 |
| Projection of section upon base has one focus on axis of cone | " |
| The plane of section formerly drawn at right angles to side of cone | 195 |
| The asymptotes | " |
| Foci and directrices determined by focal spheres | 196 |
| Sum or difference of focal distances constant | 198 |
| Tangent makes equal angles with focal distance and side of cone | 200 |
| Angle between two tangents to section determined | 201 |
| The minor axis a mean proportional to diameters of focal spheres | 202 |
| Latus rectum varies as perpendicular from vertex of cone to plane of section | " |
| The auxiliary circle lies on a sphere through the centres of the focal spheres | 203 |
| Product of focal perpendiculars on tangent constant | 204 |
| The discovery of the focal spheres | " |
| Product of segments of chord through any point | 205 |
| Examples 541–600 | 206 |

## CHAPTER VIII.

### CURVATURE.

|  |  |
|---|---|
| The circle of curvature cuts and touches the conic at same point, and is the circle of closest contact at that point | 214 |
| Length of focal chord of curvature at any point of a conic | 215 |
| The chord of curvature in any direction deduced | 216 |
| The same otherwise determined | 217 |
| Length of common chord of circle of curvature and conic | 220 |
| Orthogonal projection of chords of curvature | 221 |
| Huyghens on evolutes | " |
| The osculating circle named by Leibnitz | 222 |
| Examples 601–650 | " |

## CHAPTER IX.

### ORTHOGONAL PROJECTION.

|  |  |
|---|---|
| Parallel lines project in the same ratio | 229 |
| Degree and class of curve unaltered by projection | 230 |
| Areas in one plane project in constant ratio | 231 |
| Plane projection defined | 232 |
| Conjugate diameters of ellipse project into diameters at right angles in a circle | 233 |
| Area of ellipse to auxiliary circle as minor to major axis | 234 |
| Projection of hyperbola into equilateral hyperbola | " |
| Properties of polars | " |
| Locus ad quatuor lineas | 235 |
| Three osculating circles can be drawn through a given point on an ellipse | 236 |
| Parallel projection | " |
| Projection of any triangle into equilateral triangle | 237 |
| Lambert's theorem in elliptic motion | " |
| Examples 651–695 | 242 |

## CHAPTER X.

### CROSS RATIO AND INVOLUTION.

| | PAGE |
|---|---|
| Definition of cross or anharmonic ratio | 249 |
| Elementary properties of cross ratio | 250 |
| Four fixed rays cut any transversal in constant cross ratios | 251 |
| Condition for concurrence of lines | 252 |
| Condition for collinearity of points | 253 |
| Harmonic tetrads | 254 |
| Harmonic properties of quadrilateral | 255 |
| The diameter of a quadrilateral defined | 256 |
| Pappus on cross ratio | ,, |
| Involution defined | 257 |
| Coaxal circles cut any transversal in involution | 258 |
| A right angle turning about its summit generates a negative involution | 259 |
| General relations between points in involution | 260 |
| Two conjugate points with foci form a harmonic range | 261 |
| Pappus and Desargues on involution | ,, |
| Anharmonic properties of conics | 262 |
| Newton's organic description of conics | 263 |
| Maclaurin's description of conics | 264 |
| The conjugate diameters of a conic form a pencil in involution | 265 |
| Construction for four normals from given point | ,, |
| Eight lines from one point meet conic at given angle | 266 |
| Newton on the Locus ad quatuor lineas | ,, |
| Property of polars | 268 |
| A row of points and their polars homographic | 269 |
| Conjugate points and lines with respect to conic | 270 |
| Conjugate lines through a focus are at right angles | ,, |
| The reciprocal of a conic is a conic | ,, |
| Two triangles circumscribed to a conic are inscribed in a conic | 271 |
| The problem to circuminscribe a triangle to a given pair of conics is indeterminate | 272 |
| The problem to inscribe a triangle self-polar to one conic in a second is indeterminate | 273 |
| Gaskin's property of the orthocycle | 274 |
| Conics through four points have a common self-polar triangle | ,, |
| Frégier's theorem | 276 |
| Self-polar triangle of conics touching four lines | ,, |
| Theorem of Desargues on involution | 277 |
| Given four points on a conic, the polar of a fixed point passes through a fixed point | 278 |
| Harmonic section of common tangent to osculating circle and conic | 279 |
| Similar coaxal conics have double contact at infinity | ,, |
| Conics touching the same four lines subtend a pencil in involution at any point, and their orthocycles are coaxal | 280 |
| Poles of given line with respect to all conics touching four given lines are collinear | 281 |
| Confocal conics have four common tangents | ,, |
| Newton's property of the diameter of a quadrilateral | 282 |
| Centre-locus of conics through four points | 283 |
| Nine-point circle of a tetrastigm | 285 |

## CONTENTS.

|  | PAGE |
|---|---|
| Degeneration of conic into line or line-pair | 285 |
| Pascal's hexagon | 286 |
| Construction of conic through five points | 288 |
| Brianchon's hexagon | 289 |
| Construction for other tangents when **five are given** | 290 |
| Steiner's theorem on directrix of parabola deduced from Brianchon's hexagon | ,, |
| The original proof of Brianchon's theorem | 291 |
| Examples 696—800 | 292 |

### CHAPTER XI.
#### CONICAL PROJECTION.

|  | |
|---|---|
| **Definitions** | 307 |
| **Direction of** the line at **infinity indeterminate** | 308 |
| Concentric circles touch at the focoids | ,, |
| The focoids are conjugate points with respect to rectangular **hyperbola** | 309 |
| The lines joining a focus to the focoids touch the conic | 310 |
| Constant relation of figure moving in a plane to the focoids | ,, |
| Boscovich and Poncelet on continuity | 311 |
| Plücker's definition of the foci **of plane curves** | ,, |
| Figures in perspective homographic | 312 |
| Harmonic and polar properties projective | 313 |
| A line and two angles **may be projected arbitrarily** | ,, |
| Projection of pencil **in involution into rectangular pencil, and of any two** points into focoids | 315 |
| Four points or lines may be projected arbitrarily | 316 |
| Conics touching same four lines project into confocals | 317 |
| Projection of conic and point into a circle and its centre | 318 |
| Conics through four points project into coaxal circles, and conics **having double** contact into concentric circles | 319 |
| Projection of angles | ,, |
| Perspective in one plane | 320 |
| Locus of centre of perspective for different planes | 321 |
| Newton on the projection of cubics | ,, |
| Reversion defined | ,, |
| **Points at infinity reverse into points on the base line** | 322 |
| **Reversion of angles** | 323 |
| Frégier's theorem proved by reversion | 324 |
| Case of Newton's *Descriptio Organica* proved by **reversion** | 325 |
| Reversion of Steiner's theorem on directrix of parabola | 326 |
| Reversion of normals, conjugate diameters, asymptotes | 327 |
| Scholium on reversion | ,, |
| Homographic figures may be placed in perspective | 328 |
| A conic and point may be projected arbitrarily | 329 |
| Desargues on polar planes, and on transformation | ,, |
| Newton's rational transformation | 330 |
| Examples 801–850 | ,, |

### CHAPTER XII.
#### RECIPROCATION AND INVERSION.

|  | |
|---|---|
| Degree of curve equal to class of its reciprocal | 337 |
| Conics through four points reciprocate into confocals | 338 |

|   | PAGE |
|---|---|
| Homographic relation of reciprocal points and lines | 338 |
| Reciprocal relation between distances of a **point** from given lines **and of its** polar from their poles | 339 |
| Reciprocation with respect to a **point** | 340 |
| Reciprocals **of angles subtended at origin, and of** distances from origin | 341 |
| Determination of species of reciprocal conic | ,, |
| Property of the orthocycle proved by reciprocation | 342 |
| Circle reciprocates into conic with origin **as** focus | ,, |
| **Eccentricity,** latus rectum, centre and directrix of reciprocal of circle | 343 |
| **Reciprocation of** coaxal circles into confocal conics | 344 |
| **The discovery** of the principle of duality | 346 |
| **Properties of** minor directrices may be proved **by reversion** | ,, |
| **Or by** reciprocation | 347 |
| **Double reciprocation and reversion** | 348 |
| **Applications of reciprocation** | 349 |
| **Further properties of the minor directrices** | 352 |
| **Inversion** | 354 |
| **Nine-point circle touches inscribed and escribed circles** | 355 |
| **The cardioid** | 356 |
| **Osculating circles invert into osculating circles** | 357 |
| **Examples 851—1000** | 358 |

INDEX.

# PROLEGOMENA.

## SECTION I.

### GEOMETRY BEFORE EUCLID.

N.B. *References within square brackets* [ ] *occurring in the Prolegomena are to the pages of the present work.*

§ 1. The science of Geometry, as its name suggests, was developed from the art of land surveying, to which ancient testimony likewise refers us for its origin. The practice of systematic land measurement is said to have been forced upon the Egyptians by the annual rise and fall of the river Nile, which from time to time left portions of land that had been high and dry submerged, or *vice versa*, so that the owners were unable to distinguish what belonged to each. Thus writes Hero* the elder, of Alexandria, and to the same effect Herodotus (II. 109), Diodorus Siculus and Strabo, as cited by Bretschneider in his excellent monograph on the history of geometry before Euclid.† Whether the Nile altogether played the part attributed to it in the advancement of science is matter of question, but it may be conceded to the concurrent testimony of ancient writers that the Egyptians had laid the foundation of concrete fact upon which the superstructure of Greek abstract geometry was to be reared.

---

\* Heronis Alexandrini *Geometricorum et Stereometricorum reliquiæ*, p. 138 ed. Hultsch (Berlin 1864). He flourished within the period B.C. 285–222, or later.

† *Die Geometrie und die Geometer vor Euklides* (Leipzig 1870). See also Dr. Allman's paper on Greek Geometry from Thales to Euclid, in *Hermathena* vol III. 160–207 (Dublin 1877).

xviii    PROLEGOMENA.

*Eudemus*
*B.C.*
*340—320.*

§ 2. Eudemus of Rhodes (a disciple of Aristotle and an immediate predecessor of Euclid) was the primary authority on the early history of mathematics;* but his writings on the history of geometry and astronomy, which appear to have been composed in a philosophical spirit, now no longer survive, except so far as they are embodied in the still extant works of his successors. The important list of early geometers given by Proclus Diadochus† (A.D. 412–485) in his commentary on the first book of Euclid's elements is not unreasonably thought to have been derived from Eudemus. The following is the substance of the passage, of which the original Greek with a German rendering may be found in the above mentioned work of Bretschneider, pp. 27–31. It is taken from lib. II. cap 4 of the commentary, which was written in four books:—

"Geometry is said by many to have taken its rise from the measurements rendered necessary by the obliteration of landmarks by the Nile. And it is nothing strange that this and other sciences should have arisen from practical needs, since there is a general tendency in things from imperfection to perfection, in accordance with which law we pass naturally from perception to reflection and thence to intellectual insight. As then the Phœnicians were led on from trade and barter to systematic arithmetic, so the Egyptians discovered geometry in the manner aforesaid.

First Thales went to Egypt and brought over this science to Greece. He made many discoveries himself and suggested the beginnings of many to his successors, apprehending some things more in the abstract but others in a limited and perceptional way. Next Ameristus, brother of the poet Stesichorus, became famed in geometry, as Hippias of Elis relates. Pythagoras, who succeeded them, transformed it into a liberal science, investigating its first principles and regarding theorems from the immaterial and intellectual standpoint. He it was who

---

\* His contemporary Theophrastus also wrote something about mathematics, amongst a multitude of other subjects, according to the statement of Diogenes Laertius (lib. v. cap. 2).

† Notice the editions mentioned on p. [82], and the Latin edition of Barocius (Patavii 1560).

discovered the theory of irrational quantities and the construction of the regular solids. After him came Anaxagoras of Clazomenæ and Oenopides of Chios, who are mentioned by Plato as famed in mathematics. After them Hippocrates of Chios, who quadrated the lunule and was the earliest writer on the Elements, and Theodorus of Cyrene became eminent in geometry. Plato, who succeeded Hippocrates, greatly encouraged the study of mathematics and geometry by the frequent use of mathematical considerations in his philosophical writings. To this age also belong Leodamas of Thasos, Archytas of Tarentum and Theætetus of Athens. Younger than Leodamas were Neocleides and his disciple Leo, who added much to the work of their predecessors. Leo also composed a work on the Elements characterised by the greater number and importance of its propositions, and he assigned the limits within which a construction was possible. Eudoxus of Cnidus, an associate of the school of Plato and somewhat junior to Leo, increased the number of general theorems, added three new proportions to the three already known, and developed Plato's doctrine of the section (of a line), making use in his investigations of the method of geometrical analysis. Amyclas of Heraclea, Menæchmus (a pupil of Eudoxus and contemporary with Plato) and his brother Dinostratus made geometry as a whole still more complete. Theudius of Magnesia, a writer on the Elements, and Athenæus of Cyzicus were greatly distinguished especially in geometry. These lived and worked together in the Academy. Hermotimus of Colophon carried on the discoveries of Eudoxus and Theætetus, and also wrote some things upon loci. Philip of Mende was led by Plato to study mathematics in relation to the Platonic philosophy. Thus far do the writers on the history of geometry bring the science.*

Analysis.

Loci.

Not much junior to the above was Euclid, who compiled the Elements, putting in order many discoveries of Eudoxus,

---

* If the history of Eudemus breaks off before Aristaeus, whose writings preceded Euclid's, we may conjecture that it was completed before 320 B.C. It is impossible to determine the precise dates of the early geometers.

completing many of Theætetus, and replacing the former lax demonstrations by incontrovertible proofs. He lived in the reign of the first Ptolemy, in answer to whom he is reported to have said that there was no royal road to geometry.* He was therefore younger than the disciples of Plato, but elder than the contemporaries Eratosthenes and Archimedes. Being a Platonist, he made the construction of the Platonic bodies (or regular solids) the goal of his work upon the Elements."

*Thales B.C. 640—548.*

§3. Of the above mentioned early geometers Thales, Pythagoras, Hippocrates, Menæchmus and some others deserve to be noticed more particularly. Thales of Miletus, of Phœnician ancestry and the founder of the Ionian school of philosophy, was the first to naturalise the study of geometry amongst the Greeks. Visiting Egypt† as a trader, he brought back thence late in life to his native place such knowledge of geometry and astronomy as he had been able to pick up from the priests. He was born about the commencement of the 35th Olympiad, and died (according to one account) at the great age of 90 years, or upwards. His reputation was made once for all by the prediction—to what degree of accuracy we know not—of an eclipse of the sun, which duly came to pass (28th May, 585 B.C.)‡; and this well attested fact corroborates the statement of Diogenes Laertius (lib. I. cap. 1) that he first came to be styled σοφός in the archonship of Damasias. Although he is said by Proclus in general terms to have made many discoveries in geometry, the following alone are expressly attributed to him.‖ (1) The circle is bisected by its

---

\* The saying referred to (Bretschneider p. 163) is also attributed to Menæchmus, who is said to have replied to Alexander: "In the country, O king, there are roads ἰδιωτικαί καί βασιλικαί, but in geometry there is one road for all."

† The foreign travels of the early Greek philosophers are however sometimes thought to be attested by insufficient evidence. Cf. Renouf's *Hibbert Lectures* Lect. VI. p. 246 (London 1880).

‡ The Egyptians had doubtless supplied him with the facts on which his calculation was based. Diogenes Laertius states that they had observed more than 1200 eclipses of the sun or moon. See Bretschneider pp. 39, 52.

‖ See Proclus on Euclid I. def. 17 and props. 5, 15, 26 (Thos. Taylor's Proclus vol. I. 165; II. 54, 96, 143); Diogenes Laertius lib. I. cap. I. §§ 3, 6.

diameter. (2) The angles at the base of an isosceles triangle are equal to one another. (3) When two straight lines cut one another the vertical angles are equal. (4) A method of determining the distance of a ship at sea from the land, implying the knowledge of a theorem equivalent to Euclid I. 26. (5) The angle in a semicircle is a right angle.* And (6) a method for determining the heights of the pyramids from the lengths of their shadows, viz. at the moment when the sun is at an elevation of 45° above the horizon.†

The fact that a theorem was attributed to Thales by his successors does not altogether exclude the supposition that he had himself received it from the Egyptians; and accordingly it has been thought that the second only of the above theorems was in reality discovered by him. The theorem (5) may have been arrived at by the Egyptian geometers by supposing first a square and then any rectangle inscribed in a circle to be turned about within it; and it is impossible to lay much stress on (1) or (3). The method (4) if actually known to Thales was probably discovered by him, but if (as has been conjectured) he was acquainted only with the case of the right-angled triangle,‡ his knowledge of this, as also of (6), may very well have been derived from the Egyptians. On the whole we may conclude that he probably made some advance towards that abstraction by which the Greek geometry, in contrast with the Egyptian, was to be characterised; but more than this cannot safely be affirmed until we are better informed as to the "many things" which he is said to have discovered for himself. Thales was acquainted with the globular form of the earth, which was held by his school to be at the centre of the world.

---

\* This is of course the meaning of the statement that he was the first *to inscribe a right angled triangle in a circle*.

† Plutarch in his *Symposium* states the method in a form requiring a knowledge of similar triangles and applicable at any time of the day. The most trustworthy part of the story is that the method in its simpler form was used *in Egypt*. It would serve for an obelisk, but scarcely for a pyramid.

‡ This view is taken by Bretschneider (p. 43), who attributes (2) only to Thales himself. For a more appreciative estimate of his contributions to geometry see *Hermathena* vol. III. 173.

*Pythagoras  
B.C.  
580—500.*

§ 4. The name of Pythagoras of Samos next arrests our attention. Although the date of his birth and his age at his death are variously given, he was doubtless a young man when Thales, of whom he was regarded as the successor, died. At the instigation of Thales he visited Egypt, where he resided many years, learned the Egyptian language and received instruction from the priests. After this he is said to have visited or been carried captive to Babylon. Returning to Samos at a time when Ionia had lost her independence, he migrated thence to Crotona in Magna Graecia, where he gathered round him his exclusive brotherhood and became the founder of the famed Italian school; but in course of time he was banished by the democratic party, and died shortly afterwards at Metapontum. We proceed to notice some of the chief discoveries in mathematics attributed to him, remarking however that it is impossible to distinguish with certainty between the discoveries of the master and his scholars, since the doctrines of the sect were in the first instance communicated only to its members, and when they came at length to be divulged it was the practice to attribute everything to Pythagoras himself. Hippasus, who offended against this rule, was lost at sea for his impiety (Iamblichus *Vit. Pythag.* cap. 18). He had taken credit to himself for the construction of the sphere circumscribed to a regular dodecahedron (τὴν ἐκ τῶν δώδεκα πενταγώνων), whereas everything belonged to Him (εἶναι δὲ πάντα Ἐκείνου), "for so they call Pythagoras, and not by his name."

*a. The square on the hypotenuse of a right angled triangle is equal to the sum of the squares on the sides containing the right angle.*

In honour of this great discovery, as also on some other occasions, Pythagoras is related to have offered a sacrifice. There is no evidence to support the conjecture that the theorem was known in its generality to the Egyptians, although it must be allowed to partake of an Egyptian character,* and may have

---

* The Egyptian geometry had very little that was of an abstract or general character, but consisted mainly in the computation of areas or volumes, and in such special constructions as are required for geometrical drawing. Cf. Eisenlohr's edition of the Rhind papyrus, published under the title *Ein mathematisches Handbuch der*

been first proved by an application of the Egyptian method. The Egyptians were acquainted with the fact that the triangle whose sides contain 3, 4, and 5 units of length respectively is right angled (Plutarch *De Iside et Osiride* cap. 56), which is a special case of the theorem of Pythagoras; and they must also have been familiar with the still simpler case in which the right angled triangle is isosceles. To prove it for this case, let one square be supposed to be inscribed symmetrically in another. Then it is easily seen that the four triangles at the corners may be fitted together so as to form two squares, the sum of which is equal to the area of the inscribed square; whence the theorem at once follows for the case in question.

By some such method of dissection of figures the general theorem also was perhaps arrived at; and that it was not in the first instance proved by the method of Euclid might have been taken for granted, even without the express statement of Proclus in his comments upon Euclid I. 47. To prove the theorem generally,* let one square inscribed in another divide each side of the latter into segments equal to $a$ and $b$ respectively, and let the side of the inscribed square be equal to $h$. Then the whole figure, being made up of $h^2$ and the four triangles, is evidently equal to $h^2 + 2ab$. Next, by considering the figure of Euclid II. 4 (omitting the diagonal), we see that the outer square may also be cut up into two rectangles, each equal to $ab$, and two squares equal to $a^2$ and $b^2$ respectively. Hence it follows that

$$h^2 + 2ab = a^2 + b^2 + 2ab,$$

and therefore $h^2$ is equal to $a^2 + b^2$, or the square on the hypotenuse of one of the triangles is equal to the sum of the squares upon its sides. Thus the theorem is shewn to be true for any right angled triangle. Pythagoras added a rule for finding triads of integers $a$, $b$, $h$ satisfying the relation $a^2 + b^2 = h^2$. The problem of the three squares would naturally suggest an analogous problem relating to cubes; and to a special case of the

---

*alten Aegypter* (Leipzig 1877), and accompanied with a German translation in a separate volume.

\* The proof here given is taken from Bretschneider's *Die Geometrie &c.* pp. 81—2.

latter, the Duplication of the Cube, we shall see that the further progress of mathematics was indirectly to a very great extent due.

*b. The three angles of a triangle are together equal to two right angles.*

Eudemus, according to a statement of Proclus on Euclid I. 32, ascribes the discovery of this theorem to the Pythagoreans, together with a general proof of it not unlike that given by Euclid. But since Eutocius,* on the authority of Geminus, asserts that the ancients were accustomed to prove it separately for the equilateral, isosceles and scalene triangles, whilst only the later geometers proved it generally once for all, it has been conjectured that its truth may have been known even to the Egyptians, the general proof only being Pythagorean. Although the fact that the area about a point can be filled up by equilateral triangles, squares or regular hexagons, and by no other regular figures, is said by Proclus (on Euclid I. 15, Cor.) to be a Pythagorean discovery, the positive part of it must have been observed by the Egyptians, who must therefore have known that the three angles of an equilateral triangle are together equal to two right angles. They may also have inferred the same for any right angled triangle regarded as the half of a rectangle; and it would then remain only to observe that an isosceles or scalene triangle may be divided into two right angled triangles. By some such process the theorem (by whomsoever discovered) may have been first arrived at; or it may have been shewn experimentally that the six angles of any two triangles exactly fit into the area about a point.

*c. The regular polyhedra.*

We have seen that the construction of the regular solids was attributed to Pythagoras by Proclus, doubtless upon the authority of Eudemus. Of these five figures the tetrahedron, the cube and the octahedron were known to the Egyptians and occur in their architecture; but it does not appear that

---

* Halley's Apollonius p. 9; Bretschneider's *Die Geometrie &c.* p. 14.

they were acquainted with the icosahedron or the **dodecahedron**. In the construction of the last mentioned solid the **regular pentagon** is **required**; and with this the Pythagoreans were familiar, **since they** used the **starred** pentagram,* formed by producing its alternate sides **to meet**, as a secret token of recognition symbolical of ὑγίεια, the letters υ, γ, ι, θ (= ει), α being written at the five angles of the figure. Moreover we have seen [p. xxii] that the knowledge of the dodecahedron was said to have been possessed and divulged by the Pythagorean Hippasus. The regular solids were also called the "**cosmic** figures," the dodecahedron being taken to represent the material world, and the remaining four its elements of earth, air, fire, and water.†

*d. The application of areas.*

The παραβολή or **application** of areas is attributed in **general terms to the Pythagoreans (Proclus on** Euclid I. 44), and also to Pythagoras in particular, who is said to have sacrificed an ox, ἐπὶ τῷ διαγράμματι,‡ where the reference is either to this discovery or to that of the theorem of the three squares (Euclid I. 47). An area, according to Proclus, was said to be *applied* to a right line when an equal area was described upon the line as base; but the term was also used more generally to include the cases in which the base of an area placed upon a given line was in excess (ὑπερβολή) or defect (ἔλλειψις) of the line to which it was "applied." Although it has not been made out wherein consisted the importance of the discovery in the hands of the Pythagoreans, we shall see that it played a great part in the system of Apollonius, and that he was led to designate the three conic sections by the Pythagorean terms Parabola, Hyperbola, Ellipse. It is not however to be thought that Pythagoras or his school had any acquaintance with these curves, although, through a misunderstanding and consequent misreading of the term παραβολή of

---

\* On the *Polygones étoilés* see Chasles *Aperçu historique* pp. 476—87 (1875).

† See Boeck's *Platonica corporis mundani fabrica &c.* (Heidelberg 1809).

‡ Plutarch on Epicureanism, cap. 11. See Plut. *Op.* IV. 1338, ed. Dübner (Paris 1841), where the misreading περὶ τοῦ χωρίου τῆς παραβολῆς occurs.

areas, he has been supposed to have anticipated Archimedes (whose name, as it happens, follows in the immediate context) in his discovery of the quadrature of the parabola.

*e. Incommensurability and proportion.**

To Pythagoras, as we have seen [p. xix], was attributed the theory of incommensurable magnitudes, which may be regarded as a corollary from his theorem of the three squares (Euclid I. 47). He was also acquainted with the doctrine of proportion, and is related by Plutarch to have solved the problem, to describe a rectilinear figure equal to one and similar to another given figure (Euclid VI. 25), and on this occasion also to have offered a sacrifice; but whether he completed the theory of proportion by extending it to the case of incommensurable magnitudes we are unable to say. Iamblichus states that in the time of Pythagoras three kinds of proportion only were known, viz. "the arithmetic, the geometric, and in the third place the subcontrary, as it was then called, but which was afterwards called the harmonic by the associates of Archytas and Hippias." Further on he remarks of the so called "most perfect" or "musical" proportion,

$$a : \frac{a+b}{2} = \frac{2ab}{a+b} : b,$$

which combines in itself the three former, that it was said to be a discovery of the Babylonians and to have been brought by Pythagoras to Greece. To him belongs the credit of combining the Eastern science of arithmetic, which he esteemed so highly, with the Egyptian science of geometry.

*f. The circle.*

Iamblichus,† giving however no details, says that although Aristotle may not have squared the circle the problem was at any rate solved by the Pythagoreans. This problem, as we learn from the Rhind papyrus (ed. Eisenlohr vol. I. 98, 117), had already engaged the attention of the Egyptians, who estimated the circle on a diameter of nine units to be equal

---

* See Bretschneider pp. 75, 83.
† See the extract from Simplicius on Aristotle given by Bretschneider, p. 108.

to the square on a line containing eight, thus making $\pi$ equal to $3\frac{1\cdot 9}{6\cdot 1}$. Having regard to the perfect symmetry of the sphere and the circle, Pythagoras speaks of the one as the most beautiful of solids and the other of plane figures (Diogenes Laertius lib. VIII. cap. 1); but there is no ground for the statement sometimes made that he speaks of the circle as the maximum plane figure having a given perimeter and of the sphere as the maximum solid having a given surface.\*

§ 5. The further development of geometry was due in great measure to repeated attempts to square the circle, to trisect an angle and to duplicate the cube, which led to the discovery of various geometrical loci. Thus Pappus† ascribes to Dinostratus and Nicomedes the use of the quadratrix for squaring the circle; and Proclus (on Euclid I. 9) relates that Nicomedes trisected a given angle by means of the conchoid (of which he had himself discovered the genesis and investigated the properties), others used the quadratrix of Hippias or Nicomedes for the same purpose, whilst others by means of the spiral of Archimedes divided a given angle in any given ratio. The problem of the duplication of the cube, as we shall see, was solved by the intersections of parabolas or other conics, and perhaps actually led to the discovery of the sections of the cone. It is to be noticed that the construction of such a curve as, for example, the quadratrix implies the conception of the idea of a *Locus*, of which before the time of the above mentioned Hippias of Elis, a contemporary of Socrates, there is no trace, although the idea must have presented itself in a rudimentary form in the construction of a circle by the most obvious method. The earliest writer on loci was Hermotimus of Colophon, one of the successors of Eudoxus [p. xix]. Loci.

§ 6. Hippocrates of Chios is referred to by Aristotle (*Ethica* Eudem. VII. 14) in illustration of the fact that there are persons who are wanting in intelligence in some respects although not Hippocrates<br>B.C.<br>450—430.

---

\* For an actual mention of these theorems see Pappus *Collectio* lib. V. (vol. I. pp. 316, 350 ed. Hultsch).

† *Collectio* lib. IV. prop. 25 (vol. I. 251, ed. Hultsch).

in others. The geometer shewed his simplicity by allowing himself to be defrauded by the douaniers of Byzantium; or, according to Johannes Philoponus, he was robbed by pirates, went to Athens to obtain redress, there frequented the schools of the philosophers and made such progress in geometry that he ventured to attack the problem of the quadrature of the circle. It is related by Iamblichus that Hippocrates was expelled from the school of the Pythagoreans for having taught for hire.

*The Delian problem.* Hippocrates is celebrated as having reduced the problem of the duplication of the cube to the simpler form in which it was thenceforth attempted by geometers. By the duplication of the cube was signified the construction of a cube of twice the volume of a given cube: a problem which may possibly have first presented itself in architecture, or may have arisen speculatively in the course of an attempt to find an analogue in space to the Pythagorean property of squares (Euclid I. 47). Eutokius, commenting upon the second book of Archimedes *De Sphæra et Cylindro*, adduces a series of solutions of the problem, including the solution of Eratosthenes given in his letter to king Ptolemy II. together with a twofold tradition as to the origin of the problem (Archimedis *Op.* p. 144, ed. Torelli). Minos of Crete, according to one of the ancient tragedians, ordered a sepulchre for his son Glaucus, and then, deeming the proposed dimensions of the (cubical) structure inadequate, directed the architect to make it exactly twice as large. At a later period—so the story runs—the people of Delos, in time of pestilence, were commanded to construct a new cubical altar twice as large as one already existing, and accordingly at their request the philosophers of the Academy set to work to solve this "solid problem" [p. xxxiii], which was found to transcend the power of the known geometry of the straight line and circle. It involved in effect the extraction of a cube root, or the solution of the cubic equation $x^3 = 2a^3$. Hippocrates reduced it to the problem of finding a pair of mean proportionals to two given magnitudes $a$ and $b$, that is to say, of determining $x$ and $y$ so as to satisfy the relations,

$$a : x = x : y = y : b,$$

which evidently imply also the relation $x^3 = a^2b$. It does not appear that he himself carried the solution any further, but the problem was afterwards attacked in this form by geometers, and in particular it was solved by Menæchmus in two ways with the help of the conic sections, of which he was the discoverer. The problem of the duplication of the cube went by the name of the *Delian Problem* owing to the above mentioned tradition connecting it with one of the altars at Delos.

Hippocrates is also celebrated as having, in his attempts to square the circle, quadrated the lunule contained by the circumscribed semicircle of an isosceles right angled triangle and the semicircle described outwards on one of its shorter sides as diameter. By an extension of his method it may be shewn that the circumscribed semicircle of a scalene right angled triangle contains with the semicircles described in like manner on its two shorter sides two lunules which are together equal to the area of the triangle; but it does not clearly appear that the theorem in this more general form is rightly ascribed to him. In his further attempts to square the circle, he succeeded only in shewing that the problem could be solved if the lunule bounded by an arc equal to a sixth part of the circumference and the semicircle described outwards upon the chord of the arc as diameter could first be squared. All this is fully discussed in a passage of Simplicius,* a commentator on Aristotle, which is given at length by Bretschneider, pp. 100–121. Simplicius gives a long extract from Eudemus, interspersed with references of his own to Euclid, from which it appears that Hippocrates made use of the following propositions in his researches. (1) Circles are to one another as the squares of their diameters. (2) Similar segments (defined as those which are the same fractional part of the circumference) contain equal angles. And (3) similar segments are to one another as the squares of their bases. It is possible, as has been suggested, that by the angle in a segment he means the angle subtended by the chord of the segment at the middle point of its arc, not knowing that the angle subtended at any point of the arc is constant and equal to

*Quadrature of lunules.*

---

* Simplicius, on Aristotle *De physica auscultatione*, fol. 12a (Venetiis 1526).

half the angle at the centre; although *prima facie* it would appear that, knowing already so much, he must have been acquainted with this also.

*Antipho B.C. 440.*

§ 7. The same passage of Simplicius contains an account of the method by which Antipho, perhaps the well known opponent of Socrates, attempted the quadrature of the circle. He first inscribed say a square in the circle, then (bisecting each quadrant) an octagon, then a 16-gon, and so on continually, till at length he supposed a regular inscribed polygon to be arrived at, having an infinity of infinitesimal sides, which was to be regarded as coincident with the circle. Although his principles were regarded as unsound by the ancient critics, he had in fact introduced the fundamental idea of infinitesimals into the geometry of curves, and had virtually proved (1) that the areas of circles are as the squares of their diameters—as his contemporary Hippocrates had also somehow arrived at; and (2) that their circumferences are proportional to their diameters. On the quadrature by Bryso, a contemporary sophist, who regarded the circle as intermediate to an inscribed and a circumscribed $n$-gon, and then applied the method of Antipho, see Bretschneider's *Die Geom. vor Eukl.* pp. 126 ff.

*Plato B.C. 429—348.*

§ 8. Plato, although not greatly distinguished for his own discoveries in geometry, became the founder of a school which was soon to carry the science to unknown heights. He indeed devised an organic solution of the problem of the two mean proportionals, depending upon a double application of a property of the right angled triangle, and gave a rule of his own for constructing right angled triangles having their sides commensurable (Proclus on Euclid I. 47); but he rendered far greater service to geometry by his systematic treatment of its definitions and primary ideas, and by the impulse which he gave to the study amongst his disciples by insisting upon a knowledge of it as a prerequisite for metaphysical speculation, writing up (as it is said) before his vestibule, μηδεὶς ἀγεωμέτρητος εἰσίτω μοῦ τὴν στέγην. To Plato are attributed the propositions Euclid VIII. 11, 12. One of his disciples Theætetus, who had been led by him to the study of incommensurable quantities in connexion with proportion, became the author of the propositions Euclid x. 9, 10. Another, Menæchmus, developed the germs of stereometry received from him and was led, in what way we can only conjecture, to the discovery of the conic sections, which is sometimes erroneously attributed to Plato himself, owing to a misunderstanding of the term τὴν τομήν, *the section*, in a passage quoted above from Proclus [p. xix], where it refers not to the cone but to the right line. Archytas of Tarentum, a contemporary of Plato, propounded a solution of the Delian problem, and is said to have been the first to apply the method of organic description to geometrical figures; a method which Plato (notwithstanding his own application of it as above mentioned) condemned, as tending to materialise geometry and bring it down from the region of eternal and incorporeal ideas. It was one of his sayings, τὸν θεὸν ἀεὶ γεωμετρεῖν, which Plutarch discusses in his *Quæst. Conviv.* lib. VIII. q. 2. Plato is said (Diogenes Laert. lib. III. cap. 1) to have introduced the method of geometrical analysis, and to have communicated it to Leodamas of Thasos. Analysis.

§9. Menæchmus,* a hearer of Eudoxus and contemporary with Plato, is expressly said by Proclus (on Euclid I. def. 4), upon the authority of Geminus, to have been the discoverer of the conic sections, which were accordingly at first named after him the "Menæchmian triads" [p. 194]. He also applied them in two ways to the solution of the problem of the two mean proportionals [pp. 45, 189], to which the Delian problem had been reduced by Hippocrates of Chios. It remains to consider whether he in the first instance regarded the curves in question as plane loci or as sections of a cone. In favour of the former view it may be urged that, as geometers before and after him were led to the discovery of the quadratrix, the conchoid and other plane loci in their attempts to square the circle or trisect the angle, so Menæchmus may have dis- Menæchmus<br>B.C.<br>350—330.

---

\* The anecdote which brings Menæchmus into relation with Alexander the Great [p. xx] is consistent with the supposition that he was a younger contemporary of Plato.

covered his triad of curves in considering by means of what loci* the construction of a pair of mean proportionals to two given magnitudes might be effected. This implies the use of the method of geometrical analysis, which was said to have been discovered by Plato [p. xxxi]; and accordingly we find that Eutokius, who gives in detail the two solutions by Menæchmus of the problem of the two means (Archimed. *Op.* pp. 141–2, ed. Torelli), represents him as having employed the method in both cases. But it is more important to notice that it was used by Eudoxus, of whom Menæchmus was a hearer [p. xix].

(1) The problem being to find the two magnitudes $x$ and $y$ which with two given magnitudes $a$ and $b$ constitute the continued proportion

$$a : x = x : y = y : b,$$

it was seen that the relations $x^2 = ay$ and $y^2 = bx$ were to be satisfied. Being then, as we have seen reason to conclude, already familiar with the idea of a locus, Menæchmus had virtually discovered the parabola regarded as the plane locus determined by the relation $x^2 = ay$, and it was evident that by the intersection of two such curves the required construction could be effected [p. 45].

(2) In his second solution of the problem he makes use of a parabola and a rectangular hyperbola [p. 189], the latter curve being regarded as possessing the property that the product of the distances of any point on it from the asymptotes is constant; whence it is inferred by Bretschneider (*Die Geom. vor Eukl.* p. 162) that the asymptotes of the hyperbola must have been discovered very soon *after* the curve itself became known. But when we consider that the assumed relations,

$$a : x = x : y = y : b,$$

are evidently equivalent to $xy = ab$ and $x^2 = ay$, it commends itself as a not less simple hypothesis that, having already formed the conception of the curve $x^2 = ay$, Menæchmus was further led by the conditions of the problem to attempt the construction

---

* We have seen, from his acquaintance with the quadratrix, that Dinostratus, the brother of Menæchmus—not to mention Hippias of Elis in the preceding century—must have been familiar with the idea of a locus [p. xxvii].

of the curve satisfying the relation $xy = ab$. This at any rate seems to be the only property of the hyperbola with which he can be safely assumed to have been acquainted. The ellipse does not occur in either of his solutions. To construct his loci he may now have had recourse in the first instance to the organic methods reprehended by Plato,* not at first perceiving that they could be more simply constructed by cutting the right cone by planes. It is less natural to suppose† that *after* the discovery of their genesis from the cone Menæchmus, or his followers, should have thought it necessary to trace them by mechanical appliances, of such a nature as to be almost immediately rejected and forgotten. But even if he never so traced them, he may still have discovered them as plane loci. Their actual description‡ was felt to be a difficulty many centuries later.

§ 10. The conic sections, in whatsoever way first discovered, soon came to be regarded as "solid loci," and problems which required them for their solution were called "solid problems." The first writer on the subject was Aristæus the Elder, who distinguished the three conics as the sections of the acute-angled, right-angled and obtuse-angled right cones respectively by planes drawn at right angles to their sides [p. 195]. He is said by Pappus‖ to have written five books of Conic Elements, and five (in continuation?) upon Solid Loci,§ thus preparing the way for the work of Euclid on Conics. He also instituted a comparison of the regular polyhedra,¶ to which Euclid may have been indebted in the thirteenth and last book of his Elements. We assign to Aristæus the date B.C. 320, to indicate that he was intermediate to Euclid and Menæchmus.

Aristæus
B.C.
320.

---

\* Plutarch *Quæst. Conviv.* lib. VIII. q. 2; *Vita* Marcelli, cap. 14.
† Bretschneider *Die Geom. vor Eukl.* p. 143.
‡ Eutocius (on Apollonii *Conica* I. 20, 21) remarks that it was often necessary, διὰ τὴν ἀπορίαν τῶν ὀργάνων, to describe a conic by points, and that this might be done by means of the relations $y^2 = px$, &c.
‖ *Collectio* lib. VII. § 29 &c. (vol. II. 672—6, ed. Hultsch).
§ Viviani [p. 221], in a *Secunda divinatio &c.* (Florent. 1701), attempted to restore the *Loca Solida* of Aristæus.
¶ See prop. 2 of the so-called 14th book of Euclid's Elements.

c

## SECTION II.

### FROM EUCLID TO SERENUS.

*Euclid
B.C.
300.*

§ 1. The birthplace of **the geometer** Euclid, sometimes confounded with **his namesake of** Megara, is unknown. He lived under the **first** Ptolemy (B.C. 323—284), about two centuries after the death of Pythagoras; and we **find him** established at Alexandria, "etwa im Jahre 308, als den ersten Mathematiker seiner Zeit."* Of the various lost works attributed **to** him we may mention (1) his treatise on *Conics*, which formed **the** nucleus of the great work of Apollonius, **and** (2) **the** three books of *Porisms*, to which we shall again refer in speaking of Pappus. His Στοιχεῖα or *Elements* was **written in** thirteen books, to which **a fourteenth and a fifteenth (by** Hypsicles of Alexandria) **are sometimes appended.** The books 1–6 are too **well known to need description.** Books 7–9 are on the pro**perties of numbers;** book 10 on incommensurable magnitudes; **and books** 11–13 on stereometry. Book 10 commences with the proposition, that *If from the major of two given magnitudes more than its half be taken away, and from the remainder* **more** *than its half, and so on continually; a remainder will at length be arrived at which is less than* **the** *minor given magnitude.* Since the minor given magnitude may **be assumed** to be as **small as we please,** the proposition is seen to embody the idea of convergent **series and** the principle **of the** "method of exhaustions." **The book** ends with the proposition that the diagonal **and the side of** a square are incommensurable. The 12th **book** contains applications of the method of exhaustions **to** plane and solid figures, and **it** is shewn that the areas of

---

\* See Moritz Cantor's *Euclid und sein Jahrhundert* p. 2 (Zeitschrift. f. Math. u. Physik. Suppl I. 1867).

circles are as the squares of their diameters (prop. 2), every cone is the third part of a cylinder having the same base and altitude\* (prop. 10), and the volumes of spheres are as the cubes of their diameters (prop. 18). In the 13th book, which is a sequel to the 4th, it is shewn that there are but five regular polyhedra, such as can be inscribed or circumscribed to a sphere. The *editio princeps* of the Στοιχεῖα was published at Basel in 1533 [p. 82]: the Arabic version at Rome in 1594. The Oxford Græco-latin edition, by David Gregory, of the extant works of Euclid was issued in 1703: it contains the Elements, Data, Introductio Harmonica, Sectio Canonis, Phænomena, Optica, Catoptrica, De Divisionibus Liber, De Levi et Ponderoso fragmentum. Notice also Peyrard's *Les Œuvres d'Euclide, en grec, en latin et en français* (Paris 1814–18).

To what extent Euclid was himself a discoverer we are unable to say, but in his Elements he is to be regarded mainly as a compiler. His system as a whole must however have been more or less original in its conception; and the best testimony to its superior method and completeness is the subsequent neglect and disappearance of the cognate works of his predecessors. But his work, although the most ancient on the Στοιχεῖα still surviving, must not be supposed always to preserve the most ancient methods of proof. Thus the theorem of Thales (Euclid I. 5) cannot have been first proved in the manner of Euclid;† whilst Proclus expressly states that the theorem of Pythagoras was not originally proved as in Euclid I. 47. It was also perhaps first shewn more briefly than by Euclid that circles are as the squares and spheres as the cubes of their diameters.

§ 2. Archimedes of Syracuse was born in the year 287 B.C.‡ According to Plutarch (*Vita* Marcelli cap. 14) he was related

Archimedes
B.C.
287—212.

---

\* This theorem, as we shall see, was discovered by Eudoxus [p. xxxviii].

† The more direct way of deducing it from prop. 4 is mentioned by Proclus, in connexion with the name of Pappus.

‡ Notices of the life and works of Archimedes (and of Apollonius) are contained in Cantor's *Euclid u. s. Jahrhundert*. See also Heilberg's article on his knowledge of the *Kegelschnitte* in the *Zeitschr. f. Math. u. Physik* (April 1880).

to king Hiero, whilst Cicero on the other hand speaks of him as "humilem homunculum" (*Tusc. Disp.* lib. v. cap. 23). He was a master not only of geometry, but also of theoretical and applied mechanics. By his scientific conduct of the defence of Syracuse against the Romans the siege was protracted for two years, till at length the city was carried by a surprise from the land side, and Archimedes fell by the hand of a soldier (212 B.C.). His grave was marked by the figure of a sphere inscribed in a cylinder, in commemoration of his most cherished discoveries, and by that sign it was recognised by Cicero in the course of his quæstorship in Sicily. His works, according to the Græco-latin edition of Torelli (Oxon. 1792), are as follows:

(1) *De Planorum Æquilibriis.*\* Two books, with the tract *Quadratura Paraboles* placed between them (pp. 1–60).

(2) *De Sphæra et Cylindro.* Two books (pp. 61–201).

(3) *Circuli Dimensio* (pp. 203–216).

(4) *De Helicibus* (pp. 217–255).

(5) *De Conoidibus et Sphæroidibus* (pp. 257–318).

(6) *Arenarius* (pp. 319–332).

(7) *De iis quæ in Humido vehuntur.* Two books, in Latin only (pp. 333–354).

(8) *Lemmata*, translated from the Arabic (pp. 355–361).

(9) *Opera mechanica*, ut cujusque mentio ab antiquis scriptoribus facta est (pp. 363–370).

We learn also from one of the scattered notices of Archimedes in the *Collectio* of Pappus (lib. v. § 34 vol. I. p. 352, ed. Hultsch) that he discovered thirteen semi-regular polyhedra, bounded by regular but not similar polygons—one of them, for example, by 20 triangles and 12 pentagons, another by 30 squares, 20 hexagons and 12 decagons. But his greatest achievements in geometry were his approximate quadrature and rectification of the circle, his quadrature of the parabola, and his applications of the method of exhaustions to the quadrics of revolution.

---

\* To this treatise and to (2) and (3) are appended the commentaries of Eutocius of Ascalon (540 A.D.).

## SECTION II.

**§3.** In the introduction to his treatise Τετραγωνισμὸς Παραβολῆς, as later scribes have entitled it, Archimedes remarks that none before him, so far as he knows, has attempted to quadrate the segment cut off by a right line from the "section of a right-angled cone," for so he calls the parabola [p. 195]. His theorem, which was first arrived at by mechanical considerations and afterwards proved by pure geometry, is stated as follows:

*The segment contained by any right line and the section of a right-angled cone is equal to four-thirds of the triangle which has the same base and altitude as the segment.*

*a.* In the mechanical proof he shews first that a triangle $CDE$ suspended from a lever of equal arms $AB$ and $BC$, so as to have its side $DE$ vertical and in a line with the fulcrum $B$, is balanced by an area equal to one-third of its own suspended from $A$ (prop. 7); and that the parallel sided trapezium cut off from the triangle $CDE$ by two vertical lines drawn at horizontal distances $h$ and $k$ from $B$ is balanced by an area suspended at $A$ intermediate to $\frac{h}{BC}$ and $\frac{k}{BC}$ of the trapezium (prop. 13). Lastly, supposing a parabolic segment on $BD$ as base to be suspended with its vertex downwards, he arrives at the required quadrature by successive applications of the foregoing theorems after the manner of the method of exhaustions.

*b.* The following is a summary of his second and purely geometrical proof of the same theorem (props. 20–24). If $P$ be the vertex [fig. p. 58] and $QQ'$ the base of a segment of a parabola, the triangle $QPQ'$ is greater than half the segment. Take away this triangle from the segment, and from the remaining segments $PQ$ and $PQ'$ take away their corresponding triangles, and from the four remaining segments their corresponding triangles, and so on continually. Thus at length (Euclid x. 1) we arrive at a remainder less than any assignable magnitude.* Now the sum of all the above mentioned triangles

---

* By this continual subtraction the area of the segment is at length *exhausted*. Hence the term "method of exhaustions."

is $\left(1+\frac{1}{4}+\frac{1}{4^2}+\ldots+\frac{1}{4^n}\right)\Delta PQQ'$, and the limit of this when $n$ is infinite is $\tfrac{4}{3}\Delta PQQ'$, which is accordingly the area of the segment. Notice that at the end of prop. 3 he alludes to an existing treatise on Conics: "These things are proved ἐν τοῖς Κωνικοῖς Στοιχείοις," as he does again in *De Conoid. et Sphæroid.* props. 3, 4. There is no reason to think, as some have done, that he is referring to a treatise of his own.\*

*The sphere.*

§4. In the introductions to some of his treatises, Archimedes refers to what had been done by earlier geometers. Thus in the introduction to the *Quadratura Paraboles* (p. 18), having stated as his primary lemma,† that *the excess of one magnitude over another may be continually added to itself till the sum exceeds any assigned magnitude*, he remarks that it had been applied by those before him, viz. to prove that circles and spheres are as the squares and the cubes respectively of their diameters, and that any pyramid or cone (Euclid XII. 7, 10) is the third part of the prism or cylinder having the same base and altitude. In the introduction to *De Sphæra et Cylindro* lib. I. (p. 64), he gives the important information that the cubatures of the pyramid and the cone (Euclid XII. 7, 10) were discovered by Eudoxus.‡

*Eudoxus B.C. 365.*

These properties preexisted in the figures, but (though many notable geometers lived before Eudoxus) no one had discovered them. In like manner, none before Archimedes had discovered that the surface of a sphere is equal to four times the area of one of its great circles (prop. 35); the volume of a sphere to two thirds of the circumscribed cylinder having the same altitude, and its surface to two thirds of that of the cylinder (prop. 37); the surface of any segment of a sphere to the area of the circle whose radius is the line from the vertex of the segment to any point on its base (props. 48–9); and the volume of the solid sector determined by any segment to the cone whose base and

---

\* Compare the introductory remarks of Eutocius on Apollonii *Conica* (p. 8, ed Halley).

† See also *De Helicibus* (p. 220).

‡ Eudoxus of Cnidus flourished in the 103rd Olympiad, and died about 357 B.C., according to Bretschneider *Die Geom. vor Eukl.* p. 163.

altitude are severally equal to the surface of the segment and the radius of the sphere (prop. 50).

In lib. II. it is proved that the volume of any segment of a sphere is equal to that of a cone of certain altitude described upon its base (prop. 3); and the theorems now established are applied to solve the problems: to find a plane area equal to the surface of a given sphere (prop. 1), to describe a sphere equal to a given cone or cylinder (prop. 2),* and to divide a given sphere into segments whose surfaces or volumes shall be in a given ratio (props. 4, 5).

§ 5. In the *Circuli Dimensio* it is shewn, that any circle is equal to the right angled triangle whose sides about the right angle are equal to the radius and the circumference of the circle (prop. 1); a circle is to the square on its diameter as 11 to 14, approximately (prop. 2); and that the circumference exceeds thrice the diameter by a fraction of it less than $\frac{1}{7}$ and greater than $\frac{10}{71}$ (prop. 3). These last results are obtained by regarding the circumference of a circle as intermediate in length to those of its circumscribed and inscribed 96-gons. Thus we see that Archimedes treated the problem both as a quadrature and a rectification of the circle; and he shewed, not only that $\pi$ is nearly equal to $3\frac{1}{7}$, but that it is *less than* $3\frac{1}{7}$ *and greater* $3\frac{10}{71}$. We may therefore fairly say that his approximation was exact to three places of decimals, since the mean of his two limits gives $\pi$ equal to 3.1418 &c. The approximation in the Rhind papyrus makes it greater than 3.16 [p. xxvi].

<small>Circuli dimensio.</small>

§ 6. In the treatise *De Helicibus* he defines his helix or spiral ($r = a\theta$) as generated by the double motion of a point, which moves uniformly outwards from a fixed origin, in the direction of a radius vector which itself rotates uniformly about that origin. Supposing the generating point to start from the

<small>Spiral of Archimedes.</small>

---

\* A solution of the problem of the two mean proportionals being here presupposed, Eutocius (pp. 135–149) gives the methods of Plato, Hero, Philo of Byzantium, Apollonius, Diocles, Pappus, Sporus, Menæchmus, Archytas, Eratosthenes and Nicomedes, rejecting that of Eudoxus (pp. 135, 149), perhaps for insufficient reason (Bretschneider *Die Geom. vor Eukl.* p. 166).

origin $O$ and to arrive at the point $A$ after $n$ complete revolutions **of the radius vector,** he shews that the intercept made by the tangent **at $A$ upon the** radius vector at **right** angles to $OA$ is equal **to $n$ times the circumference** of the circle described with $OA$ as radius (prop. 19), thus effecting the *rectification* **and quadrature of** the circle with **the help of his** spiral; and in **prop. 20** he proves the corresponding theorem for any other **position** $OR$ of the radius vector.* The quadrature of the **spiral is** determined in props 24–28. From the introduction **to this** treatise we learn that **there were** other able geometers in **the time of** Archimedes, with **whom he** was in correspondence; and that **there were also** pretenders addicted **to** claiming more **than** their due, **for** whose discomfiture he propounded false theorems, of which examples **are** given (p. 218).

His spiral **affords** the simplest illustration **of the generation of** curves **by** an angular compounded with **a linear motion,** according to the idea **of** Plato, **who** "establishing **two** most simple and principal species **of** lines, **the** right and the **circular,** composes all the rest from **the mixture of these"** (Proclus on Euclid I. **def. 4).** Desargues (1639 A.D.) threw out the suggestion that **a conic might be** thus described, **but** assigned no **law of** movement.† On Roberval's rule for drawing the *tangent* to a curve at any point, regarded as the line of the **resultant of all the** movements of the point, see Chasles' *Aperçu historique* **p. 58** (1875).

**The conoids.** §7. The book *De Conoidibus et Sphæroidibus* contains various theorems on the cubature of the quadrics of **revolution,** the sphere having been already dealt with **in** a separate work. The figure generated by the rotation of a "section of the right angled cone" about its axis, **that is to** say, the paraboloid of revolution, is called the right angled conoid; the hyperboloid of revolution is called the **obtuse angled** conoid; but the "acute angle conoid," as it should be called, is more briefly termed the

---

* Thus in effect he determines the trigonometrical tangent of the angle between $OR$ and the tangent to the curve at $R$.

† Poudra *Œuvres de Desargues* vol I. 227, 298.

spheroid. It is shewn that the **areas of** ellipses **are as the** products of their axes (prop. 7); that an infinity of **right cones or cylinders can be** drawn so as to contain a **given ellipse** (props. 8–10); **and that** the plane sections of the conoids **and** and **spheroids are** conics (props. 12–15). The book concludes with **a series of propositions on** their cubature (props. 21–34), which **are proved by a process** closely **related to** the method of integration.

§ 8. The method of exhaustions employed **by Euclid** and *Method of exhaustions.* Archimedes involved a tedious *reductio ad absurdum*, **and was** perhaps first elaborated as a means of *verification* **rather than** of discovery. The idea of regarding a curve as a limiting form of **polygon was propounded, for the** case of the circle, **by Antipho [p. xxx]**, in the fifth century B.C.; and the fact that **circles are as the squares of their diameters was thus rendered intuitive,** presupposing **only a well known relation between the areas of** similar rectilinear figures. **As regards this property of circles** and the analogous property of **spheres, the proofs given by** Euclid may be supposed merely to have established **more rigidly** what had been already divined by a summary process; **but the use of** the method of exhaustions was more apparent in the **actual evaluation of** volumes and areas. Granted, for example, **that a curvilinear plane area** might be regarded as divided into **rectilinear elements by an infinity of** consecutive ordinates, the **summation of its elements could not well have been** effected directly **before the invention of some form of algebraical** calculus. Instead of regarding the **small elements of a curve as** ultimately rectilinear, **the ancients would (in the case** supposed) have proceeded somewhat as follows. Project every ordinate upon the next before and the next after it by parallels to the axis **of** abscissæ: thus two sets of parallelograms are constructed, to which the area of the curve is intermediate: suppose the difference between the two sets to be indefinitely diminished by increasing the number of ordinates, and then apply the method of *reductio ad absurdum*, **as above** mentioned. For actual cases of the subdivision **of** surfaces by parallel planes, and their cubature by this method, see *De Conoid. et Sphæroid.* props. 21–24. The stereometrical work of Archimedes was revived and continued by Kepler, whose *Nova Stereometria* prepared the way for the modern forms of the infinitesimal calculus.

Apollonius
B.C.
247—205.

§ 9. Apollonius of Perga was born in the reign of Ptolemy Euergetes (247–222 B.C.), studied at Alexandria under the successors of Euclid, and flourished in the reign of Ptolemy Philopator (222–205 B.C.). Of his various works* the most famous was the Κωνικά, which gained for him (according to Geminus)† the title of the Great Geometer. In the account of this work given by Pappus,‡ it is divided into two tetrads of books, the former founded on Euclid's four books of *Conics*, and the latter supplementary to them. Apollonius in like manner, in his introductory letter to Eudemus, draws a distinction between books 1–4, which he describes as elementary, and the remainder, which were περιουσιαστικώτερα, at the same time pointing out that the former also contained very much that was new. The Oxford edition by Halley (1710) contains books 1–4 with the commentary of Eutocius, in Greek and Latin (pp. 1–250); and in a second part, books 5–7 translated from the Arabic and lib. VIII. "restitutus" (pp. 1–171). The volume concludes with the two books of Serenus on the Cylinder and the Cone, in Greek and Latin (pp. 1–88). The contents of the several books of the Conics of Apollonius are specified below. The most striking evidence of his geometrical power is afforded by the fifth book, in which he solves the problem of drawing normals to a conic from an arbitrary point in its plane, and evaluates the coordinates of what we call the Centre of Curvature at any point of a conic. To have worked out such results with the means at his disposal is an achievement not unworthy of the greatest of geometers in any age.

(*a*) Book I. A conical superficies is defined as the surface generated by an infinite right line, which passes through a fixed vertex and moves round the circumference of a given

---

\* See the notices in the *Collectio* of Pappus; and cf. Cantor's *Euclid u. s. Jahrhundert* pp. 44–64.

† Halley's Apollonii *Conica* p. 9. Geminus lived about 150 B.C. (Cantor p. 52).

‡ *Collectio* lib. VII. § 30 (vol II. p. 672, ed. Hultsch).

## SECTION II.                                                xliii

circle: the term Cone is used specially of **the finite portion of
the superfices between the vertex and the circle or base. The**
*Axis* **is the line from the vertex to the centre of the base. The**
plane through **the axis at right angles\*** to the base cuts **the**
cone and its **base** in a triangle, which is called "the triangle
through **the axis;" and every** chord of the cone at right angles
to the **plane of this triangle is bisected by it** (prop. 5). **Any
plane at right angles to the plane of the triangle** through the
**axis meets the conical surface in general in one** of the three
**curves formerly** distinguished as the sections **of the** right, obtuse
**and** acute angled cones respectively. These **names** being thus
found to be inappropriate, others have to be suggested in their
place. The new **names may** be briefly **explained as follows**
[p. 82]. The Parabola is so called because **at every point of it,**
if $p$ be the parameter, $y^2$ is *equal* **to** $px$; the Hyperbola because
$y^2$ is *greater* **than** $px$; and the Ellipse† because $y^2$ is *less* **than**
$px$ (props. 11–13). **He is now** practically independent **of the**
cone, and **starts** afresh **from the relation** between the ordinate
and the abscissa. See also props. **20–21**. It is shewn later in
the book, that the tangent to a conic at any point and the
ordinate of the point to any diameter divide the **diameter**
harmonically (props. 34–38); and lastly a construction **is given**
for describing **two** conjugate hyperbolas with a given pair of
conjugate diameters (prop. 56).

In the use of coordinates‡ by **the ancients,—as for example**   Application
by Apollonius in this book, and **in a more striking** way in his   of areas.
fifth book—the form of procedure •was strictly geometrical
throughout. Hence we see more clearly the importance of the

---

\* After prop. 5 this plane is called briefly "a plane through the axis." There is
the same laxity of statement in def. 10, where the right line bisecting a system of
parallel chords of any curve line (πάσης καμπύλης γραμμῆς) in one plane is defined
as a diameter; whereupon Eutocius remarks that he rightly adds *in plano*, to exclude
the cylindrical helix and the sphere.

† If the three conics were first discovered in the order in which Apollonius
(perhaps following Euclid and Aristæus) here introduces them, this tends rather to
support the conjecture that they were discovered *in plano* [p. xxxii], since the
contemplation of the cone, which was regarded as a finite figure (Euclid XI. def. 18),
would have revealed the ellipse first instead of last. Geminus (Proclus on Euclid I.
def. 4) called the ellipse θυρεός, from its shape. Cf. cissoid, conchoid, cardioid.

‡ The term *ordinate* was derived by translation from the Greek.

παραβολή of areas [p. xxv.]. In its simplest form this amounted to finding the line which in conjunction with a given line determines a rectangle, or other parallelogram (Euclid I. 44), of given area, which corresponds to the algebraical division of a given product by one of its factors. A further use of the term occurs in the determination of the foci of a central conic which Apollonius calls "the points arising ἐκ τῆς παραβολῆς," *puncta ex applicatione facta*. Here the problem is to divide the axis into segments whose product is equal to the fourth part of "the figure" [p. 82], or to determine $x$ and $y$ from the relations $x \pm y = 2a$ and $xy = b^2$. The application to a given line of a parallelogram *deficient* or *exceeding* by a parallelogram similar to a given one is the subject of the propositions Euclid VI. 27—29. For an extension of the method to an indefinite *series* of magnitudes, see Archimedes *De Conoid. et Sphaeroid.* prop. 3. Thus the "application" of *areas*, so far as it went, was to the ancient geometry what algebra, which deals with *products* and factors, is to the geometry of Descartes.

(*b*) Book II. The asymptotes are thus defined: on the tangent to a hyperbola at any point $P$ take $PT$ and $PT'$, each equal to the parallel semi-diameter; then the lines $CT$ and $CT'$, and these alone, being produced to infinity, do not meet but approach indefinitely near to the curve (props. 1, 2, 14). Through a given point a hyperbola can be drawn so as to have a given pair of lines for asymptotes (prop. 4). The opposite intercepts made on any straight line by the curve and its asymptotes are equal to one another, and the product of two adjacent intercepts is equal to the square of the parallel semi-diameter (props. 8—11). The product of the distances of any point on the hyperbola from its asymptotes is constant (prop. 12). A line parallel to an asymptote meets the curve in one point only (prop. 13). The tangents to conjugate hyperbolas at the extremities of any two conjugate semi-diameters meet on one or other of the asymptotes (prop. 21). The diameter through the point of concourse of any two tangents to a conic bisects their chord of contact (props. 29—30). Supplemental chords of a hyperbola are parallel to conjugate diameters (prop. 37).

Given a central conic, to find its centre and axes (props. 45–8); and to draw a tangent making a given angle with the axis (prop. 50), or with the diameter through its point of contact (props. 51–3).

(c) Book III. The diameters through any two points of a conic contain equal triangles with the tangents at those points (prop. 1). The rectangles contained by the segments of any two intersecting chords of a conic are as the squares of the parallel tangents (props. 16–23). *Any chord through the intersection of two tangents to a conic is cut harmonically*\* *by their point of concourse and their chord of contact* (props. 37–40): the special case of any chord through the intersection of a tangent and an asymptote is treated separately in props. 34–36. Thus a substantial contribution is made to the theory of polars, afterwards completed by Desargues. Any three tangents to a parabola cut one another proportionally (prop. 41). Two tangents being drawn at the extremities of any diameter, the product of their segments by any third tangent is equal to the square of half the conjugate diameter (prop. 42). The tangent to a hyperbola cuts off a constant area from the space between the asymptotes (prop. 43). The foci of a central conic, or "puncta ex applicatione facta," are determined and their principal properties proved in props. 45–52 [p. 111]; but since the process of "application" fails when the axes become infinite, *he does not detect the existence of the focus of the parabola*.

This third book is said by Apollonius, in his preface to the entire work (p. 8), to contain many wonderful theorems, for the most part new; and he adds that Euclid was not able to construct the *Locus ad tres et quatuor lineas* generally,† but only some special case of it, and that indifferently; for in fact it was not possible to complete the construction "without our further discoveries," where the allusion is doubtless to props. 16–23 [p. 266]. From the extant works of Apollonius we learn nothing about the nature of this *Locus*, and even the com-

---

\* The expression "harmonically" is however not used by Apollonius. On polars with respect to a circle see § 13 (*f*).

† The proof for the case of the circle presents no difficulty [p. 235].

mentator Eutocius cannot explain the allusion (p. 12). Pappus however informs us what the problem really was, and states clearly that the locus in question is a conic. He then speaks of the locus analogously related to more than four given lines (*Collectio* p. 680, ed. Hultsch). Considering the distances of a point from *six* given lines, we may say that the solid contained by three of the distances varies as that contained by the remainining three; but we cannot go on to more than six given lines, and say: "if the ratio of the content of four of the distances to the content of the remainder be given,—since there is not anything that is contained by more than three dimensions. Nevertheless, men a little before our time have allowed themselves to interpret such things, signifying nothing at all comprehensible, speaking of *the product of the content of such and such lines by the square of this or the content of those*. These things might however be stated and shewn generally by means of compounded proportions &c." These predecessors of Pappus, who were not to be confined to three dimensions, were evidently algebraic geometers, who considered lines not directly as such, but only in their numerical relations to a unit of length.

<small>Germ of algebraic geometry.</small>

(*d*) Book IV. No two conics can have more than four points of concourse (props. 25, 36, 40–4, 53), or two of concourse and one of contact (props. 26, 45–8, 54), or two of contact (props. 27, 38, 49–51, 55). Two parabolas can only touch one another in one point (prop. 28).

If the earliest writers on conics had not dealt with the subject of this book (p. 217), and if the principal part of book 3 was also new, we may conclude that the Conics of Euclid contained little or nothing that is not to be found in books 1 and 2 of Apollonius and that the earlier Elements of Aristaeus were meagre, or "somewhat concisely written." Their treatises would have contained elementary propositions on the *right cone*, the relation of the *ordinate* to the abscissa in each section, the property of a *diameter*, some construction of a *tangent* with the determination of its intercept on any diameter, and the leading properties of the *asymptotes*; but nothing about foci, or normals, or the metric relations of conjugate diameters, or of intersecting chords of the general conic drawn arbitrarily.

(e) Book v. Under the title *De Maximis et Minimis*, this book treats of the four *Normals* (regarded as greatest or least lines) that can be drawn to a conic from a given point in its plane, and establishes the complete analogue of Euclid III. 7, 8 for the general conic. If $P$ be any point on a conic, $N$ its projection upon the axis, $A$ the nearer vertex, $AL$ parallel to $NP$ and *equal to half the latus rectum*, and $Q$ the point in which the diameter $CL$ meets $PN$, then

$$PN^2 = 2 \text{ trapezium } ANQL.$$

From this relation between the coordinates $PN$ and $AN$ the following results are elaborated.

If the abscissa $AN$ be less than $AL$, the least right line from $N$ to the curve is $NA$, and the greatest is $NA'$ the remainder of the axis (props. 4–6); but if $AN$ be greater than $AL$, the least line from $N$ to the curve is such that its projection upon the axis is equal to $AL$ in the case of the parabola, and in other cases to $\frac{b^2}{a^2} CN$ (props. 8–10), which is the property of the subnormal. The greatest lines from given points on the minor axis of an ellipse to the curve are then considered (props. 16–22), and it is shewn that the intercepts upon them between the curve and the points in which they meet the major axis are the least lines that could be drawn from those points to the curve (prop. 23). All such greatest and least lines meet the conic at right angles (props. 27–33); and if $O$ be any point on one of them and $N$ be the point at which it meets the curve normally, then $ON$ is also a greatest or least line from $O$ to the curve (props. 12, 21, 34). Four *normals* (as we shall now call them) to a semi-ellipse, or three normals to an elliptic quadrant, cannot meet in one point (props. 47–8). If $O$ be any point in the plane of a conic whose abscissa $AN$ is not greater than $AL$, no normal can be drawn to the conic from $O$ so as to fall within the angle $AON$ (props. 49, 50); but if $AN$ be greater than $AL$, then according as $ON$ is greater than, equal to, or less than a certain length $\lambda$ no normal, or *one*, or *two* can be drawn to the conic from $O$ so as to fall within the angle $AON$ (props. 51–2).

To determine the length λ, divide the semi-axis $CA$ in $H$ so that

$$CH : NH = CA : AL;$$

between $CH$ and $CA$ find two mean proportionals* $CL$ and $CK$, so that

$$CH : CL = CL : CK = CK : CA;$$

and lastly, supposing $CK$ to be an abscissa measured towards $A$, and $P$ the point on the curve (on the opposite side of the axis to $O$) having $CK$ for its abscissa, take λ to $PK$ in the compound ratio of $CN$ to $CH$ and $HK$ to $CK$, so that

$$\lambda : PK = CN . HK : NH . CK.$$

Hence, writing $a$ and $b$ for the semi-axes, we find that

$$CN = \frac{a^2 - b^2}{a^2} . CH = \frac{a^2 - b^2}{a^4} . CK^3;$$

and

$$\frac{\lambda}{PK} = \frac{a^2 - b^2}{b^2}\left(1 - \frac{CH}{CK}\right) = \frac{a^2 - b^2}{b^2}\left(1 - \frac{CK^2}{CA^2}\right),$$

or

$$\lambda = \frac{a^2 - b^2}{b^4} PK^3;$$

and therefore $CN$ and λ are equal to the coordinates of the centre of curvature at $P$, which is here virtually regarded as the *point of the ultimate intersection of consecutive normals*, since if the ordinate of $O$ be diminished however slightly, $O$ at once becomes a point from which two normals can be drawn. The locus of $O$ is the evolute of the conic [p. 221].

When $ON$ is less than λ, he determines a certain point Σ having $CH$ for its abscissa and a certain point $I$ on $PK$, and through $I$ he describes one branch of a *rectangular hyperbola* (pp. 39, 40), having for asymptotes the parallels through Σ to the axes of the given conic. This semi-hyperbola intersects the conic in two points $X$ and $Y$, the normals at which are $OX$

---

* The Arab interpreter gives a construction (p. 40) for two mean proportionals identical with that of Ex. 530 [p. 189].

and $OY$ [p. 265].* To determine the positions of $\Sigma$ and $I$, produce $ON$ to a point $M$ such that

$$OM : MN = a^2 : b^2;$$

through $M$ draw a parallel to the transverse axis, meeting the ordinate through $H$ in $\Sigma$, and $PK$ in $G$; and upon $PG$ take the point $I$, such that $GI.G\Sigma = \Sigma M.MO$. The rectangular hyperbola may then be described. Afterwards he considers other positions of the point $O$ from which the normals are to be drawn, making use of the same hyperbola of construction (props. 55, 58–63). In prop. 75 (the last but two) it is shewn, that if the normals at three points $PQR$ on the same side of the axis of an ellipse cointersect in $O$, the normal $OP$ nearest to the vertex remote from $O$ is the longest line from $O$ to the semi-ellipse, and the normal $OR$ nearest to the vertex $A$ adjacent to $O$ is the longest line that can be drawn from $O$ to the arc $AQ$.

(*f*) Book VI. Similar conics being those in which corresponding ordinates and abscissæ are proportional, it is **shewn** that all parabolas are similar (prop. 11); as also are **central** conics the figures upon whose axes are similar (props. 12, 13). At the end of the book it is shewn how to cut a section of given form and magnitude from a given right cone (props. 28–30), and conversely, how to draw a right cone similar to a given one through a given conic (props. 31–3).

(*g*) Book VII. In props. 6, 7, use is made of supplemental chords drawn from the vertices. Cf. lib. II. 37. The sum or difference of the squares of conjugate diameters is constant (props. 12, 13); and in the equilateral hyperbola conjugate diameters are equal (prop. 23). The conjugate **parallelogram** is equal to the rectangle contained by the axes (prop. 31). The relative magnitudes of conjugate diameters in various special cases are then discussed.

(*h*) Book VIII. Of this book there is only a conjectural restitution. Thirty-three propositions are given, containing

---

* HALLEY, in a Scholion on Serenus *De Sect. Coni* prop. 39 (p. 69), gives a construction for the **three normals** to a parabola from a given point, by means of a certain circle through the vertex [p. 224].

various special constructions, such as: given the axis and the latus rectum of a central conic, to draw a pair of conjugate diameters whose ratio, or sum, or difference is given (props. 7—12).

*Menelaus
A.D.
80.*

§ 10. In continuation of the account of the most brilliant period of ancient geometry, the century of Euclid, Archimedes and Apollonius, recourse must again be had to the *Collectio* of the much later writer Pappus, for information about the lost three books of Porisms of Euclid. But two other names meanwhile demand at least a passing allusion. In the *Sphærica* of Menelaus, a geometer and astronomer of the first century A.D., is found the theorem (lib. III. lemma 1 p. 83, Oxon. 1758): If the sides $ag$, $gd$, $da$ of a plane triangle be met by any transversal in the points $erb$ respectively, then

$$ge : ea = gr.db : rd.ba,$$

or *the product of three non-adjacent segments of the sides of the triangle by any transversal is equal to the product of the remaining three.* This was also extended to spherical triangles, and served as a basis for the spherical trigonometry of the ancients. But the property of the six segments *in plano* is here noticed on account of the great results to which it led long after, especially in the hands of Desargues. See also Chasles *Aperçu historique* Note VI. p. 291, 1875; *Les Porismes* p. 107. Menelaus is mentioned in the fourth and sixth books of the *Collectio* of Pappus (pp. 270, 476, 600—2 ed. Hultsch).

*Ptolemy
A.D.
125—139.*

§ 11. Claudius Ptolemæus was "le plus célèbre, sans contredit, mais non le plus véritablement grand astronome de toute l'antiquité." Thus writes Delambre in the *Biographie Universelle* (vol. 36. Paris 1823). In a work on the three dimensions of bodies, Ptolemy introduced the idea of determining the position of a point in space by referring it to three rectangular axes of coordinates (ibid. p. 272). His chief work, which he called a mathematical Σύνταξις, was further described by his admirers as ἡ μεγάλη, and by the Arabs as *Almagest* (ἡ μεγίστη).

SECTION II.

In it* (*Dictio Prima*, **cap. 12.** fol. 9*b*. Venet. 1515) he reproduces the theorem of the six segments (§ 10), which has accordingly been ascribed to Ptolemy—and founds upon it a system of trigonometry, plane and spherical. For a full account of his works see tome II of Delambre's *Hist. de l'Astronomie ancienne* (Paris 1817), comparing the *Præfatio* to the third volume of Hultsch's Pappus.

§ 12. The Συναγωγή of Pappus was formerly best known in the Latin version of Commandinus, but a complete Greek text of its *Reliquiæ* (with a Latin rendering) has at length been edited by Hultsch.† It is a miscellany of mathematics and mathematical history, to which we here refer chiefly to supplement our account of Euclid by some notice of his great work, the lost three books of Porisms. It is customary to place Pappus near the end of the fourth century of our era; but Hultsch, following Usener (*Rheinisches Museum* vol. XXVIII. 403), considers him to have flourished under Diocletian, 284–305 A.D. A general account of the Porisms is given in lib. VII (pp. 636, 648–60), where it is said, that the three books were an exceedingly skilful compilation, serving for the solution of the more difficult problems: the doctrine of porisms was subtle and general, and very delightful to persons of insight and resource: nothing had been added to what Euclid wrote upon them, except that some dull persons had given their second redactions of a few of his propositions. Twenty-nine genera of porisms are specified, and it is stated that the three books contained 38 lemmas and 171 theorems. The 38 lemmas constitute props. 127–164 of lib. VII of the *Collectio* (vol. II. pp. 866–918). Their enunciations are curt and unfinished, being like private memoranda of the writer rather than complete statements, and the whole doctrine of porisms long remained an impenetrable secret. The first great step towards their inter-

Pappus A.D. 300.

---

\* For his property of a trapezium *abgd* inscribed in a circle, viz. *ag.bd* = *ab.gd* + **ad.bg**, see cap. 9, fol. 5*b*.

† Pappi Alexandrini *Collectionis* quæ supersunt, *e libris MSS. edidit &c.* Frid. Hultsch (Berol. 1876–8). Lib. I. is lost, but portions of II.–VIII. remain. The edition is in three volumes, in which the text has one pagination throughout. On earlier editions see Pref. to vol. I.

pretation was made by Simson (*Phil. Trans.* May 1723), and the most complete work upon the subject is Chasles' *Les trois livres de* **Porismes** *d' Euclide rétablis...conformément au sentiment de R. Simson sur la forme des énoncés* (Paris 1860), which contains a historical *résumé*, an analysis (pp. 73–84) and enunciation (pp. 87–98) of the lemmas, and a conjectural restitution of the Porisms.

§ 13. Passing by the lemmas 18, 20, 21, 29–33, 36–38, we may group the remainder as below, adhering (except in one particular) to the classification of Chasles.

(*a*). *Four fixed radiants cut any transversal in a constant cross ratio.*

Lemmas 3, 10, 11, 14, 16, 19. Props. 129, 136, 137, 140, 142, 145.

Lemma 3 (p. 870) affirms, that *if any two straight lines* $ABCD$ *and* $AB'C'D'$ *be drawn across three straight lines* $OB$, $OC$, $OD$ [p. 253], then

$$AB.DC : AD.BC = AB'.D'C' : AD'.B'C'.$$

Here we have, not quite directly stated, the theorem (*a*). In lemmas 10 and 16 it is shewn conversely, that if $\{ABCD\} = \{AB'C'D'\}$, the line $DD'$ passes through $O$ the intersection of $BB'$ and $CC'$. Lemma 19 is simply, that if $\{ABCD\} = 1$, then $\{AB'C'D'\} = 1$. Lemmas 11 and 14 follow from 3 and 10 respectively by taking one of the two transversals, as $aBb$ in Art. 103 [p. 251], parallel to one of the three radiants.

These Lemmas are used in the proof of lemmas 12, 13, 15, 17.

(*b*). *The opposite sides and the two* **diagonals of** *any quadrilateral* **meet** *any transversal in three* **pairs of** *points in involution.*

Lemmas 1, 2, 4–7. Props. 127, 128, 130–3.

Lemma 4 (p. 872) will serve as an example of the obscure enunciations of Pappus. The statement is as follows:

*The figure* $ABCDEFGHKL$, *and as* $AF.BC$ *to* $AB.CF$ *so let* $AF.DE$ *be to* $AD.EF$. (*I say*) *that the line through the points* $HGF$ *is straight.*

This is in reality a converse of (*b*) [p. 261]. In lemmas 1 and 2 the transversal is parallel to a side of the quadrilateral.

In lemma 5 it is drawn through the points of concourse $A$ and $C$ of the opposite sides of the quadrilateral, and the line $AC$ is divided *harmonically* by the two diagonals, or if parallel to one of them (lemma 6), is bisected by the other.

(c). *Theorem reciprocal to the above.*
Lemma 9. Prop. 135.

Lemma 9 (p. 878) is that, if $AD$ and $AE$ be drawn from the vertex $A$ to the base $BC$ of a triangle, $FG$ a parallel to the base meeting $AB$ in $F$ and $AC$ in $G$, and $H$ a point on $DE$ such that
$$BH : HC = DH : HE,$$
the lines $FH$, $GH$ meet $AB$ and $AC$ respectively in points $K$ and $L$ lying on a parallel to $BC$.

This is the converse of a special case of the theorem, that *the three pairs of summits of a quadrilateral $FGKL$ subtend a pencil in involution at any point $A$ in its plane;* the point $H$ in the case supposed being the centre of the involution in which $BE$ and $CD$ are segments.

(d). *If a hexagon be inscribed in a line-pair, its* **three pairs of** *opposite sides meet in three points lying in a straight line.*

Lemmas 8, 12, 13, 15, 17. Props. 134, 138, 139, 141, 143.

Lemma 13 (p. 886) is to the effect, that if $AEB$ and $CFD$ be triads of points on a straight line, the three intersections
$$(AF, CE), (FB, ED), (BC, DA),$$
are in a straight line. The figure $AFBCED$ may be regarded as a hexagon inscribed in a line-pair. Lemma 12 is the case in which $AB$ and $CD$ are parallel. Lemmas 15 and 17 are converse forms of 12 and 13.

Lemma 8* (p. 878) is thus enunciated:

Let $ABCDEFG$ be a βωμίσκος, and let $DE$ be parallel to $BC$, and $EG$ to $BF$. Then $DF$ is parallel to $CG$.

That is to say, if $BC$ be the base of a triangle, $DE$ (terminated by the sides through $B$ and $C$) a line parallel to $BC$, and $EG$, $BF$ a pair of parallels terminated by $BD$ and $CE$ respectively or their complements, then $DF$, $CG$ are parallel.

---

* This lemma is isolated in Chasles' classification (p. 78).

In other words, *if FBCGED be a hexagon inscribed in a line-pair BDG, CEF, the intersections* (*FB, GE*), (*BC, ED*), (*CG, DF*) *are in the case supposed at infinity, and in general in one straight line.*

(*e*). *Harmonic section of a right line.*

Lemmas 22–27, 34.   Props. 148–153, 160.

These Lemmas are on the metric relations of the segments of a harmonic range; but the term **harmonic**, although coined long before [p. xxvi], is not employed.

(*f*). *Property of polars with respect to a circle.*

Lemmas 28, 35.   Props. 154, 161.

These Lemmas (pp. 904, 914) are to the effect, that any chord of a circle drawn through a fixed point *without* or *within* it is divided harmonically by the point and a certain fixed straight line. Of this proposition, which in its entirety is the foundation of the theory of polars with respect to a circle, the former part only was extended by Apollonius to the conics (lib. III. 37).

*Focus and directrix*

§ 14. To Pappus we are further indebted for the earliest trace of a focus of the parabola, and of a directrix of any conic. In the *Collectio* lib. VII. prop. 238 (p. 1013) is the theorem, that the locus of a point *in plano*, whose distance from a fixed point varies as its perpendicular distance from a fixed straight line, is a conic. Thus one focus of the parabola is at length found; but it was reserved for Kepler to complete the theory of the real "foci" of conics, and to give them their name.

*Serenus.*

§ 15. The two books of Serenus of Antissa *De Sect. Cylindri* and *De Sect. Coni* respectively form a sequel to the Conics of Apollonius in Halley's edition. Serenus was also a commentator on Apollonius, and he lived before Marinus, a disciple of Proclus.* Many geometers in his day imagining that the sections of the cylinder were not identical with the elliptic sections of the cone, he sets to work to remove this misapprehension. (*De Sect. Cyl.* props. 16–18). He then shews

---

* The date 450 A.D. for Serenus may serve as a conjectural lower limit. Baldi, *Cronica de Matematici* p. 59 (Urbino 1707), boldly assigns to him the precise date 462. Suter *Gesch. der math. Wissenschaften* I. p. 92 (Zürich 1873) prefers the date 200–100 B.C.

how to construct a cylinder and a cone on **coplanar bases, so as to be cut by one plane in the same ellipse or in similar ellipses** (props. 19—23); and further, given a cylinder and **a plane cutting it, he describes a cone having the same base and altitude, which is cut by the given plane in a** section similar to that of the cylinder (prop. 25). Props. 26—30 shew how to cut a cylinder or cone in subcontrary pairs of similar ellipses. The remaining propositions *De Sect. Cylindri*, although of still greater interest and importance, are sometimes overlooked. The property of a harmonic pencil, indirectly stated, is applied *in space* to prove that all the tangents to a cone from one point have their points of contact on two generating lines (props. 33, 34), and the idea of projection by rays emitted from a luminous point is suggested and illustrated by a simple case (prop. 35). In the book *De Sect. Coni* he breaks (as he tells us) new ground, in thoroughly discussing the triangular section determined by an arbitrary plane through the vertex. Thus he makes a step towards the generalisation of Desargues, who drew his planes of section without reference to the fixed "triangle through the axis."

§ 16. The writings of Serenus suggest an answer to the *Perspective.* question (Chasles *Aperçu historique* p. 74, 1875), Was the method of perspective\* known and used by the ancients? Certainly not by those who doubted whether the sections of a cylinder were also sections of a cone. But Serenus now shews that the property of a harmonic range may be transferred by central projection from plane to plane, and hence that any tangent to a conic section and its point of contact project into a tangent and its point of contact on any plane. The principle of perspective had thus been laid down, as the modern reader clearly sees; but if the ancients had still (as in the time of Apollonius) no complete theory of polars with respect to a conic, and if they had not learned to look upon parallel straight lines as concurrent (Chasles *Les Porismes* p. 104), the method could not have been applied by them to much effect, had it been even more distinctly formulated than by Serenus.

---

\* For some information about perspective see Poudra's *Histoire de la Perspective ancienne et moderne* (Paris 1864).

# SECTION III.

## KEPLER. DESARGUES. NEWTON.

§ 1. In the interval between the decadence of the Greek school of mathematics and the revival of learning in Europe, the Arabs preserved and commented upon the works of the old masters in geometry, and also applied themselves with effect to the new science of **Algebra**,* which was so greatly to enlarge the domain of geometry itself in the hands of the followers of Descartes. Referring for supplementary information from this point onward to Montucla's *Histoire des Mathématiques*† and Chasles' *Aperçu historique des Méthodes en Géométrie*, we pass at once to the astronomer **Kepler**, who by his contributions to the doctrine of the infinite and the infinitesimal and his firm grasp of the principle of continuity is entitled to the foremost rank amongst the founders of the modern geometry.

<small>Kepler 1571—1630.</small> § 2. Kepler was born at Weil in the duchy of Würtemberg (whither his parents had migrated from Nürnberg) on the 27th of December 1571, and died at Ratisbon in the sixtieth year of his age in 1630. A full account of his life is appended to Frisch's edition of his collected works‡ (vol. VIII. 669—1028). A famous contemporary's description makes him a man of varied ability but superficial and never incubating long over one discovery, "ingenii optimi, nec uni tantum rei dediti, sed universim plura complectentis, ut et pluribus sese tradidit. Neque diu et constanter, plures ob causas, tanquam ovis gallina

---

\* For the literature of this subject see *Elementary Algebra with brief notices of its history* (Lond. 1879) by Mr. R. Potts, the editor of Euclid's *Elements*.

† A work in two volumes 4to (Paris 1758), afterwards expanded by De la Lande into four (Paris 1799—1802).

‡ Joannis Kepleri astronomi *Opera Omnia* (Francof. a. M. 1858—70).

uni invento incubare ipsi licuit; inde fieri potuit id quod in proverbio dicitur, *pluribus intentus, minor ad singula sensus*; et sic quædam, præsertim autem Archimedearum demonstrationum vim minus accurato judicio perpendisse videri possit" (Kepleri *Op.* IV. 647). It is interesting to pass from this to the evidence of his profound **insight** into the abstract principles of geometry, and the **indomitable** perseverance with which he established the laws of **planetary motion** that have immortalised his name. Though his work might not be recognised in his lifetime, it could afford to wait "**centum** annos" for an interpreter.*

§3. *The principle of Analogy.*

The work of Kepler entitled *Ad Vitellionem*† *paralipomena quibus Astronomiæ pars Optica traditur* (Francofurti **1604**) contains a short discussion *De* **Coni** *sectionibus* (cap. IV. §4 pp. 92—6) from the point of view of **analogy or continuity.** The section of a cone by a plane "aut est **Recta,** aut **Circulus,** aut **Parabole** aut **Hyperbole** aut **Ellipsis.**" Of all hyperbolas "obtusissima est linea recta, acutissima **parabole;**" and of all ellipses "acutissima est parabole, obtusissima **circulus.**" The parabola is thus intermediate in its nature to the hyperbola and "**recta**" (or line-pair) on the one hand, and the closed curves the ellipse and the circle on the other; "infinita enim & ipsa est, sed finitionem ex altera parte affectat." He then goes on to speak of certain points related to the sections, "quæ definitionem certam **habent,** *nomen* **nullum,** nisi pro nomine definitionem aut proprietatem aliquam **usurpes.**" The lines from these points to any point on the curve make equal **angles with the tangent** thereat: "Nos lucis causa & oculis in Mechanicam **intentis ea** puncta **Focos** appellabimus." He would have called them *centres* if that term had not been already appropriated. In the circle there is one **focus,** coincident with the centre; in the ellipse or hyperbola **two,** equidistant from the centre: in the parabola one **within the** section, "*alter vel extra vel intra sectionem in* **axe**

The foci named.

---

* See lib. V. *Harmonices Mundi,* with Chasles' account of the work (Aperçu historique p. 482).

† *Opticæ* **Thesaurus.** ALHAZENI *Arabis* **libri** VII. *item* VITELLIONIS *libri* **X** (Basil. 1572).

*fingendus est infinito* **intervallo a** *priore remotus, adeo ut educta HG vel IG\* ex illo cæco foco* **in** *quodcunque punctum sectionis G sit axi DK parallelos.*"

In the **circle the focus recedes as far as** possible from the **nearest part of the** circumference, in the ellipse somewhat less, **in the parabola** much less; whilst **in the** line-pair the "**focus,**" **as he still calls** it to complete the analogy, falls upon **the line itself.** Thus in the **two** extreme cases of the circle and **the line-pair** the two foci coincide. He then goes on to compare the **latus rectum** and its intercept on the axis, or as he calls them the *chorda* and *sagitta*, in the several sections, concluding **with the case** of the line-pair, in which the **chord** *coincides with its arc*, "**abusive** sic dicto, **cum recta linea sit.**" But our geometrical expressions must be subject to **analogy,** "*plurimum namque amo analogias, fidelissimos* **meos magistros,** *omnium naturæ arcanorum conscios.*" And especial regard is **to be** had to these analogies in geometry, since they **comprise,** in however paradoxical terms, **an** infinity of cases lying **between opposite** extremes, "totamque **rei alicujus** essentiam luculenter ponunt ob oculos." **Lastly he shews how to describe** an ellipse by means **of a string fixed at the foci,** without the use of the clumsy **compasses [p. 178],** "quibus aliqui cudendis admirationem hominum venantur," and gives the corresponding constructions for the hyperbola and the parabola.

**Continuity.** (1) Hereupon be it remarked, that the principle of Analogy on which he insists so fervently is the archetype of the principle of Continuity. The one term expresses the inner resemblance of contrasted figures $A$ and $B$, which **are connected by** innumerable intermediate forms; whilst **the other** expresses the possibility of passing through those intermediate forms from $A$ to $B$, **without any change** *per saltum*. Geometry was *not* indebted to Algebra for the suggestion of the law of continuity.

Second focus of parabola. (2) **Having traced the transition** from the line-pair to the circle through **the three standard forms of** conics, he completes

---

\* The figure indicates that the line from the further **focus may be considered to** lie either within or without the parabola.

the theory of the points henceforth named Foci by the discovery of the "cæcus focus" of the parabola, which is to be taken at infinity on the axis either *without* or *within* the curve. The parabola may therefore be regarded indifferently as a hyperbola, having (relatively to either of its branches) one external and one internal focus, or as an ellipse, having both foci within the curve.

(3) **The further focus of** the parabola being taken at infinity on the axis in either direction, the two **opposite** extremities of **every infinite** straight line are thus **regarded as** coincident or consecutive points—a conception which has been **shewn** to conduct logically to the **idea** of imaginary points [p. 311]. Opposite infinities adjacent.

(4) Every straight line **from** the "cæcus focus" of the parabola to a point on **the curve** being said to be parallel to the axis, the idea of the concurrence of parallel lines at a point at infinity has at length been formed and announced.* It is to be noticed **that the new** doctrine of parallels **is here** presented in relation to one plane, and not as springing **out of** the consideration of figures in perspective in space. Parallels concurrent.

Such were Kepler's most original contributions to pure **geometry,** although he is better known by his continuation **of the work of** Archimedes in stereometry.

## §4. *Nova Stereometria doliorum vinariorum.*

Of this work (anno 1615, Lincii. *Op.* IV. 545–646 ed. Frisch) we notice chiefly **the former part, which contains a new** and abbreviated redaction of the **work of Archimedes on the** circle and in stereometry, followed **by** *Supplementum ad Archimedem* (p. 574). The circuitous method **of** exhaustions is **here** transformed into the method of infinitesimals. Thus in theor. 1, on **the** approximation to $\pi$, he treats an infinitesimal arc as **a** straight line: "Licet autem argumentari de *EB* ut de recta,

---

\* It was in the course of an attempt to trace the origin of the term Focus of a conic that I came upon the passage quoted from the *Paralipomena*. Chasles (*Les Porismes* p. 104) attributes the discovery of the concurrence of parallels to Desargues; and when he says (*Aperçu* p. 56. Cf. pp. 15, 16, 61) that Kepler first introduced "l'usage de l'*infini*" into geometry, he is referring " aux méthodes infinitésimales."

quia vis demonstrationis **secat circulum in arcus** minimos qui *æquiparantur rectis.*" In theor. 2 in like manner he regards the circle as an aggregation of triangular elements, having a common vertex at the centre, and their bases coincident with successive small arcs of the circumference. So the sphere is considered in theor. 11 to be made up of small cones, having their vertices at the centre and their bases, "quarum vicem sustinent *puncta*," on the surface of the sphere. He then passes on to the conoids &c., and thence to the solids generated in a certain way by conics, the generating curve being attached at right angles to a plane, which turns about one of its own points without change of place. He gives a slight *résumé* of his doctrine of the foci, mentioning the further focus and likewise the centre of the parabola, but not in such a way as bring out the idea of the concurrence of parallels (p. 577).

§ 5. Guldinus, quoted by Frisch (Kepleri *Op.* IV. 647), stoutly opposed Kepler's *æquiparatio* of an arc to a chord, as not permissible "per ullam ullius demonstrationis geometricæ vim"; precisely as it was objected to Antipho [p. xxx], who had made bold to do likewise some 2000 years earlier, that "he did not start from geometrical principles." It could not however be denied that Kepler's method was of service in discovering theorems, although by no means to be recommended as a method of proof—at least, if any better could be found, "si alia suppetant geometris jam probata media" (p. 653). Kepler had in reality grasped the idea of the infinitesimal, although a *calculus* remained still to be discovered. The law of continuity is now applied by him not only to the infinitely great but to the infinitely small. He has formed the conception of the continuous change of a variable: "crescit a quantita nulla *continue* &c," and discovered the law of its variation in the passage through a maximum value (Pt. II. theor. 16—22); thus laying a firm foundation for the fluxional calculus of Newton, better known by the name and with the notation proposed by Leibnitz.

§ 6. The name of Girard Desargues of Lyons (1593—1662) had fallen into oblivion, when early in the present century his

*Desargues 1593—1662.*

genius was recognised by Brianchon (1817) and Poncelet (1822). A further appreciative notice in the *Aperçu historique* of his apparently lost works was followed by Chasles' own discovery of the chief of them, *Brouillon Proiect etc.*, which with others afterwards discovered was published by Poudra in his *Œuvres de Desargues* (Paris 1864). This edition contains a biographical notice of Desargues, his recovered works with an analysis of each, and an analysis of those of his pupil Abraham Bosse, themselves founded upon the ideas of the master. He proclaims himself not in the first instance a pure mathematician, avowing that he had never a taste for study or research except with a view to some practical application, "au bien et commodité de la vie." He was an architect and engineer, and in the latter capacity served under Cardinal Richelieu at the siege of La Rochelle (1628). After the war he retired to Paris, where he devoted himself to geometry and its applications, frequenting a weekly gathering of savants for the discussion of mathematical topics, which preceded the foundation of the *Académie des Sciences* (*Œuvres* I. 14). He was esteemed by the ablest of his contemporaries as a geometer second to none, but virulently attacked by some important persons[*] of smaller calibre, who were confounded by the novelty and abstraction of his ideas. The subsequent neglect of his works was due partly to the form in which they were written, but in far greater measure to the counter-attraction of the algebraic geometry of his contemporary and friend Descartes. For full information about his works, which include Perspective, Coupe des Pierres, Gnomonique, a fragment on gravitation (I. 239) &c., we can only refer to Poudra's excellent edition; but it will be seen from the following slight account of some portions of them that the *Géométrie Projective* of the present day is in fact the geometry of Desargues.

§ 7. In his earliest work, *Méthode Universelle de mettre en Perspective &c.* (1636), he notices the cases in which concurrent lines are seen as parallels on the *tableau* (tome I. pp. 83, 94),

---

[*] The hostile critique of M. de Beaugrand, secrétaire du Roi, "est le premier écrit qui a servi au général Poncelet à reconnaître le mérite de Desargues" (*Œuvres* II. 353).

and concludes with the problem, to find the lines in a conic which correspond to the axes of its projection. Purposing to return to this work shortly, we pass on to the BROUILLON PROIECT *etc.* [p. 261], or *rough sketch* of a theory of the intersection of a cone by a plane (pp. 97–242, and 243–302). He commences with the new doctrine of infinity.* The opposite points at infinity on a right line are coincident, parallel lines meet at a point at infinity, and parallel planes on a line at infinity (pp. 103–6, 229, 245–6). A straight line may be regarded as a circle whose centre is at infinity (pp. 108, 224). The theorem of the six segments found in the Almagest and elsewhere is stated in a converse form (p. 256). The theory of *Involution de six points*, with its special cases, is fully laid down, and the projective property of pencils in involution is established† (pp. 246–61). The theory of polar lines is expounded, and its analogue in space suggested (pp. 263–6, 271–7, 214, 291). A tangent is a limiting case of a secant (pp. 262, 274, 277), and an asymptote is a tangent at infinity (pp. 197, 210). The joins of four points in a plane determine three couples in involution on any transversal (p. 266), and any conic through the four points determines a couple in involution with any two of the former (p. 267). The points of concourse of the diagonals and the two pairs of opposite sides of any quadrilateral inscribed in a conic are a conjugate triad with respect to the conic, and when one of the three points is at infinity its polar is a diameter (pp. 188–9); but he does not explain the case in which the quadrilateral is a parallelogram, although he had formed the conception of a straight line wholly at infinity (p. 265).

"Mais voicy dans une proposition comme un assemblage abregé de tout ce qui precède" (pp. 195, 277). Thus he introduces the general theory of projection, which is the main subject of the *Brouillon*. Given any conic $O$ and a cone through it, let $O'$ be any section of the cone. Through the vertex $V$

---

* He must have been acquainted with Kepler's theory of the foci. Notice his use of Kepler's term "foyers" (pp. 210, 222).

† Notice his form of expression (p. 104), "*Rangée de points alignez*" [p. 249].

[p. 314] draw a plane parallel to that of $O'$ meeting the plane of $O$ in $ab$, and let $P$ be the pole of $ab$ with respect to $O$. Then the system of planes through $VP$ determine the *diameters* of $O'$, the centre being considered, as we should say, to be the pole of the line at infinity; and any two such planes drawn through points on $ab$ conjugate with respect to $O$ determine *conjugate diameters* of $O'$, the tangents at the extremities of which, "à distance ou finie ou infinie" are also known (p. 197). Since a point at which $O$ meets $ab$ coincides with its own conjugate, an asymptote of $O'$ (besides being a tangent at infinity) is a double or self-conjugate diameter. He concludes: "Comme entr' autre, que sur la quelconque de ces coupes de rouleau peut estre construit un rouleau qui sera coupé selon quelconque espèce de coupe donnée" (p. 198). The ancients had always taken a circle for the base of their cone, and had drawn all planes of section at right angles to one and the same fixed plane.

The Foci of a conic are determined *in plano* as the intersections of the axis with a certain circle, which may have for diameter the intercept on any tangent (or on an asymptote, p. 288) by the tangents at the vertices, in accordance with Apollonii *Conica* III. 45 [p. 111]. He determines the axes and foci of a conic in the cone by a process (pp. 215–23, 293) which Chasles summarises as follows:[*]

The line $ab$ being drawn as above, take any point $t$ upon it, and let the chord of contact of the tangents from $t$ to $O$ meet $ab$ in $t'$. Also let $rr'$ be any segment of $ab$ which subtends a right angle at $V$. The two sets of points $tt'$ and $rr'$ constitute two involutions, having one segment $cc'$ in common. The polars $X$, $X'$ of the points $c$, $c'$ correspond to the Axes of $O'$. The tangents to $O$ from the points $r$ and the lines from $r'$ to their several points of contact determine upon $X$ an involution, whose double points correspond to the Foci of $O'$, since every tangent and its normal are harmonic conjugates to the focal distances of the point of contact.

---

[*] *Rapport sur les progrès de la Géométrie* p. 305.

§ 8. A sequel to **Desargues'** *Perspective* of 1636, found in the *Perspective* of Bosse (1648), contains some explanations of the principles of the former work.

*a. Proposition fondamentale de la pratique de la Perspective.*

The statement and proof by Desargues (*Œuvres* I. 403—7) are analysed by the editor, who reduces his fundamental proposition to the anharmonic property of a pencil of four rays (p. 425), which cut any transversal in a constant cross ratio.

*b. Figures in homology.*

Three other geometrical propositions are given, which embody the principle of perspective in one plane (pp. 413—22, 430—5). The first is on triangles in homology [p. 307, 321], the second and third on quadrilaterals in homology. On the second he remarks generally that a like reduction of the figure to one plane may be used "en semblable cas" (p. 417); and in the third he gives a metrical relation between a system of corresponding segments (p. 435). Notice that he passes from solid to plane figures in the manner afterwards used by the school of Monge (Chasles' *Aperçu historique* p. 87).

School of Desargues.

§ 9. Although the roughly sketched essays of Desargues themselves fell into neglect, his ideas were preserved by an illustrious school of disciples, numbering amongst its members **Bosse, Pascal** and **De la Hire**. The writings of the engraver **Abraham Bosse** (1643—1667) are analysed by Poudra in vol. II. of the *Œuvres de Desargues*. The famous Hexagrammum Mysticum of Pascal was a corollary from what he had learned from Desargues. The theory of polars was brought into prominence by De la Hire (1685), and forthwith supposed to have been discovered by him. The reader of Brianchon's *Mémoire sur les Lignes du Second Ordre* (Paris 1817), and Poncelet's *Traité des Propriétés Projectives* will not need to be reminded how great a part of modern geometry is actually and confessedly founded on the work of Desargues.

Newton 1642—1727.

§ 10. **Newton** was born at Woolsthorpe near Grantham in 1642, the year of the death of Galileo, and died in the eighty-fifth year of his age in 1727. The first edition of his *Philosophiæ* *naturalis Principia mathematica* was published 1687, **the second,** edited by Roger Cotes, in 1713, and the third, **by Henry Pemberton** in 1726. His *Opticks* was published in English in 1704 and in Latin in 1706, in each case with an *Appendix* in Latin containing the *Enumeratio Linearum Tertii Ordinis* and a tract *De Quadratura Curvarum*. **For an account of** his life and writings see **Brewster's** *Memoirs of the Life etc. of Sir Isaac Newton* (Edinburgh **1855), and Edleston's** *Correspondence of* **Sir Isaac Newton and Professor** *Cotes* (London **1850**); and for **the works themselves see** Horsley's five volumes, 1779–85 **[p. 264], and the** *Newtoni Opuscula* (3 vols.) **of** Joh. Castilloneus (Laus. et Genevæ 1744). **Presupposing** a general acquaintance with his geometrical discoveries, we **shall** confine **our attention to a few particulars.**

**§ 11.** In the fourth and fifth sections of the first book of the *Principia* he solves various forms of the problem, to describe a conic subject **to the equivalent of five conditions, (1) when a focus is given, and (2) when** neither **focus** is given. **It will** suffice to allude briefly in passing to the former case. **The title** of lib. I. sect. 4 is *De inventione orbium ellipticorum, parabolicorum & hyperbolicorum ex umbilico dato*. In it he makes much **use of the simple property** (lemma 15), that **the perpendicular from one focus of a conic to any tangent** intercepts a length equal **to the axis on the further focal distance** of the point of contact. The section concludes with the construction of an orbit of which **one focus and three points are given, a** problem which had been solved, " **Methodo haud multum dissimili** " **by** de la Hire, *Sect. Conic.* lib. **VIII. prop. 25. In this** construction and in prop. 20 Newton assumes **that the focal** distance of a **point on a** fixed conic varies as its distance from **the** directrix, a theorem proved in **the** *Arithmetica Universalis* prob. 24 (Cantab. 1707), and sometimes attributed to Newton as its first discoverer, **although** in reality known **to** Pappus.

**§ 12.** *Inventio orbium ubi umbilicus neuter datur.*

A.

The 5th section of the **first book** of the *Principia*, under the above title, treats with the utmost generality of the *point-*

*e*

*properties* and the *tangent-properties* of conics. It commences with the problem of the *Locus ad quatuor lineas* (lemmas 17—19), of which no geometrical solution was extant [p. 266]. Then follows a theorem (lemma 20) which may be thus stated: If $AB\ PC$ be four fixed points on a given conic, the chords from $B$ and $C$ to a variable point on the curve meet the parallels through $P$ to $AB$ and $AC$ respectively in points $T$ and $R$, such that $PT$ varies as $PR$, and conversely. From a limiting case of this lemma he deduces his organic description of a conic by means of two rotating angles* (lemma 21), giving somewhat later (prop. 27 Scholium) his construction for the centre and asymptotes of the conic thus generated. By means of the above mentioned lemmas he shews how to describe a conic when five points on it are given,† or four points and a tangent, or three points and two tangents (props. 22—4). Next follows lemma 22, *Figuras in alias ejusdem generis mutare*, in which it is shewn that any curve may be transformed into another of the same order by substitutions of the form $X = \dfrac{ab}{x}$ and $Y = \dfrac{ay}{x}$ [p. 330], and two applications of the lemma follow. (1) In order to describe a conic passing through two given points and touching three given lines, he transforms two of the given lines into parallels, and the third given line and the join of the given points into parallels (prop. 25); and (2) to describe a conic passing through a given point and touching four given lines, he transforms the four lines into the sides of a parallelogram (prop. 26).

### B.

The lemmas next following lead up to some important properties of the tangents to conics, the discovery of which by Newton is commonly overlooked. First it is shewn that if $AC$ and $BD$ be lines given in position, terminated at $A$ and $B$, and having a given ratio to one another, the locus of the point which divides $CD$ in a given ratio is a straight line‡ (lemma 23).

---

\* Cf. Ex. 855 [p. 558]. The equation to the locus in lemma 21 is given in the *Arithmetica Universalis* prob. 53 (Cantab. 1707).

† A solution of this case is found in Pappi *Collectio* (p. 1077 ed. Hultsch).

‡ It also divides $AB$ in the same ratio, since $AC$ and $BD$ vanish together.

## SECTION III.        lxvii

Next, if two given parallel tangents **viz.** at $A$ and $B$ to a conic be cut by any third tangent in $M$ and $I$ respectively, the semi-diameter parallel to the two tangents is a mean proportional to $AM$ and $BI$ (**lemma 24**). From this, which is identical with **lib. III. prop. 42** of the *Conics* of Apollonius, **he** deduces two corollaries: (1) if any fourth tangent meet $AM$, $BI$, $MI$ in $F$, $Q$, $E$ respectively, then

$$AM : BQ = AF : BI = MF : IQ = ME : EI;$$

from the first of which proportions it follows (2) that $FI$ and $MQ$ intersect upon $AB$.* The next lemma and its **corollaries are** of **peculiar** importance **in relation** to the modern geometry.

### Lemma XXV.

If $ML$, $LK$, $KI$, $IM$ be the sides of a parallelogram touching a conic in $A$, $C$, $B$, $D$ respectively, and if any fifth tangent cut them in $F$, $H$, $Q$, $E$ respectively, then by lemma 24 cor. 1,

$$ME : EI = AM : BQ = BK : BQ;$$

or $\qquad ME : MI = BK : BK + BQ = BK : KQ.$

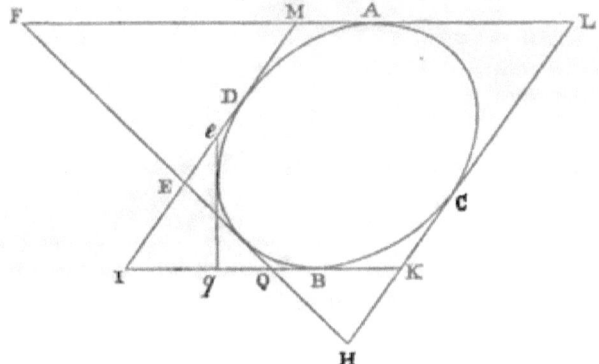

In like manner

$$KH : HL = BK : AF = AM : AF,$$

or $\qquad KH : KL = AM : AF - AM = AM : MF.$

---

* It is easy to generalise this result by transforming the parallel tangents into non-parallels by Newton's method. Cf. Art. 121 Cor. [p. 276].

e 2

### Corollary 1.

If the parallelogram and the conic be fixed and the fifth tangent variable, then

$$KQ.ME = MI.BK = \text{a constant,}$$

and $KH.MF$ has the *same* constant value. The result may be expressed in words as follows: *given two tangents to a fixed conic, the product of the intercepts upon them between the diameter parallel to their chord of contact and any third tangent is constant.* The relation between the intercepts $IE$ and $IQ$ is of the form $(IM - IE)(IK - IQ) = $ a constant, or

$$a.IE.IQ + b.IE + c.IQ + d = 0,$$

which is the "tangential equation" to any conic referred to any two fixed tangents, and also expresses that *any two given tangents are cut homographically by a variable third* tangent.

### Corollary 2.

If $eq$ be any other position of the tangent $E$, it follows that

$$KQ : Me = Kq : ME = Qq : Ee.$$

### Corollary 3.

Since $QK$ is to $eM$ as $Qq$ to $eE$, it follows by lemma 23 that the middle points of $KM$ and $qE$ are in a straight line with the middle point of $Qe$. Hence, the middle point of $KM$ being the centre of the conic, *if a conic be inscribed in a quadrilateral, its centre lies on the straight line bisecting any pair of the diagonals of the quadrilateral.*[*] This theorem suggested to Brianchon and Poncelet the investigation of the centre-locus of a conic passing through four given points [p. 282], and prepared the way for the general consideration of *systems of conics subject to four conditions.*

### C.

In prop. 27 the conic touching five given lines is described,

---

[*] It may or may not have occurred to Newton that this theorem might be generalised by projection; but in any case he would not have turned aside to notice results so distantly related to his *Inventio orbium.* It may be shewn that he must have been acquainted with the theory of Polars.

its centre being first determined **as the point of concourse of the** diameters of any two of the five quadrilaterals formed **by the** given tangents. In a Scholium he remarks that an **asymptote is** a *tangent at infinity*, and also **shews how to** determine **the axes and foci of** a conic described by the organic method of lemma 21. There are also other lemmas which he might have used for the construction **of conics, such as that the locus of the** middle point **of a chord drawn through a fixed point to a conic is** a parallel conic: "sed propero ad magis utilia." In lemma 26 and a **corollary** it is shewn how to describe **a triangle of** given species **and** magnitude having its vertices severally on three given lines, **and how** to draw a transversal the intercepts upon which **by** three given lines shall be of given lengths. This lemma **is used** in prop. 28. In lemma 27 and a corollary it is **shewn how to** describe **a trapezium of given species having its vertices** severally **on four given lines, and how to draw a transversal** which shall **be cut in given ratios by four given lines, which is a special case of section in a constant** *cross ratio* [p. 296]. An application follows in prop. 29, and the section **concludes:** "Hactenus de orbibus inveniendis. Superest ut motus **corporum** in orbibus inventis determinemus."

### §13. *Curvarum Descriptio Organica.*

A well known generalisation of Newton's description of a conic by angles would certainly have been passed **by** in the *Principia* with a "propero ad magis utilia," since it merely shews how to describe a conic by **assuming that a conic is already** described. When however he is treating **of pure mathematics, he** extends his method to the utmost, applying **it not merely to cubics,** as in Ex. 760, but to curves of all orders having "double" points. In the case of a cubic, **of which a** *double point* $A$ and six other points $BCDEFG$ are given, let the angle $CAB$ turn about $A$ and the angle $ABC$ about $B$; then as the intersection $C$ of the arms $AC$, $BC$ assumes **the new** positions $DEFG$, the intersection $I$ of the other arms determines four other points, say $PQRS$. Draw the conic $APQRS$, and let $I$ move round its circumference: **then** $C$ traces the cubic as required (*Opticks*, II. 160, 1704).

If instead of the point $C$ a *tangent* $BC$ be given, the angle $CAB$ vanishes, and the curve is described by means of one finite angle and a straight line, which latter moves parallel to itself when the fixed point through which it passes is at infinity.* How was Newton led to his organic description of conics and other curves? Possibly he took a suggestion from Euclid III. 21 [p. 172], and first described a *circle* by means of an angle and a line parallel to one of its arms, or by two angles having one pair of their arms constantly parallel.

§ 14. *Proof and extension of Newton's Descriptio Organica.*

Let two angles $AOB$ and $A\omega B$ of given magnitudes turn about $O$ and $\omega$ respectively, and let the intersection $A$ trace a curve of the $n$th order. For a given position of the arm $OB$ there are $n$ positions of $A$ and therefore $n$ of $B$. When $OB$ is in the position $O\omega$ the $n$ $B$'s coincide with $\omega$, which is therefore an $n$-fold point on the locus of $B$, as is also the point $O$; and since any line through $O$ (or $\omega$) meets the locus of $B$ in $n$ other points, the locus is of the order $2n$. Its order is the same when $A\omega B$ is a zero-angle or straight line.

Let a given trihedral angle $O$ ($ABC$)—or a plane $OBC$ and a line $OA$ rigidly attached to it—turn about $O$, and let a variable plane through a fixed point $\omega$ meet $OA$ in $A$ and the plane $OBC$ in $BC$; then if the line $BC$ describes a ruled surface of the order $n$ the point $A$ describes a surface of the order $4n$.† For a given position of the line $BC$ the locus of $A$ is a conic, and when the director surface is a cone of the $n$th order every plane through $\omega$ and its vertex meets the surface which is the locus of $A$ in $n$ conics.

When the director is a plane, $BC$ must be made to pass through a fixed point or touch some curve in it, *except* in the case in which $OA$ is normal to the plane $OBC$. In the last case the locus of $A$ is a quadric, which becomes a sphere when the director is at infinity.

---

\* For some earlier essays at **Descriptio Organica** see Chasles' *Aperçu* p. 100.

† For the determination of the order of the surface described by the point $A$ I am indebted to Professor Cayley.

## SECTION III.

§ 15. Colin Mac Laurin developed Newton's theory of curves in his *Geometrica Organica, sive descriptio curvarum universalis* [p. 345]. He also wrote a treatise on *Fluxions*\* (Edinb. 1742). The work *Algebra with an Appendix etc* [p. 128] was first published, after the death of the author, in 1748, and in the same year *An Account of Sir Isaac Newton's Philosophical Discoveries*. The *Appendix* to the Algebra (pref. p. xi) was founded on Cotes' theorem of harmonic means, of which further use has been made by Poncelet and other modern geometers (Salmon's *Higher Plane Curves* § 132 1879).

*Maclaurin 1698—1746.*

§ 16. The property of the focus, directrix and determining ratio remaining buried in the *Collectio* of Pappus, modern writers looked to later works for the first notice of the focus of the parabola and the directrix of the general conic. Thus Robertson in his *Sect. Conic.* pp. 340, 363 (Oxon. 1792) refers for the focus of the parabola to an anonymous work *De Speculo Ustorio*—known to Roger Bacon, and perhaps translated from the Arabic—which was published at Louvain in 1548. Gregory St. Vincent knew the property of the directrix for the case of the parabola, and virtually arrived at it for the ellipse (*Opus Geomet.* lib. IV. prop. 139, p. 317, 1647), in the form of Ex. 16 [p. 35]; as did De la Hire for the hyperbola, measuring distances from the directrix parallel to an asymptote [p. 155 Ex. 397], in his *Sect. Conic* lib. VIII. prop. 18 (1685). In prop. 25 De la Hire introduces the directrix as the polar of the focus: in props. 23-4 he had proved that the tangents at the extremities of any focal or other chord subtend equal angles at the focus. It remained for Newton to bring the property of the determining ratio fully to light, and for Boscovich, with

*Ratio determinans.*

---

\* On the rival claim of Leibnitz to the first discovery of the differential calculus see Montucla *Hist. des Math.* vol. II. 330—343 (1758); Gerhardt's *Hist. et Origo Calc. Diff.* (Hannover 1846) and other publications; Brewster's *Life of Newton* II. 23—47 (1855); Weissenborn *Die Principien der höheren Analysis* (Halle 1859); Sloman *The claim of Leibnitz to the invention of the Differential Calculus*, translated from the German (Cambridge 1860). It is disputed to what extent Leibnitz was indebted to the letters and MSS. of Newton. Leibnitz several times discovered things already in print (*Biographie Universelle* XXIII. 627—8, 634); and it is a striking fact that the leading propositions of the PRINCIPIA reappeared under the name of G. G. L. in the *Acta Eruditorum* pp. 82—96 (cf. p. 36), 1689.

the help of it, to compose the first really complete *Sectionum Conicarum Elementa*. The term Directrix, formerly used in the case of the parabola, is now used of the general conic (p. viii.); as in the following year also (1758) by Hugh Hamilton, who added a construction for the focus and directrix in the Cone [p. 204]; but the term was still used only for the parabola by Le Seur and Jacquier (1760) in their edit. 2 of the *Principia* (I. 134, 179), and the characteristic property of the line itself in the general conic was not familiarly known even some years later, to judge from Lexell's elaborate discovery of the simplest of corollaries from it in the *Nova Acta Petropol.* I. (147), 1787.

Later works founded upon the properties of the determining ratio and the eccentric circle [p. 3] were those of Thos. Newton (Camb. 1794), G. Walker of Nottingham (Lond. 1794), and John Leslie (Edinb. 1821), who describes Thos. Newton's work as "clear, neat, and concise," whilst Walker's " though ingenious and strictly geometrical, is unfortunately so prolix and ponderous as to damp the ardour of the most resolute student." Walker was under the impression that he was the first discoverer of what he called the Generating Circle,* but Thos. Newton rightly referred it to Boscovich. Leslie's account of the work of Boscovich† is that it consisted "of only a few propositions, but drawn out into a string of corollaries." It is nevertheless a clear and compact treatise, which for simplicity, depth and suggestiveness will not readily be surpassed.

---

\* Compare Mr. S. A. Renshaw's *The Cone and its Sections* pp. 26—8 (Lond. 1875); *Messenger of Mathematics* vol. II. 97 (1873). Walker's complete treatise was to be in five books, but one only appeared.

† From the preface to vol. III. of the *Elementa Univ. Math.* (1757) we gather that Boscovich's scheme of Conics was first published "in Romano litteratorum diario ad annum 1746," in the form of a short article, "schediasma brevissimum"; and that several years elapsed before the complete work, after repeated delays, was given to the world. Leslie says that it was published in 1744, and Walker's in 1795. Walker had discovered the generating circle "near thirty years" before publication. An edition of Boscovich's *Elementa* seems to have been published at Rome in his name in 1752—3; but I have only seen the "editio prima Veneta."

## SECTION IV.

### MODERN GEOMETRY.

§ 1. ALTHOUGH the law of Continuity, the vital principle of *Continuity.* the modern geometry, had been decisively laid down by Kepler, it was not until the great discovery by Poncelet of the circular points at infinity in any plane that it came to be universally acknowledged. The principle as enunciated by Kepler was wholly independent of algebraical considerations, but its later developments were suggested by the occurrence of negative and imaginary roots in equations applied to geometry, whilst the discovery of the differential calculus gave a new zest to speculations *De infinito* and *De nihilo*. The earliest thorough and geometrical treatment of the subject with which we are acquainted is found in Boscovich's appendix* to his *Sectionum Conicarum Elementa* [p. 311], of which a slight account is given below. The complete dissertation occupies more than two-thirds as much space as is devoted to the entire subject of conics *in plano*. The writer cautiously abstains from the too bold assertion of novelty in his speculations, but remarks that the essay contains many things which "ego quidem nusquam alibi offendi," and many which, although found elsewhere, "nusquam ego quidem ad certos reperi redacta canones, et geometrica methodo pertractata."

§ 2. His first principle is, that every member of a geome- *Boscovich 1711—1787.* trically defined locus must have the same nature and properties,†

---

\* *Elementa Univ. Math.* tom. III. 228–356 (Venet. 1757).
† See Chasles' *Aperçu historique* p. 197 on Monge's use of the principle of *contingent relations*.

which are wrapped up in **the definition itself, so that** whatsoever is demonstrable **of one part of the locus should be** demonstrable in like manner **of every** other. On this principle we conclude *a priori* from the nature of the problem, to trisect a given circular arc, **that any construction must give a series of solutions, three of them geometrically** distinct **[p. 141].** In geometrical demonstrations a **determinate** configuration is present to the senses, but the reasoning applies to an infinity of cases. **This is** clearly seen when, for example, we bisect a given finite **straight** line; but it is true none the **less** in cases in which **a new** configuration seems to render **the** proof nugatory, **although some artifice may** be **required,** "ad servandam analogiam, et **retinendam** solutionis ac **demonstrationis vim**" (p. 229). Notice his **use** of the term Analogy, **by which** the idea of geometrical continuity was first expressed.

The correspondence of change of sign with change of **direction in lines carries with it the** idea of negative rectangles and squares, and thus of imaginary magnitudes (pp. 234, 308). Change of sign implies a transition through **zero** or *infinity*, and never takes place *per saltum* (p. 250). To illustrate this, take an indefinite line $AB$ **and a point** $C$ **without it, draw** $CH$ **perpendicular to** $AB$, **and let a line** turning continuously about $C$ meet $AB$ in $P$. **As** $CP$ passes through $H$, the sign of $HP$ changes, say **from** positive to negative: when $CP$ becomes parallel to $AB$, the point $P$ is at infinity on the *negative* side **of** $H$, and the next instant it is at infinity on the positive side of $H$. Thus the passage through infinity carries with it a change of sign and, like the passage through zero, is **effected** by the *continuous* rotation of $CP$, **and not** *per saltum*. The opposite extremities **of an** infinite straight line are thus in **a** manner joined, **as if the line were an** infinite circle (p. 254), whose **centre may be considered to** be at infinity on either side of the line. In illustration of the principle that opposite infinities are thus adjacent, take the case of an infinite double ordinate **to the axis or any** diameter of the parabola, regarded as a **closed curve (pp. 265,** 343).

The consideration of a circle of infinite radius leads to the **idea of a** "veluti plus quam infinita extensio" (pp. xv, 281).

Through fixed points $AC$ on an indefinite straight line $MACN$ draw a circle, and bisect its minor segment $AIC$ in $I$ and its major segment in $I'$. **If now the** centre be removed to infinity, **the** arc $AIC$ becomes the line $AC$, whilst of the arc $AI'C$ part coalesces with the infinite segments $AM$ and $CN$, and the rest **recedes to** infinity with the point $I'$, "ut nusquam jam sit;" **or as we may say,** the circle degenerates into the endless line $MACN$ together with the line at infinity* [p. 344]. Hence it is deduced that whilst the line $AC$ is bisected in a point $I$, its complement $AM \infty NC$ is bisected at a point $I'$ at infinity (p. 274); which might also have been arrived at by dividing $AC$ harmonically, and making one point of **section** coalesce with the middle point of $AC$ (pp. 6, 344).

If $AC$ be a segment of an **infinite right line,** the remainder $A \infty C$ may be regarded as its "complementum ad quendam veluti infinitum circulum" (pp. 276, 280, 292, 297 &c.). The hyperbola, regarded as a quasi-ellipse, has for its axis $A \infty A'$ the complement of $AA'$ (pp. 264, 276, 289, 292). The further developments of this idea already given [pp. 18, 22, 102, 153, 311] are in accordance with the views of **Boscovich;** but the note on Art. 13 was written, and a Scholium in **continuation planned** [p. 102], before his dissertation on continuity had been consulted.†

The change from the real to the imaginary state is contingent upon the transition of some element of a figure through zero or infinity, and never takes place *per saltum* (p. 277). Examples of imaginaries are the exterior diameters of a hyperbola, whose **squares are negative;** for the so-called "secondary" axis and **diameters have no real analogy** to the minor axis and conjugate diameters in the ellipse, although the unwary geometer may be imposed upon by the conjugate

---

* Although the arc $AI'C$ seems to lie wholly on one side of the line $AB$, it is to be remembered that *opposite infinities are adjacent.* Thus every line-circle passes through all points at infinity in its plane.

† Without reference to the idea of an infinite line-circle, I had used the term COMPLEMENT of a straight line several years before I was acquainted with any work of Boscovich. See *Oxf. Camb. Dubl. Messenger of Mathematics* IV. 140 (1867).

hyperbola, and persuaded to think that there can be a curve of the second order which a straight line meets in four points (pp. 311—317). In comparing properties related to diameters in the ellipse and in the hyperbola, we should endeavour to bring in the *squares* of the diameters, the signs of certain of which will merely have to be changed in passing from the one curve to the other (p. 320). The general method of procedure in dealing with a geometrical figure one or more of whose elements is evanescent, infinite or imaginary is summed up in eleven CANONS, formally stated and fully illustrated (pp. 284—339). The 5th relates to negative angles, and the 11th to the ratios of infinite magnitudes. He might have added a Canon 12 on the ratios of the Newtonian-nascent or evanescent quantities, but promises another volume when time permits (p. 348); whilst on p. 353 he refers to his former dissertation *De Natura et usu infinitorum et infinite parvorum* (1741). In the course of the essay now under consideration he treats of curves of all orders, their infinite branches and asymptotes, their curvature, cusps, points of inflexion, and the tangents thereat (pp. 245, 267, 270—3, 325).

§3. The general solution of the problem, *given the focus directrix and determining ratio, to find the intersections of an arbitrary line with the conic*, completely determines by implication the nature and properties of the curve (p. 286). His construction is as follows. Take a point $\omega$ exterior to the given line, instead of a point $O$ upon it [p. 10]; draw $\omega Z$ parallel to $PQ$ meeting the directrix in $Z$, and let the parallel through $Z$ to $SR$ meet the eccentric circle of $\omega$ in $p'$ and $q'$. Then the focal radii parallel to $\omega p'$ and $\omega q'$ meet the given line at its intersections with the conic (p. 39). This construction failing (1) when $PQ$ is parallel to the directrix, in which case the line $Zp'q'$ is indeterminate, and (2) when $PQ$ passes through $S$, in which case the parallels through $S$ to $\omega p'$ and $\omega q'$ coalesce with $PQ$, instead of cutting it each in one point; he shews how to meet the difficulties thus arising. The former case however leads only to a simplification when the centre of the circle is taken upon $PQ$, as in the present work. Considering the second case in relation to Art. 16 Cor. 1 [p. 29], let a fixed chord drawn through the focus $S$ be intersected in $\Sigma$ by a variable chord containing a given angle with **the axis.** Then the products of the segments of the two chords are **as** the squares of the parallel tangents, however near $\Sigma$ be taken to $S$, and therefore in the limit the products of the segments of any **two** *focal chords* **are as the squares of** the parallel tangents.

A line still meets a conic in two points, even though, for example, one of them should disappear at **infinity.** In the parabola, **any** two chords as they become infinite **are in a** ratio of equality. In the hyperbola, if from any **two points** $LL'$ parallels be drawn meeting the curve in $PP'$ **and its adjacent** asymptote in $hh'$ **respectively, then as** the **latter points recede** to infinity the intercepts $Ph$ and $P'h'$ **remain finite** [p. 146], and the ratios of $LP$ to $Lh$ and $L'P'$ to $L'h'$ tend to equality as their limit. The infinite segments $LP$ and $L'P'$ are as the distances of $L$ and $L'$ **respectively** from the asymptote $hh'$ (p. 347).

§4. It is remarkable that Boscovich enters upon **these** abstruse speculations in an elementary treatise for beginners, **and even several times** touches upon the subject of the appendix **in the text itself, as for example when** he notices that the properties **of chords of a conic may be** transferred to one of its limiting **forms, the line-pair** (p. 100). **The preface to** the volume contains an **earnest plea for the introduction** of the modern ideas into **the schools.** He had taught the appendix *viva voce* to his own tyros **with the happiest results.** The **mind** of the tyro is commonly **overwhelmed** with a multitude **of details not** reduced to any system; demonstrations are put before him **in** an unsuggestive form which gives no play **to** his inventive faculty; and thus it comes to pass that of **the many students so few** turn out genuine geometers. Let the learner be furnished with principles, and not alone with fully explained **facts, and** be continually stimulated to exertion by the intense **pleasure of finding** something left to discover for himself.

§ 5. The newly founded *Ecole Polytechnique* led the way in the geometrical revival of the nineteenth century. From this source issued first the works of Carnot and Monge, which further illustrated the principles of continuity above described. The leading conception of Carnot's *Géométrie de position* (Paris 1803) is the doctrine of quantities "dites positives et negatives" (p. ii), to which he recurs in his *Essai sur la théorie des Transversales* &c. p. (96), Paris 1806. This essay is in great measure based upon the ancient theorem of the six segments [p. l.].

Referring to the *Aperçu historique* for a good description of the works of Monge, we pass from the master to one of his most illustrious scholars, whose short incisive essays in pursuance of the ideas of Desargues, Pascal and Newton were the prelude to their fuller development by Poncelet, Steiner and Chasles.

Brianchon.

§ 6. Second in importance only to the principle of Continuity is the principle of *Duality*, of which Brianchon's hexagram (1806) occasioned the discovery [p. 290]. Noticing again the important article of Brianchon and Poncelet on the Equilateral Hyperbola [pp. 175, 282], we next come to the separate publications:*

(1) *Mémoire sur les lignes du Second Ordre, faisant suite aux recherches publiées dans les journaux de l'Ecole royale polytechnique.* Par C. J. Brianchon, capitaine d'artillerie, ancien élève de l'Ecole Polytechnique. Paris 1817.

(2) *Application de la théorie des Transversales* (Brianchon. Paris 1818).

The latter memoir consisted of *Leçons données à l'école d'artillerie de la garde royale en mars* 1818. The former, which is of greater interest, must be described in detail. A line of the *Second Order* is defined as the section of any circular cone by an arbitrary plane: the term *projection* is introduced in relation to perspective: *poles* and *polars* are defined: as also is the expression *Géométrie de la règle*. The term polar had been introduced by Gergonne as correlative to "pole," an old

---

* It would be worth while to republish Brianchon's articles and memoirs in one volume.

expression for a fixed point, which was beginning to be used in its restricted modern sense (Gergonne's *Annales* I. 337; III. 297).

Pp. 7—10. The **property** (afterwards called anharmonic) of four radiants is **enunciated, the** case of the harmonic pencil specially noticed—the term *faisceau harmonique* being introduced apparently as new—and it is noticed that the harmonic property holds "pour toutes les projections de la figure," a reference being given to Gregory.St. Vincent's *Opus Geometricum* p. 6, prop. 10 (1647). A fourth harmonic to three given points in a straight line is found "avec la règle seule" by the property of the complete quadrilateral, and a reference for this is given to De la Hire's *Sectiones Conicæ* p. 9, prop. 20 (1685).

Pp. 10—16. Any transversal is cut in involution (1) by the sides and diagonals of a quadrilateral, regarded as the *projection of a parallelogram:* and (2) by these and any circumscribed conic regarded as the *projection of a circle*. The latter theorem (§ XI) was due to Desargues and had been preserved by Pascal. The case in which a conic degenerates into a **line-pair is** noticed. In a note (p. 14) he refers, on the theory of **transversals,** to the works of the ancients—the *Almagest* for example; to Fr. Maurolycus *Opuscula Mathematica* p. 281, 1575; and to Schubert in the *Nova Acta Petropol.* tome xii. ann. 1794.

Pp. 17—28. Pascal's theorem is proved by considering the six points of concourse of a conic with any triangle, and in a note is added the property of triangles in homology. The extension of the theory of polars to quadrics is ascribed to Monge (p. 19), although in reality due to Desargues [p. 329], who would however have thought definitely only of the quadrics of revolution. The theorem that the joins of four points on a conic determine a self-polar triangle with respect to it [p. lxii] is proved, and the reciprocal property of four tangents deduced. Hence follows "une propriété bien remarquable des lignes du second ordre," viz. that *the intercepts on any two fixed tangents by* **the diameter** $MN$ *parallel to their chord of contact and a variable tangent EI have their product $EM.NI$ constant* (§ XXVIII). He does not seem **to** be aware that this is one of Newton's theorems [p. lxviii] although he refers to Newton more than once (pp. 38, 45), has a general acquaintance with lib. I. sect 5 of the *Principia* from which it is taken, and knows the third corollary of the lemma to which it is the first corollary.

Pp. 28—53. He draws conics passing through $n$ points (one or more of which may be at infinity) and touching $5-n$ lines (where $n$ is 0, 1, 2, 3, 4 or 5), referring also to Newton's methods, and in one case to Maclaurin's *Algebra*. He makes use in these constructions of his own property of the hexagram, the reciprocal of Pascal's; and from the two together deduces (p. 35), that the six summits of two triangles touching a conic lie on a conic, and conversely.

Pp. 53—60. The theorem of Desargues (§ XI) "va nous découvrir de nouvelles propriétés des coniques à branches infinies." Take four points $UXYZ$ on a conic, of which $XY$ are at infinity and $U$ variable, then any fixed chord $AB$ is met by $UX$ and $UY$ in points $C$ and $F$, such that the cross ratio $\frac{AC}{AF} : \frac{BC}{BF}$ is constant (p. 54). If $AXYZ$ be fixed points on a parabola or hyperbola, of which $Z$ only is at infinity, and $U$ a variable point on the curve, the lines $UX$ and $UY$ meet $AZ$ at distances from $A$ which are in a constant ratio. By making $U$ coincide with each of the points $XY$, he deduces Ex. 429 [p. 159] and its analogue for the parabola; as also Ex. 427, that the arms of any angle in a fixed segment of a hyperbola intercept a constant length on either asymptote. By means of these results he shews how to describe a hyperbola, having given an asymptote, and in addition three points or a point and two tangents or two points and one tangent. All that remained to bring the *anharmonic point-property of conics* fully to light was a simple application of the method of projection, which the writer had already used with such effect. Knowing so well the importance of the projective property of the anharmonic pencil, it is remarkable that he should have left it for others to take the final step.

"Toutes les proportions contenues dans ce Mémoire se rattachent au théorème XI" (p. 61). Thus the name of Desargues is brought effectually into notice. He also refers to Lambert (*Perspective* ed. 2, 1774), Blondel, Muller &c.

Pp. 61—65. He deduces from Pascal's hexagon **the properties** of similar central conics, and concludes with some new properties of the tangent cones to quadrics. If **a** conic passes through two fixed points $AB$ and touches two given lines, the chord of contact passes through a fixed point on the line $AB$ (p. 20). Hence, if **a** quadric passes through two fixed points $AB$ **and has a given enveloping cone**, the plane of contact passes through **a fixed point on the line** $AB$. A construction is deduced **for a quadric passing through** *four given points* and having a *given enveloping cone*.

§ 7. To Poncelet as a geometer belongs the double **honour** of supplying what was lacking in the **theory of** Continuity **by** his discovery of the focoids [p. **311**],* and bringing to light the principle of Duality by **his method** of reciprocal **polars [p. 346]**; whilst, like Desargues **and Archimedes** before him, he was no less a master of **the** principles and practice **of mechanics.** Born at Metz on the first of July 1788, he was allowed to **grow up** almost without instruction at St. Avold, until in his seventeenth year, at the end of 1804, he entered the Lycée imperial **de** Metz. Three years later he gained his admission to the *Ecole Polytechnique,* **was** employed in 1811 upon the fortifications of **Rammekens in the island** of Walcheren, marched in 1812 with Napoleon **to Moscow, and was taken** prisoner and interned at Saratov on the Volga until the general peace of 1814. In his captivity he set **to work,** in spite of all hindrances, to reconstruct for himself a course **of mathematics, and entered upon** those bold speculations which are the characteristic **of his famous** works **on** geometry. For a full list of **his scientific** publications see Didion's *Notice sur la Vie et les Ouvrages du Gen. J. V. Poncelet* (Paris 1869). The appearance of his *Cours de Mécanique appliquée*, dating in part from 1826, is described **as** having "fait sensation dans le monde de la science et de l'industrie" (**p. 33**). The dominating idea **of** his geometrical works was to increase the **resources** of pure geometry, to generalise its con-

Poncelet
1788—1867.

---

* Poncelet, before Plücker, spoke **of** a conic as having four foci (*Propriétés Projectives* p. 271, 1822).

ceptions and language, and thus to raise it to the level of analysis. See the Introduction to his *Traité des Propriétés Projectives des Figures* edit. 1 p. xxxiii. (1822); edit. 2 tome I p. xxii (1865–6). At the end of the year following he died (Dec. 23 1867), with his thoughts turned again to mechanics: "Ma tête est bonne, et j'espère bien pouvoir cet hiver publier ma Mécanique (p. 45)."

Steiner.

§ 8. One of the leading contributors to the further systematisation and development of geometry was Jacob Steiner, the author of numerous mathematical articles and of the works:

(*a*) *Systematische Entwickelung etc.* 1832 [p. 262].

(*b*) *Die geometrischen Konstructionen, ausgeführt mittelst der geraden Linie und eines festen Kreises* (Berlin 1833).

The work (*a*) was a first instalment of a treatise in five parts, of which no further part appeared in the author's lifetime; but his *Vorlesungen über synthetische Geometrie* were edited posthumously by Geiser and Schröter (Leipzig 1867, 1876). Of the second part of the *Vorlesungen*, containing the projective geometry of conics, the third section is on *Kegelschnittbüschel* and *Kegelschnittschaar*.

Chasles 1793–1880.

§ 9. Michel Chasles, the most famous of living[*] geometers, is perhaps best known as the author of the *Aperçu historique sur l'origine et le développement des Méthodes en Géométrie etc.* (Paris 1837, 1875), which contains an invaluable series of notes on the history of geometry from the earliest times, followed by a *Mémoire de Géométrie* (pp. 573–848) devoted to the exposition of the two general principles of Duality and Homography. Supplementary to the *Aperçu* was his *Rapport sur les progrès de la Géométrie* (1870), forming one of a series of official reports on the various branches of literature and science. He is also author of treatises on the *Géométrie Supérieure* (1852, 1880), *Porismes d'Euclide* (1860), *Sections Coniques, première partie* (1865), and of a multitude of separate articles, several of which relate to Maclaurin's theorem in attractions (*Comptes*

---

[*] These pages were already in type when the death of Chasles took place, on the 18th December, 1880.

*Rendus* v. 842. vi. 808–812, 902–915. 1837–8, &c.). At this point we may fitly offer some remarks upon the history of the "anharmonic" properties of conics and the general principle of "homography," with which the name of Chasles is so intimately associated.

(a) *The anharmonic point-property of conics.*

From the *Collectio* of Pappus we are led to infer that Euclid was acquainted with a form of the theorem (1) that the cross ratios of four fixed radiants are constant, and Apollonius with the theorem (2) that the locus $a\gamma = k . \beta\delta$ is a conic. From the union of these two at once arises the well-known "Propriété anharmonique des points d'une conique," which nevertheless remained unnoticed for upwards of 2000 years longer. Although the theorem (1) was rediscovered by Desargues and taken as his fundamental property in Perspective, whilst (2) was brought into notice by Descartes and afterwards proved synthetically by Newton, the combination of the two was not yet thought of.* The third and last stage in their history was inaugurated by Brianchon, who proved that, *if AB be a fixed chord of a conic and XY its points at infinity, the chords from a variable point on the curve to ABXY cut AB in constant cross ratios.* Chasles shewed, in course of an account of his "Transformation Parabolique," that the same is true when $X$ and $Y$ are any two fixed points on the conic; and he deduced that the locus of the point at which four given points subtend a *harmonic* pencil is a conic through the four points. See Quetelet's *Correspondance Math. et Physique* tome v. 293–4, 301 (1829): Chasles *Rapport etc.* p. 268. All that was still wanting was a familiarity with the "théorie complète des rapports anharmoniques,"† which might have been found in the Barycentrische Calcul of Möbius (1827). The property of conics now under consideration is fully stated, and its impor-

---

* Although it is convenient to deduce Newton's description of a conic by angles from the four-point property [p. 264], we ought, historically speaking, rather to reverse the process, and say that the anharmonic property is evidently contained in his *Descriptio Organica*.

† See the Preface to **Cremona's** *Géométrie Projective* p. xv (Paris 1875). In this work, originally written in Italian, the reader will find references to many of the leading treatises and historical facts of geometry.

tance pointed out in the *Aperçu historique* (pp. 80, 334–341), which was presented to the Brussels Academy in a rudimentary form in 1829, and ultimately published in 1837: it is also found in Steiner's *Systematische Entwickelung*, published in 1832 [p. 262].

(b). *The anharmonic tangent-property of conics.*

It was shewn by Newton that, if $IM$ and $IK$ be certain fixed segments of two given tangents to a conic, any third tangent cuts them in points $E$ and $Q$ respectively such that the rectangle $(IM-IE)(IK-IQ)$ is of constant magnitude; and the same theorem was reproduced by Brianchon in 1817 [p. lxxix]. Poncelet, who refers to Brianchon's *Mémoire*, proves[*] that, if the opposite sides $AB$ and $CD$ of a fixed quadrilateral circumscribed to a given conic be met by any fifth tangent in $L$ and $N$ respectively, and if $BC$, $DA$ be met by any sixth tangent in $M$ and $P$, then

$$\frac{DP}{AP} \cdot \frac{BM}{CM} = \frac{BL}{AL} \cdot \frac{DN}{CN};$$

and by fixing the tangent $MP$ he deduces that *the cross-ratio* $\frac{BL}{AL} : \frac{CN}{DN}$ *of the segments of AB and CD by any fifth tangent is constant.*[†] In the case of the parabola this cross ratio is equal to *unity* [p. 295]—a theorem which he believes to be due to Halley (p. 118).[‡] Chasles gave a second proof of Poncelet's generalisation, regarding the tangents to a conic as projections of the generators of a ruled hyperboloid, and shewed how to pass from it to Newton's theorem, which however he ascribed only to Brianchon (Quetelet's *Correspondance* IV. 364–70. Cf. Chasles *Rapport* p. 239). He afterwards proved it again in the form, that *the ratio of the products of the* distances *of the* fifth tangent from $A$, $C$ and $B$, $D$ respectively is constant (ibid. v. 289, 1829); and also shewed that the envelope of a

---

[*] *Propriétés projectives* p. 115 (1822); vol. I. p. 111 (1865).

[†] This is obvious from Newton's figure p. [lxvii] for the case of two pairs of *parallel* tangents: it then follows by projection for any two pairs of tangents, the "cross-ratio" having *different* but constant values for different planes. See Art. 153 (ii) [p. 312].

[‡] Apollonii Pergæi *De Sectione Rationis &c.* lib. I. pp. 64–5, ed. Halley (Oxon. 1706).

line cut *harmonically* by four fixed lines is a conic touching them (p. 294). The property of which this last is a special case was at length completely stated, simultaneously with its reciprocal (a), by Steiner and Chasles. It might have been deduced at once by projection from Lambert's solution of Sir Christopher Wren's problem [p. 296].

(c). *A conic regarded as the projection of a circle.*

Chasles, in his *Sections Coniques* (at the suggestion of M. J. Delbalat), defines a conic as the projection of a circle (p. 7), deduces its anharmonic properties, and founds his treatise upon them. The effectiveness of this method and the ability with which he applies it are known to all. Nevertheless, however excellent in a supplementary course of geometry, it is less suited for beginners, owing to the difficulty of proving conversely that every conic—secondarily defined by the anharmonic properties—can be placed in perspective with a circle. The problem is indeed solved concisely on p. 5, but not without references to a later paragraph and a separate work for further reasoning in justification of the construction. It naturally presents some difficulty to the tyro, being in fact a form of a problem which no geometer was able to solve generally before Desargues.

(d) *Homographic figures in two and in three dimensions.*

The general principle of "homographic"—as it was named by Chasles—is somewhat obscurely set forth in the works of Desargues, who regarded figures in homology as special cases of figures in perspective in space, at the same time taking for his *Proposition Fondamentale de la pratique de la Perspective* a form of the property of the anharmonic pencil.* The idea of transforming solid figures also is briefly hinted at by Desargues [p. 329]. Poncelet studied the relations of figures in "homology" (to use his own expression), and devoted a supplement of his *Traité des Propr. Projectives* pp. 369–416 (1822) to the projective properties of figures in space. Not the least valuable part of the *Aperçu historique* is the full exposition of the

---

\* Poudra *Œuvres de Desargues* I. 425.

principle of *Homographie*, as applied to plane and to solid figures, with which it concludes.

§ 10. The following works may be mentioned as having advanced the knowledge of the new geometry in this country. The essay on *Transversals* in the 12th edition of Hutton's Course of Mathematics, by T. S. Davies, also an editor of the journal the *Mathematician*; Salmon's compendious works on the various geometrical and other methods, to which we have so often referred; Gaskin's *Geometrical Construction of a Conic Section;* Mulcahy's *Principles of Modern Geometry;* and Townsend's two volumes on the *Modern Geometry of the Point, Line and Circle*, which within their prescribed limits are as complete an exposition of the principles of the subject as could be desired. We must not omit to notice also Prof. H. J. S. Smith's article on the *Focal Properties of Homographic Figures* in vol. II. 196—248 of the *Proceedings* of the London Mathematical Society, some of the results of which are given below, with especial reference to the case of reverse figures.

*Properties of reverse figures.*

§ 11. If $MPN$ be any angle which is EQUAL to its reverse $mpn$, it must be equal to $MON$ [p. 323]. Hence (1) for a given position of $P$ there are an infinity of angles $MPN$, each of which reverses into an equal (or supplementary) angle; and the arms of such a system of angles constitute a pencil in involution, since the points $MN$ always lie on a circle of the coaxal system through $O$ and $P$. (2) If $O'$ and $\omega'$ be the points such that the base-line bisects $OO'$ and $\omega\omega'$ orthogonally, every angle subtended at $O$ or $O'$ reverses into an equal angle subtended at $\omega$ or $\omega'$. (3) Every conic which has $O$ or $O'$ for a focus reverses into a conic having $\omega$ or $\omega'$ for a focus [p. 317 (i)]. (4) An ellipse (or hyperbola) having $O$ and $O'$ for foci reverses into a hyperbola (or ellipse) having $\omega$ and $\omega'$ for foci; their normals at reverse points $Pp$ correspond—being fourth harmonics to the focal distances and tangents at $P$ and $p$ respectively; and therefore their *centres of curvature* and their *evolutes* correspond. (5) The coaxal circles of which $OO'$ are the limiting points reverse into those of which $\omega\omega'$ are the limiting points. (6) The parallels through $\omega$ and $\omega'$ to the base-line are such that every segment of either reverses into an equal segment. And (7) on any two reverse lines $MP$ and $mp$ there are two sets of equal corresponding segments, which determine two pairs of involutions having their centres on the base-line—as may readily be deduced from the constancy of the product $MP.mp$ [p. 328] for given positions of the two lines. These results apply *mutatis mutandis* to homographic plane figures in general, however placed.

§ 12. The organic description of curves has within the last few years received developments of the greatest theoretical interest and practical importance, consequent upon the discovery (1864) by Peaucellier, an officer of engineers in the French army, of an apparatus for the *Inversion* of circular into rectilinear motion. Let $AOB$ be an angle of equal arms, and $ACBP$ a rhombus whose sides are less than the arms of the angle. Then $OCP$ is a straight line, and the product $OC.OP$, being equal to $OA^2 - AC^2$, will be **constant** if the sides of the rhombus and of the angle be constant. Let these be now replaced by bars or "links" jointed at the five points $OABCP$, then the whole linkage is called a *Peaucellier cell*.

If this linkage be moved about a fixed pivot at $O$ in any possible manner in one plane, then whatever be the locus of $C$ the point $P$ will trace its inverse with respect to $O$, on account of the constancy of the product $OC.OP$. To make $P$ describe a *straight line* we must make $C$ describe a *circle* through $O$; which is at once effected by joining $C$ to an "extra link" $CQ$, whose end $Q$ works about a fixed point at a distance equal to its own length from $O$. This apparatus may evidently be applied also to produce Parallel Motion; and we may make $P$ describe an arc of a circle of as great a radius as we please by making the distance $OQ$ sufficiently **nearly** equal to the length of the "extra link." The principle of linkages is well explained by Mr. A. B. Kempe in his concise work *How to draw a Straight Line, a lecture on Linkages* (London 1877), and references are given in it to the chief articles that had been written upon the subject. To conclude, in the words of Sylvester,* to whom Peaucellier's method of linkages owes so much of its further development: "*It is possible, by means of an apparatus consisting exclusively of rigid rods, compass joints, and pivots, to convert circular into linear, hyperbolic, elliptic, and parabolic motion; and, in general, to describe any curve of the form* $x\phi(x^2, y^2) + \psi(x^2, y^2) = 0$, *where $\phi$, $\psi$ are homogeneous forms of functions of any degree respectively in $x^2$, $y^2$.*"

---

* Educational Times *Reprint* vol. XXI. 58 (1874). Later information is to be sought in the same and other scientific periodicals.

NOTE.

*In continuation of Note † p. lxx.*

The order of the surface is thus determined by Professor Townsend. For a given position of $OA$, the plane $OBC$ envelopes a quadric cone, $2n$ of whose tangent planes pass each through a generator of the director scroll. These generators give $2n$ positions of the plane $\omega BC$ and $2n$ of the point $A$; and when $OA$ coincides with $O\omega$ all the $A$'s coalesce at $\omega$, which is therefore a $2n$-fold point on the locus. Again, every line through $\omega$ meets the scroll in $n$ points, through each of which passes a generator; and these generators severally determine $n$ conics, cutting the line through $\omega$ in $2n$ points, all of which, when the line is $\omega O$, coalesce at $O$. Thus $O$ also is a $2n$-fold point; and every line through $O$ or $\omega$ passes through $2n$ other points on the locus, which is accordingly of the order $4n$.

# THE GEOMETRY OF CONICS.

## DEFINITIONS.

A CONIC SECTION, or, briefly, a *Conic*, is a curve traced by a point which moves in a plane containing a fixed point and a fixed straight line in such a way that its distance from the fixed point is in a constant ratio to its perpendicular distance from the fixed straight line. The Conic Sections were so named from the circumstance that they are, and were originally defined as, the plane sections of a cone.

The fixed point is called the *Focus*; the fixed straight line the *Directrix*; and the constant ratio the *Eccentricity*, or the *Determining Ratio*.

A Conic is called an *Ellipse*, a *Parabola*, or a *Hyperbola*, according as its eccentricity is less than, equal to, or greater than unity.

*Similar Conics* are such as have the same eccentricity.

The *Axis* is the straight line through the focus at right angles to the directrix, and the point between the focus and the directrix in which it cuts the conic is called the *Vertex*.

When the eccentricity is either greater or less than unity, the conic cuts its axis in a second point, which is also called a vertex. In such cases the term *Axis* may denote the *finite* straight line which joins the vertices. Its middle point is called the *Centre* of the conic, and the conic is called a *Central Conic*.

The *Latus Rectum*, or, as it is sometimes called, the *Parameter*, is the chord through the focus at right angles to the axis

Other uses of these terms will be noticed in the course of the work.

A *Diameter* is the locus of the middle points of a system of parallel chords. It will be shewn that the diameters of conics are straight lines. The points in which diameters and chords meet the curve are called their ends or *Extremities*. The extremities of diameters which do not meet the curve will be defined in the chapter on Central Conics. The diameter at right angles to the axis of a central conic is called the *Minor*, or *Conjugate*, *Axis*.

Two diameters are said to be *Conjugate* when each bisects the chords parallel to the other; and two chords are said to be conjugate when they are parallel to conjugate diameters.

*Supplemental Chords* are such as join the extremities of a diameter to a point on the curve.

A *Tangent* to a conic is the limiting position of a secant, whose two points of intersection with the curve have become coincident. Thus, if $P$, $Q$ be adjacent points on the curve, and if the chord joining them be turned about $P$ until its further extremity $Q$ coincides with $P$, the chord in its limiting position will have become the tangent at $P$. Hence a tangent is said to be a straight line which passes through two *consecutive* or *coincident* points on the curve.

If the point of contact of a tangent to a hyperbola be removed to infinity the tangent will coalesce with one of two straight lines through the centre, which are called *Asymptotes*.

The *Normal* at any point of the curve is the straight line drawn through that point at right angles to the tangent.

The perpendicular upon the axis from any point is called absolutely the *Ordinate* of that point; but the ordinates of a specified diameter are the segments of the chords which that diameter bisects. The term *Abscissa* will be defined later.

The portion of the axis intercepted between the tangent at any point of the curve and the ordinate of that point is called the *Subtangent*.

The portion of the axis intercepted between the normal at any point of the curve and the ordinate of that point is called the *Subnormal*.

## DEFINITIONS.

A straight line is said to be divided *harmonically* at four points $P, S, Q, R$, when $PQ$, and $PQ$ produced, are cut in the same ratio by $S$ and $R$, so that

$$RP : RQ = SP : SQ = RP - RS : RS - RQ,$$

since, when this is the case, the lengths $RP$, $RS$, $RQ$ are in harmonical progression, the extremes being to one another as their differences from the mean. In the case in which $S$ is the middle point of $PQ$, the point $R$ is at infinity, or $PSQ \infty$ is divided harmonically.

The locus of intersection of the tangents at the extremities of a chord which passes through a fixed point, or *Pole*, is called the *Polar* of the point. It will be shewn that the polar of any point with respect to a conic is a straight line; and that the polar of an external point coincides with the chord of contact of the tangents to the conic from that point.

If about any point in the plane of a conic, other than the centre of the conic, a circle be described, such that the ratio of its radius to the perpendicular distance of its centre from the directrix is equal to the *Eccentricity*, the circle may be called the *Eccentric Circle of the Conic with respect to that point*, or, briefly, the *Eccentric Circle\** *of the point*. It is evident that the circle will cut, touch, or fall short of the directrix, according as the conic is a hyperbola, a parabola, or an ellipse.

The circle which is described according to the same law of magnitude about the centre of an ellipse or hyperbola is called the **Auxiliary Circle** of the curve. This latter is commonly defined as the circle described upon the axis as diameter, but it will be seen that the two definitions are coincident. The circle described upon the *Minor Axis* as diameter is called the *Minor Auxiliary Circle*.

In a central conic, the locus of intersection of tangents at right angles to one another is a circle, which is called the

---

\* The properties of this circle form the groundwork of the treatise of Boscovich, *Sectionum* **Conicarum** *Elementa nova quadam methodo concinnata*, contained in his *Elementa Universæ Matheseos*, VENETIIS, 1757. Boscovich gave no name to his circle, but some later writers have called it the *Generating Circle*, since it affords a ready means of tracing a conic whose elements are given.

*Director Circle.* The corresponding locus in the parabola is the directrix.

The *Order*, or *Degree*, of a curve is determined by the number of points in which it can be met by a straight line; and the *Class* of a curve by the number of tangents which can be drawn to it from a point. Thus, a curve of the second order, or degree, is one which a straight line meets generally in two, and never in more than two, points; and a curve of the second class is one to which generally two, and never more than two, tangents can be drawn from a point.

# CHAPTER I.

## DESCRIPTION OF THE CURVE.

**1.** *To trace* a conic whose focus, directrix, and eccentricity are given.

Let $S$ be the focus,* $MM'$ the directrix, and $X$ the point in which the axis meets the directrix. In $SX$ take a point $A$

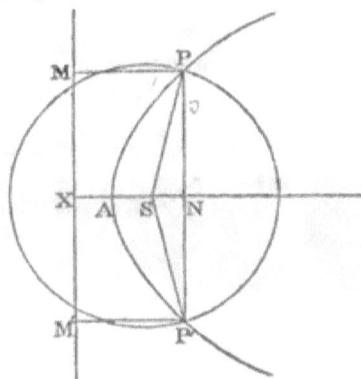

such that the ratio of $SA$ to $AX$ may be equal to the eccentricity. Then $A$ is the vertex.

Let a straight line cut the axis at right angles in $N$. About $S$ as centre, with radius $SP$, such that
$$SP : NX = SA : AX,$$
describe a circle cutting the straight line in $P$, $P'$. From these points draw perpendiculars $PM$, $P'M'$ to the directrix. Then evidently
$$SP : PM = SA : AX,$$

---

* The planets **describe** approximately ellipses about the sun in one focus. For this reason the first letter of SOL is here used, as by Newton, to denote the *Focus*, or as he called it, the *Umbilicus*.

or $P$ is a point on the curve. And in like manner $P'$ is a point on the curve.

If now we suppose the straight line $PNP'$ to move parallel to itself, the points $P$, $P'$ upon it will trace out the entire curve. From this construction it is evident that the curve is symmetrical with respect to its axis, since its points are always determined in pairs, as $P$, $P'$, which are symmetrically situated with respect to the axis. It appears also that the tangent at the vertex is at right angles to the axis, since when the point $N$ coincides with the vertex, $SP = SA = SP'$; that is to say, the points $P$, $P'$ coalesce at $A$, and the chord joining them, which is always at right angles to the axis, becomes the tangent at $A$.

In order that pairs of real points may be determined by the above construction, it is necessary and sufficient that $SN$ should be less than $SP$, and therefore
$$SN : NX < SA : AX,$$
a condition which enables us to discriminate between the three species of conics as follows:

(i) *The Parabola.*

If the eccentricity be equal to unity, we must have $SN < NX$, a condition which is satisfied by taking $N$ anywhere in $XA$ produced. The point $N$ therefore may be supposed to start from $A$, and to move in the direction $AS$ to infinity, so that the extremities of the chord $PP'$ trace out a single infinite branch.

(ii) *The Ellipse.*

If the eccentricity be less than unity, the curve will have a second vertex $A'$ in $XA$ produced, and in order that the condition
$$SN : NX < SA : AX$$
may be satisfied, it may be shewn that the point $N$ must be taken between $A$, $A'$. Hence the ellipse consists of one oval branch, as in the figure of Art. 3.

(iii) *The Hyperbola.*

If the eccentricity be greater than unity, the curve will

have a second vertex $A'$ lying in $AX$ produced **beyond the** directrix; and the point $N$ may lie anywhere in $AA'$ produced **either** way, but not between $A$, $A'$. Hence, the hyperbola consists of two infinite branches situated on opposite sides of the directrix.

A point is **said to** lie WITHIN a conic when **it lies** between the extremities of a chord perpendicular to the axis; and all other points in the plane of the conic, with the exception of those which are upon the curve itself, are said to lie WITHOUT the conic.

Let $ON$ be the ordinate of an internal point $O$, and let $NO$ be produced to meet the conic in $P$, then evidently

$$SO : NX < SP : NX,$$
$$< SA : AX.$$

Next let $O$ be an external point. Then if $N$, the foot of its ordinate, fall within the curve, it may be shewn in like manner that

$$SO : NX > SA : AX.$$

But if $N$ fall without the curve, then

$$SN : NX > SA : AX,$$

and *a fortiori* $\quad SO : NX > SA : AX.$

Hence, in every case, *a point will lie within or without a conic according as the ratio of its focal distance to its perpendicular distance from the directrix is less or greater than the eccentricity.*

### SCHOLIUM.

THE CIRCLE is the limiting form of an ellipse whose eccentricity is indefinitely diminished, and whose directrix is removed to an infinite distance from the focus. For if, in **the next** figure, $PM$, $P'M'$ be perpendiculars on the directrix from any two points $P$, $P'$ on a conic, and if the distance of the directrix from $S$ be increased indefinitely whilst $SP$, $SP'$ remain finite, then (i) the ratio $SP : PM$ is diminished indefinitely; and (ii) the ratio $PM : P'M'$ tends to equality. But

$$SP : SP' = PM : P'M'.$$

Therefore ultimately $SP : SP'$ is a ratio of equality, and the conic becomes a circle about $S$ as centre.

**2.** *The focal distances of all points on a conic are to one another as their parallel distances from the directrix.*

Let $S$ be the focus; $P$, $P'$ any two points on the curve;

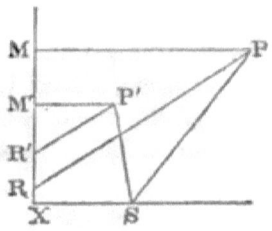

$M$, $M'$ their projections upon the directrix. Then from the definition

$$SP : PM = SP' : P'M'.$$

From $P$, $P'$ draw a pair of parallels meeting the directrix in $R$, $R'$. Then by similar triangles,

$$PM : PR = P'M' : P'R'.$$

Therefore $\qquad SP : PR = SP' : P'R'.$

Hence the focal radii $SP$, $SP'$ are to one another as the parallels $PR$, $P'R'$; and, whatever be the position of the point $P$ on the curve, the ratio of $SP$ to $PR$ will be constant if $PR$ be drawn to meet the directrix at a constant angle.

**3.** *A conic is a curve of the second order.*

For if $P$, $Q$ be any two points on a conic, as in the figure of Art. 4, and if the straight line joining them meet the directrix in $R$, then, drawing perpendiculars $PM$, $QN$ to the directrix, we have

$$SP : SQ = PM : QN$$
$$= PR : QR.$$

Hence $SR$ makes equal angles with $SP$, $SQ$; and, conversely, if $P$ be a point on the curve, and $SQ$ be drawn meeting $RP$, and equally inclined with $SP$ to $SR$, then $Q$ will be a point on the curve.

It is evident from this construction that no third point can be found on the conic in the same straight line with $P$, $Q$.

Hence a straight line which meets a conic will in general meet it in two points, and no straight line can meet a conic in **more points than two**. It is for this reason that conics are **called** curves of the second order, or of the second degree.

Let a straight line *parallel to the axis* meet the directrix in $M$ and **the curve in** $P$. Make the angle $MSR$ equal to $MSP$, and let $RS$ meet $MP$ in $Q$; then, from above, the point $Q$ **lies on the curve.** In the **case of the** *ellipse* the

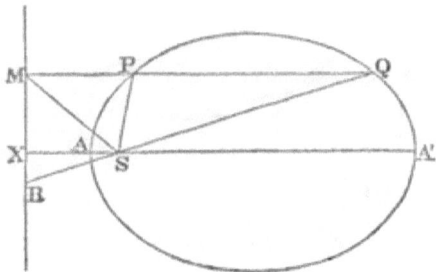

points $P$, $Q$ will lie on the same side of the directrix; **for,** since $SP$ is less than $PM$, the angle $SMP$ is less than $MSP$, and therefore the alternate angle $MSX$ is less than $MSP$ or $MSR$. Hence the straight line $SR$ falls without the angle $MSX$, and meets $MP$ **on the same** side of the directrix with $P$. By similar reasoning it may be shewn that a straight line **parallel to the axis of a** *hyperbola* intersects the curve in two **points on opposite** sides **of the directrix.** In the case of the *parabola*, $SR$ coincides **with the axis, to which** $MP$ is parallel. Hence a straight line parallel **to the axis of a parabola** meets the curve in one point only.

4. *To describe a conic of given focus, directrix, and eccentricity by means of the eccentric circle of any given point.*

Describe the eccentric circle of any point $O$ in the plane **of the conic,** and let a straight line through $S$ meet the circle in $p$ and the directrix in $R$. Let $RO$ meet the focal radius parallel to $pO$ in $P$, and let $OD$, $PM$ be the perpendiculars from $O$, $P$ to the directrix.

# DESCRIPTION OF THE CURVE.

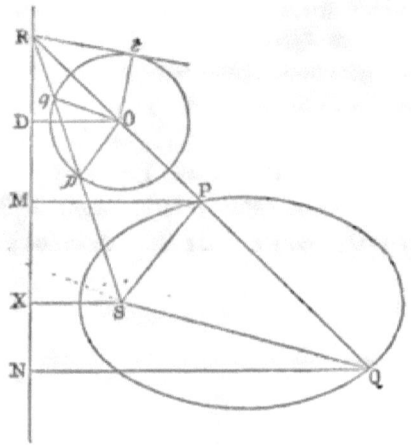

Then by parallels,
$$SP : Op = PR : OR$$
$$= PM : OD,$$
or $$SP : PM = Op : OD$$
$$= \text{the eccentricity.}$$

Hence, as $p$ moves round the circle, $P$ traces the conic which was to be described.

In the case of the hyperbola* it may be seen that the directrix divides the circle into two parts, each of which corresponds to one branch of the curve.

5. *To determine the points in which a given straight line intersects a conic of given focus, **directrix**, and eccentricity.*

Let the given straight line meet the directrix in $R$. Describe the eccentric circle of any point $O$ on the straight line, and let it cut $SR$ in $p$, $q$. Let the focal radii parallel to $pO$, $qO$

---

* Since the **locus of** $p$ **is a** continuous curve, the conic, which is the locus of $P$, is also to be regarded as in all cases a continuous curve. In the case of the hyperbola, as the point $p$ crosses the directrix, the point $P$ passes from infinity on one side of the axis to infinity on the other side of the axis. Hence the two branches of the hyperbola may be conceived of as connected *diagonally* at infinity.

meet the straight line in $P$, $Q$. Then, as above, if $OD$, $PM$ be perpendiculars on the directrix,

$$SP : PM = Op : OD$$
$$= \text{the eccentricity},$$

or $P$ is a point on the **conic**.

Similarly it may be shewn that $Q$ **is** a point on the conic.

From this **construction it** follows that **a** conic is a curve of the **same order as the circle**; that is **to say, it** is a curve of **the second order, as was shewn in Art. 3**.

**6.** *A conic is a curve of the second* **class.**

If the points $p$, $q$ become coincident, the points $P$, $Q$ likewise become coincident, since $Op$, $Oq$ are always parallel to $SP$, $SQ$ respectively; that is to say, if $SR$ touches the **circle**, $RO$ touches the conic.

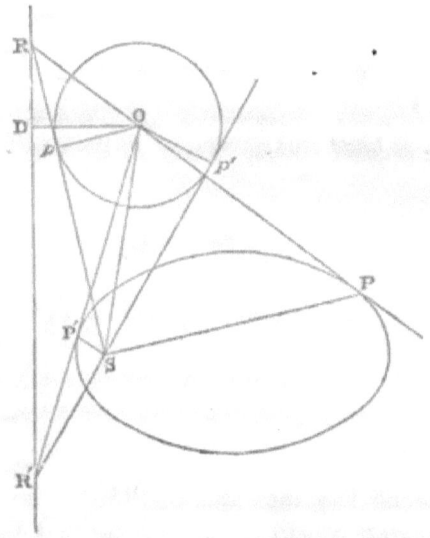

Hence the problem of drawing tangents to a conic **from** a point $O$ is reduced to that of drawing tangents from $S$ **to** the **eccentric** circle of $O$; for if the tangents from $S$ to the circle meet the directrix in $R$, $R'$, then $RO$, $R'O$ will be the required tangents to the conic.

Since the same number of tangents can be drawn from $O$ to the conic as from $S$ to the circle, it follows that a conic is a curve of the same class as the circle; that is to say, it is a curve of the second class.

In order that two real tangents to the conic may be determined by the above construction, it is necessary and sufficient that $S$ should lie without the circle. The point $O$ must therefore be so situated that $SO$ may be greater than the radius of the circle, and therefore

$$SO : OD > Op : OD$$
$$> \text{the eccentricity,}$$

where $D$ is the projection of $O$ upon the directrix.

When $O$ is on the curve the circle passes through $S$, and the two tangents coalesce.

No tangent can be drawn to a conic from any point between the curve and its axis, since at every such point

$$SO : OD < \text{the eccentricity.}$$

Hence no tangent can pass between the curve and its axis,* and the curve is therefore concave at all points to its axis.

## EXAMPLES.

1. If an ellipse, a parabola, and a hyperbola have the same focus and directrix, the ellipse will lie wholly within the parabola, and the parabola wholly within the hyperbola; and no two conics which have the same focus and directrix can intersect one another.

2. If parallels from the focus and any point $P$ on a conic meet the directrix in $D$, $R$, and if $L$ be equal to half the latus rectum, then

$$SP : PR = L : SD.$$

---

* In particular it is to be noticed that no tangent can be drawn to either branch of a hyperbola from any point within the other branch.

3. If the focus and two points of a conic be given, the directrix will pass through one of two fixed points.

4. If $PN$ be the ordinate of any point $P$ on a conic, then
$$SP \pm L : SN = \text{the eccentricity.}$$
Hence shew that if $PSP'$ be any focal chord, then
$$\frac{1}{SP} + \frac{1}{SP'} = \frac{2}{L}.$$

5. **Determine the** condition that the chord of a conic may **be greater** than, equal to, or less than the diameter of the eccentric circle of its middle point.

6. In the figure of Art. 4, if $OpSP$ be a quadrilateral formed by drawing through $O$, $S$ a pair of parallels, and a pair of straight lines which intersect on the directrix, then $p$ will lie without or within the eccentric circle of $O$, according as the ratio of $SP$ to $PM$ is greater or less than the eccentricity. Prove also by means of this construction that a tangent at any point to a conic cannot meet the curve in any other point.

7. If $p$ be made to describe a series of circles about $O$ as centre, $P$ will describe a series of conics having a common focus and directrix; and the eccentricities of the conics will be to one another as the radii of the circles.

8. **If $p$ be made to describe a curve of any** degree, $P$ will **describe a curve of the same degree;** and the corresponding arcs of the **two curves will subtend equal angles at the** points $O$, $S$ respectively.

9. If $pm$ be the perpendicular from $p$ to the directrix, then $PM.pm = OD.SX$. Hence shew that the sum of the reciprocals of the segments of a focal chord of a conic is constant, and any focal chord is divided harmonically by the focus and the directrix. Shew also that if $OP = OQ$, then $RqSp$ is divided harmonically.

10. Shew from the construction of Art. 6 that the tangents $OP$, $OP'$ subtend equal angles, and that $RP$, $RP'$ subtend right angles, at the focus.

# CHAPTER II.

## THE GENERAL CONIC.

In this chapter we shall prove some of the principal properties which are common to the Parabola, the Ellipse, and the Hyperbola, reserving for future consideration the properties which are distinctive of the three species of conics.

### PROPERTIES OF TANGENTS.

#### PROPOSITION I.

**7.** *The tangents to a conic from any point on the directrix subtend right angles at the focus.*

Let $P$, $Q$ be adjacent points on the curve, and let $PQ$ produced meet the directrix in $R$. Then, as in Art. 3,

$$SP : SQ = PR : QR,$$

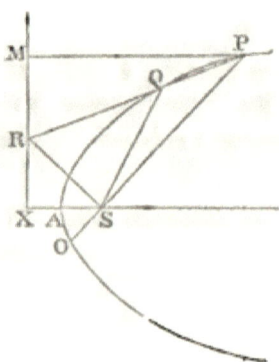

and $SR$ bisects the angle which $SQ$ makes with $PS$ produced.

Let $PS$ produced meet the conic in $O$. Then since the angles $RSQ$, $RSO$ are always equal, therefore in the limit, when $SQ$ coincides with $SP$, each of these angles becomes a right angle, and $RP$, which becomes the tangent at $P$, subtends a right angle at $S$.

Hence, (i) to draw the tangent to a conic at a given point $P$ on the curve, make $PSR$ a right angle, and draw $PR$ to the point in which $SR$ meets the directrix; and (ii) to draw tangents to a conic from a given point $R$ on the directrix, draw the focal chord $OSP$ at right angles to $SR$, and join $RP$, $RO$.

### Corollary.

Hence it appears that the tangents at the extremities of any focal chord $PO$ meet at a point $R$ on the directrix; and conversely, if tangents be drawn from any point $R$ on the directrix their chord of contact $PO$ will pass through the focus.

**The Directrix is therefore the Polar of the Focus.**

### PROPOSITION II.

**8.** *If from any point $T$ on the tangent at $P$* **perpendiculars** *$TL$, $TN$ be drawn to $SP$ and* **the** *directrix respectively,* **then** *$SL : TN =$ the eccentricity.*\*

(i) For if the tangent at $P$ meet the directrix in $R$, **and**

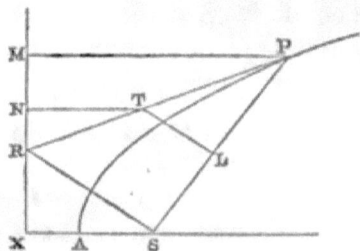

if $PM$ be a perpendicular to the directrix, then, since $SR$ is at right angles to $SP$, and is therefore parallel to $TL$, we have
$$SL : SP = TR : PR$$
$$= TN : PM.$$

---

\* It will be shewn at the end of the chapter that this theorem, which, with its applications as in the text, was discovered by Prof. Adams, is the geometrical analogue of the polar equation between $ST$ and its inclination to the axis.

Therefore
$$SL : TN = SP : PM$$
$$= SA : AX,$$
where $A$ is the vertex, and $X$ the foot of the directrix.

(ii) **It appears** from the above proof, that **this proposition may be** regarded as a **corollary from the** preceding; but **the two may** be proved at once, **as** follows, if we **consider the tangent to** be defined *mutatis mutandis* after the **manner of** EUCLID.

Let $P$ be a point on the curve, and $R$ a point on the directrix, such that $PR$ subtends at right angle **at** $S$. Take any point $T$ in **the** same straight line with $P$, $R$, and let fall the perpendiculars $PM$, $TN$ on the directrix, and the perpendicular $TL$ on $SP$. Then, as before,
$$SL : TN = SA : AX.$$
Hence $\qquad ST : TN > SA : AX,$

and the point $T$ lies without the curve in every case except that in which it coincides **with** $P$. **The straight line** $PR$ is therefore the tangent at $P$.

### *Corollary.*

It is evident that if $L$, $N$ be the projections of a point $T$ upon a fixed focal chord and the directrix respectively, and if
$$SL : TN = SA : AX,$$
the point $T$ will lie on the tangent at one or other of the extremities of the fixed focal chord.

Hence, a second construction **analogous to that of Art. 6**, for drawing tangents to a conic from **a given point** $T$. About $S$ describe a circle **equal to the eccentric circle of** $T$, and draw $TL$, $TM$ touching the circle at $L$, $M$; then $SL$, $SM$ will pass **through the points of contact** of the two tangents which **can be drawn** to the conic **from** $T$. There is an apparent **ambiguity in this** construction, since each of the focal chords through $L$, $M$ meets the conic in two points; but to determine the actual tangents, draw $SR$ at right angles to $SL$ to meet the directrix, and join $RT$; and draw $SR'$ at right angles to $SM$ to meet the directrix, and join $R'T$.

THE GENERAL CONIC.

## PROPOSITION III.

9. *The two tangents which can be drawn to a conic from any external point subtend equal or supplementary angles at the focus.*

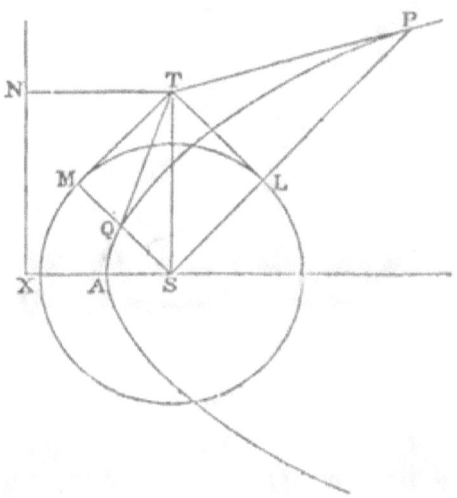

For if $TP$, $TQ$ be the two tangents, and $TL$, $TM$, $TN$ be perpendiculars upon $SP$, $SQ$, and the directrix respectively, then since $T$ lies on the tangent at $P$,

$$SL : TN = SA : AX.$$

In like manner

$$SM : TN = SA : AX,$$

since $T$ lies on the tangent at $Q$. Therefore in the right-angled triangles $STL$, $STM$, the sides $SL$, $SM$ are equal; and the hypotenuse $ST$ is common to the two triangles; therefore

$$\angle TSL = TSM.$$

**Now** (i) if $TP$, $TQ$ touch the same branch of the conic, the angles which they subtend at $S$ will be either *equal* to $TSL$ and $TSM$, as in the above figure, or *supplementary* to $TSL$ and $TSM$. In either case $TP$, $TQ$ will subtend EQUAL angles at $S$.

But (ii) if **TP**, $TQ$ touch opposite branches of a hyperbola, so that one, and **one** only, of the radii $SL$, $SM$ has

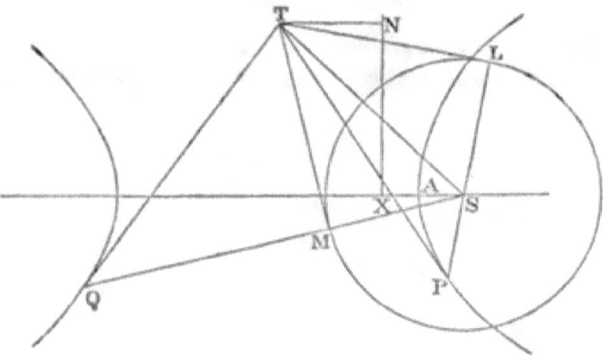

to be produced backwards to *P* or *Q*, then the angles *TSL*, *TSM* being equal as before, the tangents *TP*, *TQ* will subtend SUPPLEMENTARY angles at *S*.\*

### Corollary 1.

If the chord of contact *PQ* of a pair of tangents *TP*, *TQ* meet the directrix in *R*, then *ST*, *SR* bisect supplementary angles

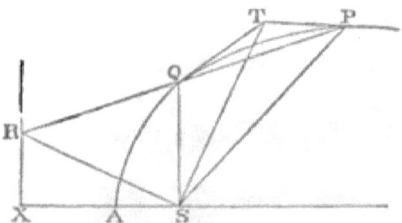

at *S*, and are therefore at right angles to one another. And the chord of contact *PQ* is divided internally and externally in the same ratio *SP* : *SQ*, that is to say, it is divided harmonically, at the points at which it meets *ST* and the directrix. Since the straight line *ST* is evidently the polar of *R*, it follows that the chord *PQ* is cut harmonically by the point *R*, and the polar of *R*.

---

\* But in this case also, we may say that they subtend EQUAL angles, if, in accordance with the principle of the Note on Prop. VII, we regard *TQ* as subtending at *S*, not the angle *TSQ*, but its supplement.

## Corollary 2.

If $O$ be any point on $PQ$, or $PQ$ produced, and $M$ the projection of $O$ upon the directrix, and if the perpendicular from $O$ to $ST$ meet $SP$, or $SQ$, in $L$; then, since this perpendicular is *parallel to SR*, it follows, precisely as in Prop. II., that

$$SL : OM = \text{the eccentricity.}$$

## THE NORMAL.

### PROPOSITION IV.

**10.** *If the normal at $P$ meet the axis in $G$, then $SG : SP = $ the eccentricity.*

For if the tangent at $P$ meet the directrix in $R$, the circle on $PR$ as diameter will pass through $S$, since the angle $PSR$ is a right angle; and likewise through $M$, the projection of $P$ upon the directrix; and $PG$, which is at right angles to $PR$, will touch the circle.

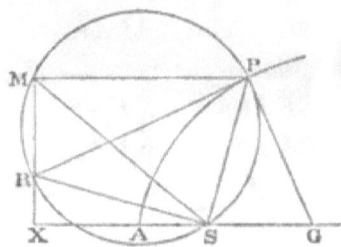

Therefore $\angle SPG = SMP$, in the alternate segment.
Also $\angle PSG = SPM$, by parallels.
Hence the triangles $SGP$, $PSM$ are similar, and
$$SG : SP = SP : PM = SA : AX.$$

Conversely, if in $AS$ produced a point $G$ be taken such that
$$SG : SP = SA : AX,$$
then will $PG$ be the normal at $P$.

This suggests an obvious method of drawing a normal to a conic at a given point on the curve, or from a given point on the axis.

## PROPOSITION V.

11. *The perpendicular let fall upon the focal radius to any point of a conic from the foot of the normal at that point meets the focal radius at a distance equal to half the* **latus** *rectum from its extremity.*

Let $G$ be the foot of the normal at a point $P$ whose ordinate is $PN$, and let a perpendicular $GK$ be drawn to $SP$. Then

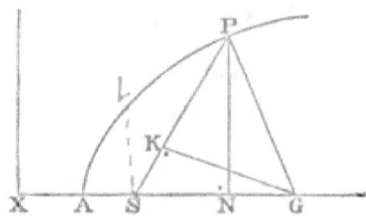

by similar right-angled triangles
$$SK : SG = SN : SP.$$
Therefore by the preceding proposition
$$SK : SA = SN : AX.$$
But, from the definition of the curve,
$$SP : SA = NX : AX.$$
Therefore $SP \sim SK : SA = SX : AX.$

Therefore $SP \sim SK$, or $PK$, is constant, and equal to half the latus rectum.

## ANGLE PROPERTIES OF SEGMENTS.

### PROPOSITION VI.

12. *The chords containing the angles in a focal segment of a conic intercept on the directrix lengths which subtend right angles at the focus.*

Let $PSp$ be a focal chord, and $PQp$ an angle which it subtends at the circumference. Let $PQ$, $Qp$ meet the directrix in $R$, $r$ respectively. Produce $QS$ to $q$.

Then since $\qquad SP : SQ = PR : QR,$

and $\qquad\qquad Sp : SQ = pr : Qr,$

# THE GENERAL CONIC.

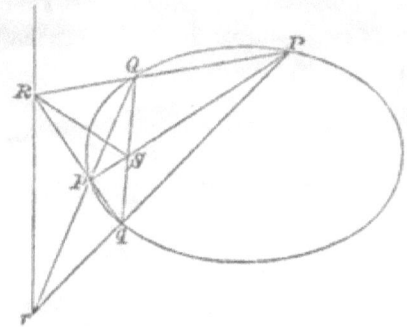

therefore $SR$ and $Sr$ bisect the supplementary angles which $Sp$ makes with $Qq$, and consequently the angle $RSr$ is **a right angle,** or $Rr$ subtends a right angle at $S$.

### Corollary.

**The** opposite sides of a quadrilateral whose vertices are **at** the ends of a pair of focal chords $PSp$, $QSq$ intersect upon the directrix, and the portion of the directrix which they intercept subtends a right angle at the focus. For, proceeding as above, we see that each of the straight lines $PQ$, $qp$ meets the directrix on the bisector of the angle $pSQ$; and each of the straight lines $Pq$, $Qp$ meets the directrix on the bisector of the supplementary angle $pSq$; that is to say, the two pairs of opposite sides of the quadrilateral intersect upon the directrix at points $R$, $r$, such that $Rr$ subtends a right angle at $S$.

### PROPOSITION VII.

**13.** *The chords containing the angles in* **a fixed segment of a conic** *intercept on the directrix lengths which subtend constant angles at the focus, the constant angles being equal or supplementary to half the angle which the chord of the segment subtends at the focus.*

Let $PQ$ be a fixed arc **of a conic,** and $PRQ$ a variable angle at the circumference. Let $PR$, $QR$ meet the directrix in $p$, $q$ respectively.

Then since $\qquad SP : SR = Pp : Rp,$

the straight line $Sp$ bisects the angle $RSP$, and likewise the straight line $Sq$ bisects the angle $RSQ$, externally or internally.

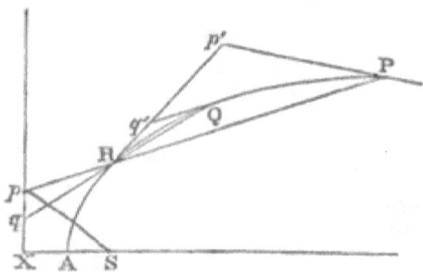

Hence, by addition or subtraction, as the case may be, the angle $pSq$ is equal or supplementary* to $\tfrac{1}{2}PSQ$. For example, in the figure drawn,

$$\angle RSq = \tfrac{1}{2} \text{ supplement of } RSQ,$$

and $\qquad \angle RSp = \tfrac{1}{2}$ supplement of $RSP$;

whence, by subtraction,

$$\angle pSq = \tfrac{1}{2}PSQ.$$

*Corollary.*

In like manner it may be shewn, by successive applications of Prop. III., that if the tangents at $P$, $Q$ meet the tangent at $R$ in $p'$ and $q'$, the angle $p'Sq'$ will be equal or supplementary to $pSq$, or $\tfrac{1}{2}PSQ$.

SCHOLIUM A.

THE ANGLE PROPERTIES of conics comprise some simple generalisations of fundamental theorems in the geometry of the circle, as may be seen by removing the directrix to infinity, when, as has been already shewn, the conic becomes a circle about $S$ as centre.

(i) Removing the directrix to infinity, we have, referring to the figure of Art. 8, the tangent $PR$ parallel to $SR$, and therefore at right angles to $SP$. That is to say, at any point $P$ on a circle the tangent is at right angles to the radius.

---

* The theorem appears to fail when $P$, $Q$ are on opposite branches of a hyperbola, in which case $\angle pSq =$ complement of $\tfrac{1}{2}PSQ$. But in this case the chord of the segment is not the finite straight line $PQ$, which lies without the conic, but the portion of the unlimited straight line through $P$, $Q$ which falls within the conic. The angle subtended at $S$ by the chord of the segment is therefore not $PSQ$, but the **supplement** of $PSQ$.

# THE GENERAL CONIC.

(ii) Removing the directrix to infinity in Art. 12, we have $PQ$ parallel to $SR$, since $R$ is at infinity; and $Qp$ parallel to $Sr$, since $r$ is at infinity. Therefore $PQ$, $Qp$ contain an angle equal to $RSr$, or the angle in a semicircle is a right angle.

(iii) Proceeding similarly with reference to Art. 13, we have $PR$ parallel to $Sp$, since $p$ is at infinity; and $QR$ parallel to $Sq$, since $q$ is at infinity. Therefore the angle $PRQ$, or its supplement, is equal to $\frac{1}{2}PSQ$. Hence, by varying the positions of the points upon the circumference, **we come** to the properties of the circle, that **the angle at** the centre is double of the **angle** subtended by **the same arc at the** circumference; that angles **in the** same segment **are equal to one another; and** that the **opposite** angles of an **inscribed** quadrilateral **are** together equal **to** two right angles. Lastly, by making $R$ coalesce with $P$, we deduce that the tangent at $P$ makes with a chord $PQ$ an angle equal to $PRQ$ in the alternate segment.

## DIAMETERS.

### PROPOSITION VIII.

14. *The locus of the middle points of any* **system of parallel** *chords of a conic is a straight line which meets* **the directrix** *on the straight line through the focus at right angles to the chords.*

Let $PQ$ be any one of a system of parallel chords, and $V$ the point in which the focal perpendicular upon them meets the **directrix. Let $PQ$** meet $SV$ in $Y$, and the directrix in $R$.

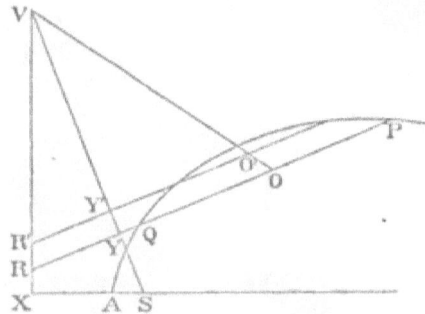

Then since $SP : PR = SQ : QR$;

therefore $SP^2 \sim SQ^2 : PR^2 \sim QR^2 = SP^2 : PR^2$,

or, subtracting $SY^2$ from each of the magnitudes $SP^2$ and $SQ^2$,

$PY^2 \sim QY^2 : PR^2 \sim QR^2 = SP^2 : PR^2$.

24 THE GENERAL CONIC.

But, if $O$ be the middle point of $PQ$, the sum of $PY$ and $QY$ will be equal to $2OY$, and their difference to $PQ$, or *vice versa*.

Therefore $PY^2 \sim QY^2 = 2OY.PQ$,

and in like manner $PR^2 \sim QR^2 = 2OR.PQ$.

Therefore $OY : OR = SP^2 : PR^2$,

which, by Art. 2, is a constant ratio for all parallel chords.

Hence the locus of $O$ is a straight line through $V$.*

### Corollary 1.

The tangents at the extremities of diameters are parallel to the ordinates of those diameters, since a bisected chord as $PQ$ may be supposed to move parallel to itself until its segments vanish together, and its extremities coalesce, viz. at the end of its diameter. Hence the diameter through the point of contact of any tangent meets the directrix at a point $V$ such that $SV$ is perpendicular to the tangent. If a diameter meets the curve in two points, the tangents at those points are parallel to one another, and to the ordinates of that diameter. Conversely, the chord of contact of any two parallel tangents is a diameter.

### Corollary 2.

If $POQ$, $poq$ be double ordinates of a given diameter $Oo$, then since $PQ$, $pq$ are both bisected by the same diameter,

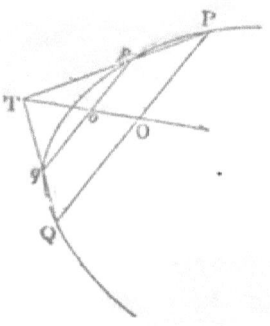

---

\* This may also be proved by means of the eccentric circle of $O$. For in Art. 16, if $OP$ be made equal to $OQ$, then $Sp : Sq = pR : qR$, or $RpSq$ is divided harmonically;

the directions of $Pp$, $Qq$ will intersect at some point $T$ on that diameter. Hence, making $pq$ coalesce with $PQ$, so that $PT$, $QT$ become the tangents at $P$, $Q$, we see that *the tangents at the extremities of any chord meet upon the diameter which bisects the chord;* and conversely, that the diameter through an external point bisects **the chord of contact of the** tangents from that point.

*Corollary* 3.

If the chord $PQ$ be parallel to the **axis**, so that $SY$ the focal perpendicular upon it is parallel to the directrix, then, proceeding as before, and supposing $PQ$ to meet the directrix in $M$, we have

$$OY : OM = SP^2 : PM^2;$$

and, the ratio of $OY$ to $OM$ being thus constant, the locus of

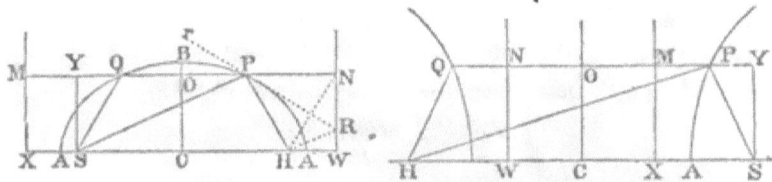

$O$ is a straight line perpendicular to the axis. Let it meet the axis in $C$, which (Def. p. 1) is the *Centre* of the conic. Then, evidently, $CO$ divides **the curve symmetrically, since it** bisects every chord $PQ$ **to which** it is at right **angles; and the** conic has therefore a second focus $H$, and directrix $NW$, which are the exact counterparts **of the** original focus **and directrix** with reference to which the curve was considered **to be described.**

From the symmetry of the curve, it is manifest that every chord through the centre is bisected at that point, and hence that all diameters pass through the centre.* Other immediate

---

and therefore the focal perpendicular $SY$ is the polar of $R$ with respect to the circle, and $OY.OR$, being equal to the square of the radius, is in a constant ratio to $OR^2$, if the inclination of $PQ$ to the directrix be invariable. Therefore $OY : OR$ is a constant ratio, and the locus of $O$ is a straight line through $V$

* A diameter is sometimes defined as a straight line through the centre.

consequences of the twofold symmetry of Bifocal Conics will be assumed as self-evident in the course of the work.

In the **case of the parabola**, since $SP^2 : PM^2$ is a ratio of equality, $OY : OM$ and $CS : CX$ are likewise ratios of equality. Hence the parabola may be regarded as *a conic whose centre is at infinity*. Its diameters are straight **lines parallel to the axis, since** they all co-intersect at the **infinitely distant** point $C$ on the axis; and conversely, every **straight** line parallel to the axis is a diameter.

*Corollary* 4.

**In a central** conic, if one diameter bisect chords parallel to a second, **the** second will bisect chords **parallel to** the former. For if the two diameters meet the directrix in $V$, $V'$, and if $SV$ be perpendicular to $CV'$; then, $CS$ being perpendicular to $VV'$, the focus is the **orthocentre of the** triangle $CVV'$, or $SV'$ is perpendicular to $CV$. That is to say, if $CV$ bisects chords parallel to $CV'$, then $CV'$ bisects chords parallel to $CV$.

If $CV$, $CV'$ be thus related, it is easily seen that

$$VX.V'X = CX.SX.$$

## THE SEGMENTS OF CHORDS.

### PROPOSITION IX.

15. *The semi-latus rectum is a harmonic mean between the segments of any focal chord.*

Let a focal chord $PSQ$ meet the directrix in $R$, and let $PM$, $SX$, $QN$ be perpendiculars to the directrix.

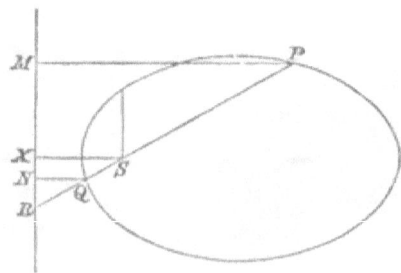

Then $\quad SP : SQ = PM : QN$
$\quad\quad\quad\quad\quad = PR : QR,$

or $PQ$ is divided harmonically at $S$ and $R$.

But by parallels, and from the definition of the curve, if $L$ be the semi-latus rectum,

$$PR : SR : QR = PM : SX : QN$$
$$= SP : L : SQ.$$

And, from above, $PR$, $SR$, $QR$ are in harmonical progression. Therefore also $SP$, $L$, $SQ$ are in harmonical progression.

*Corollary.*

This result may also be written in the forms

$$\frac{1}{SP} + \frac{1}{SQ} = \frac{2}{L},$$

and $\quad L.PQ = L(SP + SQ) = 2 SP.SQ.$

Hence, if $PQ$, $pq$ be any two focal chords,

$$PQ : pq = SP.SQ : Sp.Sq,$$

or *focal chords are to one another as the rectangles contained by their segments.*

PROPOSITION X.

**16.** *A chord of a conic being divided at any point, to determine the magnitude of the rectangle contained by its segments.*

Let $O$ be any point on a chord $PQ$ of a conic, or on the chord produced; it is required to determine the magnitude of the rectangle $OP.OQ$.

Let the chord, produced if necessary, meet the directrix in $R$, and let $OD$ be a perpendicular to the directrix. Describe the eccentric circle of $O$, and let it cut $SR$ in $p$ and $q$. Then, as in Art. 5, the radii $Op$, $Oq$ are parallel to $PS$, $QS$ respectively.

Therefore $\quad OP : Sp = OR : Rp,$

and $\quad\quad\quad OQ : Sq = OR : Rq.$

Hence $\quad OP.OQ : Sp.Sq = OR^2 : Rp.Rq,$
$\quad\quad\quad\quad\quad\quad\quad = OR^2 : Rt^2,$

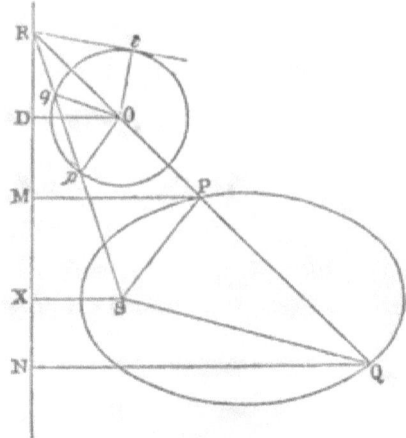

if $Rt$ be a tangent from $R$ to the circle; **or (Euclid III., 35)** if it be a semi-chord at right angles to the diameter through $R$, in the case in which $R$ falls within the circle.*

In this result it is to be noticed **(i)** that **the** magnitude $Sp.Sq$ depends **only upon the** *position* **of** $O$, since when $O$ is given, its eccentric **circle being given,** $Sp.Sq$ is constant; and (ii) **that the ratio** $OR^2 : Rt^2$ **depends only** upon the *direction*† of $PQ$, **since when the angle** $ORD$ **is** given, $OR^2$ varies as $OD^2$, **and therefore as** $Ot^2$, and therefore as $OR^2 \sim Ot^2$, or $Rt^2$.

### Corollary 1.

If through any other point $O'$ there be drawn a chord $P'Q'$ *parallel to* $PQ$, and if $p'$, $q'$ be the points corresponding to $p$, $q$, viz. on the eccentric circle of $O'$, then, the ratio $OR^2 : Rt^2$ being the same for any two parallel chords, **it** follows that,

$$OP.OQ : Sp.Sq = O'P'.O'Q' : Sp'.Sq',$$

where the consequents **depend** only upon the positions of $O$, $O'$. If therefore any **second pair of** parallel chords be drawn through the same points $O$, $O'$, we have the general theorem that:

---

* This happens **when** $P$, $Q$ are on **opposite** branches **of a hyperbola**, since $p$, $q$ then lie on **opposite sides of** the directrix.

† This follows most readily in the case of the parabola, since then the circle touches the directrix in $D$, and the ratio in question becomes that of $RO^2$ to $RD^2$, which is constant for a given inclination of the chord.

*The ratio of the rectangles contained by the segments of any two intersecting chords of a conic is equal to that of the rectangles contained by the segments of any other two chords parallel to the former, each to each.*

Taking special cases, we see that this ratio is equal to that of the parallel focal chords (**Prop. IX., Cor.**); and to that of the squares of any pair of tangents parallel to the chords; and, in a central conic, to the ratio of the squares of the semi-diameters parallel to the chords.

Hence also, any two intersecting tangents are to one another in the subduplicate ratio of the parallel focal chords; and, in a central conic, they are in the ratio of the semi-diameters to which they are parallel.

Lastly, to take a case which will be made use of in Prop. XII., if $OTO'$ touch a conic in $T$, and if $OPQ$, $O'P'Q'$ be a pair of parallel chords, then

$$OT^2 : O'T^2 = OP.OQ : O'P'.O'Q'.$$

### Corollary 2.

If a circle and a conic intersect in four points, their common chords will be equally inclined, two and two, to the axis of the conic.* For if $POQ$, $pOq$ be one of the three pairs of common chords of a circle and a conic, the rectangles $PO.OQ$ and $pO.Oq$ will be as the focal chords parallel to $PQ$, $pq$; and the same rectangles will be equal to one another, by a property of the circle. Therefore the focal chords will be equal, and therefore equally inclined to the axis.

### Corollary 3.

Let the conic be a parabola,† so that the eccentric circle touches the directrix in $D$; and let $SD$ meet the circle again in $Z$. Then, for a given inclination of the chord, the rectangle $OP.OQ$ varies as $SD.SZ$. Let $V$ be the extremity of the

---

\* That is to say, each pair of chords will form an isosceles triangle with the axis; but they will not be parallel to one another, except when they are parallel or perpendicular to the axis.

† Another proof will be given in the chapter on the Parabola.

diameter through $D$. Then, since $OZ$ is equal to $OD$ and parallel to $VS$, it is easily seen that

$$OV : SZ = VD : SD = SD : 2SX.$$

Hence $OP.OQ$ varies as $2SX.OV$, and is equal to $F.OV$, where $F$ is the focal chord parallel to $PQ$.

This may also be deduced as a special case from Cor. **1**, by regarding any two diameters as chords $V\infty$ and $V'\infty$, whose further extremities are at infinity; for, if the parallel chords $PQ$, $P'Q'$ meet the two diameters in $O$ and $O'$, then

$$OP.OQ : O'P'.O'Q' = OV.O\infty : O'V'.O'\infty = OV : O'V',$$

since it may be shewn that $O\infty : O'\infty$ is a ratio of equality; and therefore $OP.OQ$ varies as $OV$.

## POLAR PROPERTIES.*

### PROPOSITION XI.

**17.** *If a chord of a conic pass through a fixed point, the tangents at its extremities will intersect on a fixed straight line; and conversely, if pairs of tangents be drawn to a conic from points on a fixed straight line, their chords of contact will pass through a fixed point.*

If $O$ be any point on the chord of contact of the tangents from $T$ to a conic, and if $TL$ be a perpendicular to $SO$, and

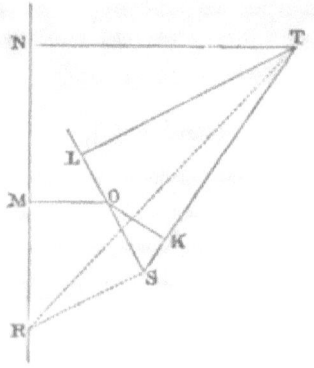

---

\* The theory of *Polars*, although the name is of later origin, was known to **Desargues.** See Poudra's *Œuvres de Desargues*, vol. I., p. 263, (PARIS, 1864).

$OM$, $TN$ be perpendiculars to the directrix, then will the rectangle $SO.SL$ be in a constant ratio to $OM.TN$.

Draw $SP$ to one of the points in which the chord meets the conic, and let $TL'$ be a perpendicular to $SP$.* Upon $ST$ let fall the perpendicular $OK$, and produce it to meet $SP$ in $L''$. Then, since each of the ratios $SL' : TN$, and $SL'' : OM$ (Prop. III., Cor. 2), is equal to the eccentricity, the rectangle $SL'.SL''$ is in a constant ratio to $OM.TN$.

And because the angles at $K$, $L$, $L'$ are right angles, the points $K$, $T$, $L$, $O$, and the points $K$, $T$, $L'$, $L''$, are concyclic. Therefore
$$SO.SL = SK.ST = SL'.SL'',$$
which has been shewn to vary as $OM.TN$. Hence, *if O be a fixed point*, $SL$ varies as $TN$, and the locus of $T$ becomes a straight line,† which meets the directrix at a point $R$, such that $OSR$ is A RIGHT ANGLE. Conversely, if $T$ be taken on the fixed straight line $TR$, the chords of contact will co-intersect at $O$.

When the POLE $O$ lies without the conic, its POLAR, the locus of $T$, is *the chord of contact of the tangents from O*, since these points of contact are evidently points on the locus.

### Corollary 1.

From the above investigation it is evident that, if a point $T$ lies on the polar of $O$, then $O$ lies on the polar of $T$. Take any two straight lines $A$, $B$, and let $a$, $b$ denote their poles. Then the polar of any point on $A$ passes through $a$, and the polar of any point on $B$ passes through $b$, and therefore the polar of the intersection of $A$, $B$ passes through both $a$ and $b$. That is to say, *the intersection of any two straight lines is the Pole of the straight line which joins their two Poles*.

### Corollary 2.

Since every point at infinity in the plane of a central conic is the point of intersection of a pair of tangents whose chord of

---

* See the lithographed figure, No. 1.
† See Scholium B.

contact, being a diameter (Prop. **VIII, Cor.** 1), passes through the centre, all such points at infinity are **on the** polar of the centre, **and may therefore be** regarded as lying **on a** straight line, which is called the *Straight Line at Infinity*.

### Corollary 3.

Since when $O$ is a fixed **point $SL$ varies as $TN$, the straight line which** is the locus of $T$ **is a** tangent (Art. 8), viz. **at the** point in which it meets $SO$, to a conic having the same focus **and** directrix, and whose determining ratio is that of $SL$ to $TN$; **and further,** it will be a tangent to the same conic if $O$ be **no longer fixed,** but subject only to **the** condition that the ratio of $SO$ to $OM$ **is constant.** Hence, if a point $O$ lie on a conic, the envelope **of its polar** with **respect to a** conic having the same focus and **directrix** will be **a third conic having** the same focus and directrix, **and** conversely; **and the** eccentricities of the three conics will be proportionals.

### PROPOSITION XII.

18. *All chords drawn through any point to a conic are cut harmonically by that point, and its polar with respect to the conic.*

Let $HT$, $HT'$ be a pair of tangents to a conic, and $PP'$ a chord which passes through $H$, and cuts the chord of contact $TT'$ in $K$; so that $H$ is on the polar of $K$, and $K$ on the polar of $H$. Through $P$, $P'$ draw parallels to $TT'$, and let them meet the curve in $Q$, $Q'$, and the two tangents in $O$, $O'$ and $R$, $R'$ respectively.

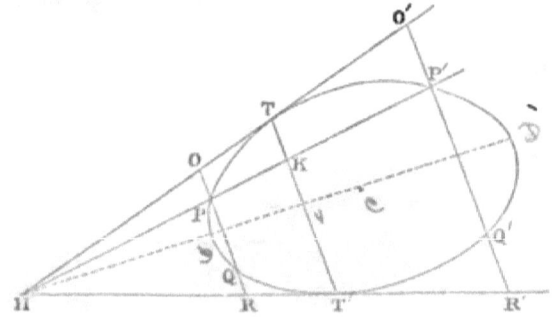

Then since the straight line which bisects $TT'$ and **passes through** $H$ bisects also $OR$ and $O'R'$; and since, by Prop. VIII., Cor. 2, the same straight line is the diameter which bisects the chords $PQ$, $P'Q'$; therefore the intercepts $OQ$, $PR$ are equal, and likewise the intercepts $O'Q'$, $P'R'$.

Hence, and by Prop. X., **Cor. 1,**

$$OT^2 : O'T'^2 = OP.OQ : O'P'.O'Q'$$
$$= OP.PR : O'P'.P'R'$$
$$= OH^2 \quad : O'H^2,$$

by **similar** triangles. That is to say, $HOTO'$ **is cut** harmonically, and therefore $HPKP'$ is cut harmonically.

### Corollary.

The diameter through $H$ **is** divided **harmonically at that point**, and the point in which **it meets the double ordinate** $TT''$. Let it meet the latter **in** $V$, **and the curve in** $D$ **and** $D'$. Then, if $C$ be the centre of the conic, and therefore the middle point of $DD'$, it follows from the nature of harmonic section that $CV.CH = CD^2$. But in the case of the parabola, if $D$ and $\infty$ be the extremities of the diameter through $H$, then $HV$ is divided harmonically at $D$ and $\infty$, and therefore $HV$ is bisected at $D$.

#### SCHOLIUM B.

In Prop. XI, having **shewn that** $SL$, **the projection of** $ST$ **on a** fixed straight line $SO$, **varies as the perpendicular distance of** $T$ from another fixed straight line, the **directrix, we inferred** that the locus of $T$ was a straight line; **and that it met the** directrix at a point $R$ such that $\angle OSR = $ a right angle. **This is virtually** proved in Art. 8, where, leaving the curve out of consideration, we may regard the eccentricity as any constant ratio. In Prop. XI. there is the same ambiguity as in Prop. II, Cor., the locus of $T$ apparently **consisting of** *two* straight lines through $R$. This arises from the **circumstance** that when the *magnitude* only of the ratio $SO : OM$ is given, the point $O$ is **not** completely determined, but the choice lies between two points $O$, $O'$ collinear with the focus, each of which has its own polar. If, however, the actual position of $O$ be given, as in the proposition, then taking into consideration the *sign* of the ratio $SL : TN$ as well as its magnitude, let the direction $SO$ be regarded as positive, and that of $OS$ negative; and let perpendiculars to the directrix from its $S$-side be positive, and those from the further side negative. Then, $SL$, $TN$ being positive or negative **together,** the locus of $T$ is seen to be the single straight line $TR$.

The polar of $O$ is **the** straight line through $R$ *parallel to the ordinates of the diameter through $O$;* for if $\infty$ be the **pole of that** diameter, **and** $V$ its point of concourse **with the directrix,** then (Prop. III., Cor. 1) $S\infty$ is at right angles to $SV$, and is **therefore** (Prop. VIII.) parallel to the ordinates of the diameter $OV$. In **the case** of central conics this follows at once from Prop. VIII., Cor. 1.

If $e$ denote the **eccentricity,** then, referring to the proof of **Prop. XI.,** we see that

$$SO \cdot SL = e^2 \cdot OM \cdot TN.$$

**Hence** (i) if $SO$ be less than $e \cdot OM$, then **will** $SL$, and *a fortiori* $ST$, be greater **than** $e \cdot TN$; but (ii) if $SO$ be greater **than** $e \cdot OM$, then will $SL$ be less than $e \cdot TN$, and $ST$, which **may have** any magnitude not less than $SL$, may be either less **or** greater than $e \cdot TN$. It follows that *the polar of $O$ will cut **or not cut** the conic according as $O$ lies without or within the conic.*

### SCHOLIUM C.

THE POLAR EQUATION of a conic referred to its focus and **axis** may be seen, from Example **4, to be of the form**

$$\frac{l}{r} = 1 + e \cos\theta,$$

where $r$ denotes $SP$; $\theta$ the angle $ASP$; and $e$, $l$ denote the eccentricity **and the** semi-latus rectum. The corresponding equations of **the** Tangent, the Normal, of any Chord, and of the Polar of any **point,** may be deduced, as below, from geometrical theorems which we have already proved.

(i) *The Tangent.*

In Prop. II., let $r$, $\theta$ be the coordinates **of $T$, and let $a$ be the** angular coordinate of the point of contact $P$.

Then $\qquad SL = ST \cos TSL = r \cos(\theta - a),$

and $\qquad e \cdot TN = e(SX - ST \cos AST) = l - e \cdot r \cos\theta.$

**Hence** $\qquad \dfrac{l}{r} = e \cos\theta + \cos(\theta - a).$

(ii) *The Normal.*

In Prop. IV., let **a parallel to the axis cut** $SP$ in $Z$, and $PG$ in $Q$. Denote $\angle ASP$ by **a, and let $r$, $\theta$ be** the coordinates $Q$.

Then $\qquad ZQ = e \cdot ZP = e(SP - SZ),$

or $\qquad e \cdot SP = ZQ + e \cdot SZ;$

and
$$\frac{ZQ}{r} = \frac{\sin(\theta - a)}{\sin a}; \quad \frac{SZ}{r} = \frac{\sin \theta}{\sin a}.$$

Hence
$$\frac{e \cdot SP \sin a}{r} = \sin(\theta - a) + e \sin \theta;$$

or
$$\frac{l}{r} \cdot \frac{e \sin a}{1 + e \cos a} = \sin(\theta - a) + e \sin \theta,$$

since $SP$, $a$ are the coordinates of **a point** on the **curve**.

(iii) *Any Chord.*

From Prop. III., Cor. 2, it is easy to deduce the equation

$$\frac{l}{r} = e \cos \theta + \sec \beta \cos(\theta - a),$$

representing the chord which cuts the **conic** at the points whose angular coordinates are $a \pm \beta$ respectively.

(iv) *The Polar of any Point.*

In Prop. XI., let $r$, $\theta$ be the coordinates of $T$, **and** $\rho$, $a$ those of $O$. Then it is easily seen that

$$e \cdot TN = l - e \cdot r \cos \theta; \quad e \cdot OM = l - e \cdot \rho \cos a; \quad SK \cdot ST = \rho \cos(\theta - a) \cdot r.$$

Hence, equating $SK \cdot ST$ to $e^2 \cdot OM \cdot TN$, we deduce that

$$\left(\frac{l}{\rho} - e \cos a\right)\left(\frac{l}{r} - e \cos \theta\right) = \cos(\theta - a),$$

which is the equation of the polar of the point $(\rho, a)$.

## EXAMPLES.

11. Determine the **pole of the latus rectum of a** conic.

12. Every tangent is the polar **of its point of contact.**

13. The segments of any focal **chord subtend equal** angles at the point in which the directrix meets **the axis.**

14. **If** two conics have a common focus, their common chord or chords will pass through the intersection of their directrices.

15. The tangents at the ends of a focal chord meet the latus **rectum** at points equidistant **from** the focus.

16. The **focal distance** of any point on a conic is equal to the ordinate at that point produced to meet the tangent at an extremity of the latus rectum.

17. The directions of any two tangents to a circle are equally inclined to the diameter through their point of intersection. State this theorem in a form applicable to all conics.

18. Given the focus of a conic and a focal chord, the locus of the extremities of the latus rectum is a circle.

19. Given the focus, the length of the latus rectum, a tangent, and its point of contact, shew how to construct the conic.

20. When the focus and three points of a conic are given, shew how to construct the curve.

21. Given the focus of a conic inscribed in a triangle, determine the points of contact.

22. Given a chord of a conic and the angle which it subtends at the focus, shew that the focal radius to the pole of the chord passes through a fixed point.

23. With given focus and eccentricity construct a conic which shall pass through two given points.

24. Determine in what cases a chord of a conic will be a maximum or a minimum.

25. The portion of any tangent intercepted between the tangents at the ends of the parallel focal chord is divided at its point of contact into segments whereof each is equal to the focal distance of that point.

26. If the tangent at any point of a conic meet the directrix in $D$, and the latus rectum in $L$, then
$$SL : SD = SA : AX.$$

27. If $PM$, $QN$ be the ordinates of the extremities of a focal chord $PQ$, and if the direction of the chord meet the directrix in $R$, then will $RN$ meet $MP$ at a distance from the axis equal to $2PM$.

28. If $M$ be the projection upon the directrix of any point $P$ on a conic, then will $SM$ meet the tangent at the vertex upon the bisector of the angle $SPM$. If a focal chord of central

conic meet the tangents at the vertices in $V$, $V'$, give a construction for determining the points in which the circle on $VV'$ as diameter meets the conic.

29. Prove the following construction for drawing tangents to a conic from a given point $T$. Divide $ST$ in $t$, so that

$$St : ST = AX : TN,$$

where $TN$ is a perpendicular to the directrix; about $S$ as centre describe a circle touching the conic, and from $t$ draw tangents to the circle, and let them meet the tangent at the vertex in $V$, $V'$; draw $TV$, $TV'$, which will be the tangents required.

30. If a chord of a conic subtend a constant angle at the focus, the locus of its pole will be a conic having the same focus and directrix. Shew also that the envelope of the chord will be another conic having the same focus and directrix, and that the eccentricities of the three conics will be proportionals.

31. The vertex of a triangle which circumscribes a conic, and whose base subtends a constant angle at the focus, lies on a conic.

32. Two sides of a triangle being given in position, if the third subtends a constant angle at a fixed point, determine its envelope.

33. If a fixed straight line intersect a series of conics which have the same focus and directrix, the envelope of the tangents to the conics at the points of section will be a conic, have the same focus, and touching both the fixed straight line and the directrix of the series of conics.

34. The focal perpendicular upon any tangent to a conic is a mean proportional to the segments into which it divides the portion of that tangent intercepted between the tangents at the extremities of any focal chord.

35. If $SY$ be the focal perpendicular on the tangent at any point $P$ to a conic, and $X$ the point in which the axis meets the directrix, then

$$SY : YX = SA : AX.$$

Determine the locus of *Y*, **and shew that it is** the envelope of the circle on *SP*.

36. **If** *PN* **be the** perpendicular from any point *P* on a conic to the latus rectum, the straight line connecting *N* with **the** point in which the axis meets the directrix will pass through **the** foot of the perpendicular let fall from **the focus upon the** tangent at *P*.

37. If the diameter at a point *P* on a conic bisects the chord normal at *Q*, the diameter at *Q* bisects the chord normal **at** *P*.

38. **In Art. 10, shew that the normal** *PG* becomes equal to the semi-latus rectum when *P* coincides with the vertex of the conic.

39. The perpendicular from *G* on *SP* varies as the ordinate of *P*; and the foot of this perpendicular lies upon the straight line which passes through the foot of the ordinate of *P*, and is parallel to *SM*.

40. If *Q* be any point on the **normal at** *P*, **and** *L* **and** *M* be its **projections on** *SP* **and the ordinate of** *P*, **shew that**

$$QL : PM = SA : AX.$$

41. The perpendicular upon a focal chord from the intersection of the normals at its extremities meets the chord at a distance from one extremity which is equal to the focal distance of the other; the locus of the foot of this perpendicular is a conic; and the straight **line drawn** parallel **to** the axis through the intersection of **the normals passes** through **the** middle point of **the chord.**

42. **If** *P* **be the pole of** a normal chord which meets the directrix **in** *Q*, **shew that** the circle *SPQ* passes through an extremity **of the chord.**

43. **If a circle touch** a conic on opposite sides of its axis, it will intercept a constant length upon the focal chords through the points of contact. When the circle passes through the focus, determine the focal radii to the points **of** contact.

44. The rectangle contained by the focal perpendicular upon the tangent at any point to a conic and the portion of the normal intercepted between the curve and its axis is equal to the rectangle contained by the semi-latus rectum and the focal distance of the point.

45. If $QQ'$ be a focal chord of a conic, and if the normal at $P$ be at right angles to the chord and meet the axis in $G$, then
$$PG^2 = SQ.SQ'.$$

46. Shew also that, if a parallel to the chord be drawn through $G$ and meet the direction of $PS$ in $U$, then $PU = \frac{1}{2}QQ'$.

47. If the normal to a conic at $P$ meet the axis in $G$, and if $SY$ the focal perpendicular upon the tangent meet the directrix in $V$, shew that
$$PG : SY = SV : VY.$$

48. The ratio of the normals, terminated by the axis, at any two points of a conic is equal to that of the tangents at those points.

49. Given an arc of a conic, shew how to construct the curve.

50. The parallel diameters of two similar and similarly situated conics bisect the same systems of parallel chords. If the two conics be concentric ellipses or hyperbolas, or equal parabolas whose axes are coincident, shew that any chord of the exterior conic is divided into pairs of equal segments by the interior, and that any chord of the former which touches the latter is bisected at the point of contact.

51. The angle between any two chords of a conic is equal to the angle subtended at the focus by the portion of the directrix intercepted by the diameters which bisect the chords.

52. The arms of the angle which a focal chord of a conic subtends at any point on the circumference meet the directrix upon diameters through the points of contact of tangents at right angles.

53. The polar of any point with respect to a conic meets the directrix on the diameter which bisects the focal chord through that point.

54. The diameter through any point, and the polar of that point, meet the directrix and the axis respectively on a straight line parallel to the focal distance of the point. Hence shew that the foot of the ordinate of any point in the plane of a central conic is at a distance from the centre which varies inversely as the distance therefrom of the intersection of the polar of the point with the axis.

55. From the preceding example deduce a construction for drawing tangents to a conic from a given point.

56. The triangle whose angular points are the focus of a conic and the intersections of the tangent and the diameter at any point with the axis and the directrix respectively has its orthocentre at the point in which the tangent meets the directrix.

57. Given the focus and the directrix of a conic, shew that the polar of a given point with respect to it passes through a fixed point.

58. If the polar of a point $O$ with respect to a conic intersect a conic having the same focus and directrix in $P$, and if $SQ$ be drawn at right angles to $SP$ to meet the directrix in $Q$, the locus of the intersection of $QO$ and $SP$ will be a conic.

59. Deduce from Art. 16 that the square of the ordinate at any point of a conic varies either as the distance of the foot of the ordinate from the vertex, or as the rectangle contained by the segments into which it divides the axis.

60. A focal chord of a conic and the diameter which bisects it meet any fixed straight line perpendicular to the axis at points whose ordinates contain a constant rectangle; and the square of the ordinate of the middle point of the chord varies either as the distance of the foot of the ordinate from the focus, or as the rectangle contained by its distances from the focus and the centre of the conic.

61. If a chord of a conic passes through a fixed point in the axis, determine the locus of its middle point, and in the case of a central conic, the locus of its intersection with another chord which passes through a fixed point in the axis and is parallel to the diameter which bisects the former.

62. If a tangent be drawn parallel to any chord of a conic, the portion of it terminated by the tangents at the ends of the chord is bisected at its point of contact.

63. Two tangents being applied to a line of the second order, if from any point in one of them a straight line be drawn parallel to the other, the portion of it intercepted by the chord joining the points of contact will be a mean proportional to its segments made by the curve. Examine the case in which the secant becomes a tangent.

64. In Art. 16, investigate the case in which $O$ coincides with $S$, and shew that $SX$ is then a mean proportional to the distances of $P$ and $p$ from the directrix.

65. Shew also that, if $SZ$ be drawn parallel to $PQOR$ to meet the directrix, then
$$OP.OQ : Sp.Sq = SZ^2 : SZ^2 - L^2,$$
where $L$ denotes the semi-latus rectum.

66. If a chord of a conic subtends equal angles at the extremities of another chord, it likewise subtends equal angles at the extremities of any chord parallel to the latter.

67. If $ABC$ be a triangle whose sides touch a conic at the points $a, b, c$, then
$$Ab.Bc.Ca = Ac.Ba.Cb.$$

68. If any conic be drawn through four given points, and if a fixed straight line meet the conic in $P$, $Q$, and one of the pairs of straight lines joining the four points in $A$, $B$, then will the ratio of the rectangle $PA.AQ$ to the rectangle $PB.BQ$ be constant.

69. Any tangent to a conic is divided harmonically by its point of contact and the three points in which it meets any two other tangents and their chord of contact. Examine the cases in which two of these four straight lines become parallel.

70. If from any point on a conic parallels be drawn to two adjacent sides of a given inscribed quadrilateral figure, the rectangles under the segments intercepted by those adjacent and by the other two opposite sides will have a given ratio.

71. If $ABC$ be a **triangle inscribed in a conic,** and if from any point $O$ on the curve there be drawn a parallel to $BA$ meeting $BC$ **and the tangent at** $A$ in $P$, $Q$, **and a parallel to** $BC$ meeting $AB$, $AC$ in $P'$, $Q'$; then will $OP.OQ$ be to $OP'.OQ'$ in a constant ratio, viz. that of the **focal** chords parallel to $BA$ and $BC$ respectively.

72. If from any point on a conic pairs of perpendiculars be drawn to the opposite sides of a given inscribed quadrilateral, the rectangle contained by the one pair of perpendiculars will be in a constant ratio to the rectangle contained by the other pair.

73. The perpendicular from any **point on a conic to** a fixed chord is a mean proportional to the perpendiculars from that point to the tangents at the extremities of the chord.

74. If from any point on a conic straight lines be drawn at given angles to two adjacent sides of a given **inscribed** quadrilateral figure, the rectangle under the segments intercepted by those adjacent and by **the other two opposite sides** will have a given ratio.

75. **Hence shew that, if from a** given point $M$ **there be drawn two fixed straight lines** meeting **a** conic in $A$, $B$ and $C$, $D$; **and likewise a** variable straight line meeting the curve in $E$, $E'$, **and** the straight lines $AC$, $BD$ in $K$, $L$; then

$$EM^2 : E'M^2 = LE.EK : LE'.E'K;$$

and investigate the form which this relation assumes when the fixed straight lines become tangents to the conic.

76. Deduce from the preceding **example that, if** $A$, $B$, $C$, $D$ be any four points on a conic, the **three** straight lines joining the intersections of $AB$, $CD$; $BC$, $DA$; and $CA$, $BD$, are cut harmonically by the curve, **and** that each of these points is the pole **of the** straight line which **joins the other two.**

77. Hence **shew how to draw tangents to a conic from any** external point **with the help of the ruler** only.

78. If $PVP'$ and $QVQ'$ be any two intersecting chords of a conic, and if the circle through $Q$, $P$, $Q'$ meet $PP'$ in $R$, then

will the ratio of $VP'$ to $VR$ be equal to that of the focal chords parallel to $PP'$ and $QQ'$. Examine the cases in which two or all of the points $P$, $Q$, $Q'$ coalesce.

79. If $PQ$ be any chord of a conic, and $F$ the parallel focal chord, and if the direction of $F$ meet the tangent at $P$ in $T$, then
$$PQ.ST = F.SP.$$

80. If there be a quadrilateral figure inscribed in a conic section, and if from one of its angular points there be drawn parallels to the sides about the opposite angle; and if from the two remaining angles there be drawn straight lines to any point in the curve to meet the parallels; the intercepted portions of the parallels, estimated from their common point, will have a given ratio, wherever in the curve the fifth point be taken.

# CHAPTER III.

## THE PARABOLA.

**19.** The parabola being a conic whose determining ratio is one of equality, some of its properties may be at once deduced by equating $SA : AX$ to unity from properties of the general conic already proved; thus $SL$ becomes equal to $TN$ in Art. 8, and $SG$ equal to $SP$ in Art. 10. The semi-latus rectum of the parabola is equal to $SX$; that is, to $SA + AX$, or $2SA$.

Other properties of the parabola may be derived from those of central conics by regarding it as a conic whose centre* and second focus are at infinity, and the further extremities of whose diameters† are likewise at infinity; but in the present chapter we shall give independent proofs of such properties, commencing with the original definition of the parabola.

The portion of any diameter intercepted between the curve and the ordinate of any point with respect to that diameter is called the *Abscissa* or *Absciss* of the point; and any focal chord of a parabola is called the *Parameter* of the diameter which bisects it.

### CHORD PROPERTIES.* †

#### PROPOSITION I.

**20.** *The ordinate of any point on the parabola is a mean proportional to the abscissa and* **the latus rectum.**

---

\* See Art. 14, Cor. 3.

† See Art. 16, Cor. 3.

‡ Under this head are included such propositions only as can be proved antecedently to the definition of a tangent; but the restriction does not apply to the *Corollaries* from those Propositions.

Let $AN$ be the abscissa of any point $P$ on the curve, and $X$ the point in which the directrix meets the axis. Then, by Euclid I. 47, and from the definition of the parabola,

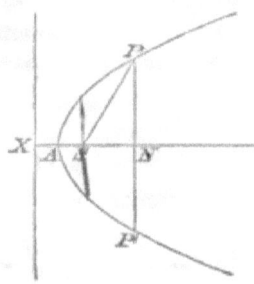

$$PN^2 + SN^2 = SP^2 = NX^2.$$

Hence $\quad PN^2 + (AN \sim AS)^2 = (AN + AS)^2.$

Therefore $PN^2 = 4AS \cdot AN$, or $PN$ is a mean proportional to $AN$ and $4AS$, which latter, by Art. 19, is equal to the latus rectum.

Conversely, if the square of the ordinate of any point $P$ vary as its abscissa, the locus of the point will be a parabola.

The above proposition suggests an obvious method of tracing the curve, since for any assumed magnitude of $AN$ the magnitude of $PN$ and the position of $P$ are determined.

### Corollary.

Hence, to find two mean proportionals between a given pair of straight lines,* with latera recta equal to the given lines describe two parabolas, having a common vertex, and their axes at right angles; then will the ordinates of either of their points of intersection be mean proportionals to their latera recta, as required; for it is evident that the ordinate in either parabola will be a mean proportional to its own latus rectum and the ordinate in the other.

---

* This problem, which is of great historical interest, was solved as above by Menæchmus, according to the statement of Eutokius. Compare Bretschneider's, *Die Geometrie und die Geometer vor Euklides,* p. 160 (LEIPZIG, 1870).

## PROPOSITION II.

**21.** *The locus of the middle points of any system of parallel chords of a parabola is a straight line parallel to the axis; and the bisecting* **line meets the directrix on the straight line through the focus at** *right angles* **to** *the common* **direction of the chords.**

Take $QQ'$, **any one of a** system of parallel **chords**, and let $M$ **and** $M'$ be the projections of its extremities upon the **directrix**.

**Let the** focal perpendicular upon the chords meet $QQ'$ **in** $Y$, **and the** directrix **in** $O$; **and through** $O$ draw a parallel to the **axis** meeting $QQ'$ in $V$. Then will $V$ be the middle point of $QQ'$.

For
$$OM^2 = OQ^2 - QM^2 = OQ^2 - SQ^2$$
$$= \overline{OY^2 - SY^2};$$

and $OM'^2$ may be shewn to have the **same value.**

Therefore $OM$, $OM'$ being equal, **the straight** line through

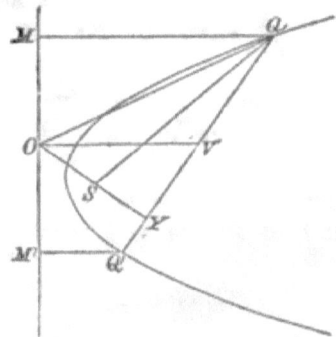

$O$ parallel to the axis bisects $QQ'$; that is **to say, it** bisects every chord which is at right angles to $OS$.

Hence it is evident that every **straight line** parallel to the axis of a parabola is **a** diameter **of the** curve, and that all diameters are parallel **to the axis and to** one another.

### Corollary.

**It** follows, as **a** particular case of the above proposition, that the direction of the focal perpendicular $SY$ on the tangent at $P$ to a parabola meets the directrix at a point $M$ such that $PM$ is parallel to the axis.

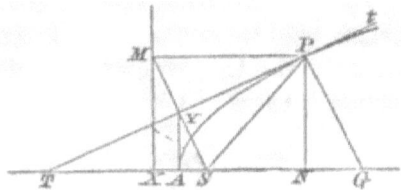

Hence it appears that the tangent at $P$ bisects the angle $SPM$, as will be otherwise proved in Art. 25; and it may also be deduced, independently of Art. 7, that the intercept on the tangent made by the curve and the directrix subtends a right angle at $S$.

PROPOSITION III.

22. *To find the length of any focal chord of a parabola.*

Let $QQ'$ be any focal chord; $M$ and $M'$ the projections of its extremities upon the directrix; and $O$ the point in which

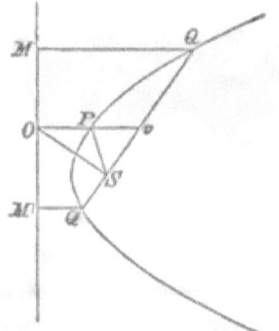

the focal perpendicular upon the chord meets the directrix. Let a parallel through $O$ to the axis meet $QQ'$ in $v$, which, by Prop. II, will be the middle point of the chord.

Hence, and from the definition of the curve,

$$QQ' = SQ + SQ' = QM + Q'M'$$
$$= 2vO.$$

And because $OSv$ is a right angle, and $SP = PO$; therefore $Pv = SP = PO$, and therefore $vO = 2SP$.

Hence $QQ' = 4SP$, or the parameter of any diameter of a parabola is equal to four times the focal distance of the extremity of that diameter. In particular, as we have already seen, the **latus rectum is equal to** $4SA$.

### PROPOSITION IV.

23. *The ordinate of any point on a parabola with* **respect** *to any diameter is a mean proportional to its parameter* **and** *the abscissa of the point.*

Let $QV$ and $PV$ be the ordinate and abscissa of any point $Q$ on the curve; let $VP$ meet the directrix in $O$, and the focal

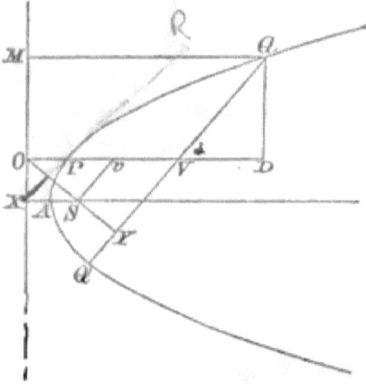

chord parallel to $QV$ in $v$; and let $OS$, which (Prop. II.) is at right angles to $Sv$ and $QV$, meet the latter in $Y$.

Then, as in Art. 21, if $D$ and $M$ be the projections of $Q$ on the diameter $PV$, and on the directrix,

$$QD^2 = OM^2 = OY^2 - SY^2.$$

And since, by similar triangles, the lengths $QD$, $OY$, $SY$ are proportional to $QV$, $OV$, $vV$, therefore, from above,

$$QV^2 = OV^2 - vV^2.$$

And since $Pv = SP = PO$, as in Art. 22, the sum of $OV$ and $vV$ is equal to $2PV$, and their difference to $2SP$, or *vice versâ*; and therefore, in either case, the difference of their squares is equal to $2SP \cdot 2PV$.

THE PARABOLA.   49

Therefore $QV^2 = 4SP.PV$, or the ordinate $QV$ is a mean proportional to the parameter $4SP$ (Prop. III), and the abscissa $PV$.

### Corollary 1.

It may be shewn that $QD^2 = 4AS.PV$; and further, that if a straight line $QD'$ be drawn *in any direction* from $Q$ to the diameter $PV$, it will be a mean proportional to the parallel focal chord and the abscissa $PV$. This follows most readily with the help of the theorem (Art. 30, Cor. 1), that *if the base of a triangle be parallel to the axis of a parabola the squares of its remaining sides will be as the parallel focal chords.*\*

### Corollary 2.

If the tangent at $P$ meet $QM$ in $R$, then the figure $PVQR$ being a parallelogram, it follows that $RP^2 = 4SP.RQ$. Hence, if $R$ be any point on the tangent at a given point $P$ to a parabola, and if the diameter through $R$ meet the curve in $Q$, then will $RP^2$ vary as $RQ$.

### Corollary 3.

On the tangent at a given point $P$ to a parabola take any two points $T$ and $R$; and let the diameters through them meet in curve in $E$ and $Q$, and let the former diameter meet $PQ$ in $F$. Then, by Cor. 2, and by similar triangles,

$$TE : RQ = TP^2 : RP^2 = TF^2 : RQ^2.$$

Hence    $TE : TF = TF : RQ = PF : PQ$,

or, *the portion of any diameter intercepted by any chord and the tangent at either extremity of the chord is divided at the curve in the same ratio as that in which it divides the chord.*

### PROPOSITION V.

**24.** *A chord of a parabola being divided at any point, to determine the magnitude of the rectangle contained by its segments.*

Let any chord $QR$ be divided internally or externally at the point $O$; and let the diameters through $O$ and the middle

---

\* This may be deduced without the help of tangent-properties from the second note on p. 28, or from Art. 22, where $QQ'$ varies as $SQ.SQ'$, that is to say, as $SO^2$, the angle $QOQ'$ being a right angle. It follows that the focal chords of a parabola vary inversely as the squares of the sines of their inclinations to the axis.

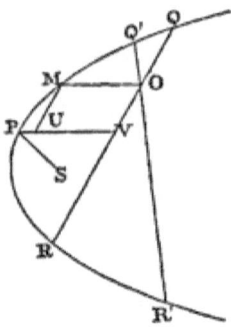

point $V$ of the chord meet the parabola in $M$ and $P$. Then by Prop. IV, and by Euclid II. 5, Cor., if $MU$ be the ordinate of $M$ with respect to the latter diameter,

$$QO.OR = QV^2 \sim OV^2 = QV^2 \sim MU^2$$
$$= 4SP(PV \sim PU)$$
$$= 4SP.MO,$$

or the rectangle whereof the magnitude was to be found varies as $MO$, which depends only upon the *position* of the point $O$; and as the parameter $4SP$, which depends only upon the *direction* of the chord $QR$.

### Corollary.

Hence, if $Q'R'$ be any second chord through $O$, and $4SP'$ the corresponding parameter,

$$QO.OR : Q'O.OR' = 4SP : 4SP',$$

or these rectangles are proportional to the focal chords parallel to $QR$, $Q'R'$, as was proved also for the general conic in Art. 16. Hence also, the squares of any two intersecting tangents are as the focal distances of their points of contact.

## TANGENT PROPERTIES.*

### PROPOSITION VI.

25. *The tangent to a parabola at any point is the bisector of the angle which the focal radius makes with the diameter produced.*

---

* See also the Corollaries in Articles 21, 23, 24.

(i) **Let the tangent at any point** $P$ **meet the directrix in** $R$, and let the diameter produced beyond the **curve meet the** directrix in $M$.

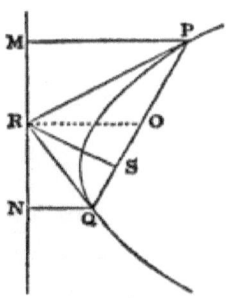

Then since **PR** subtends **a right angle at** $S$, **and since** $SP = PM$, **and** $PR$ **is common to the right-angled triangles** $SPR$, $MPR$, **therefore their angles at** $P$ **are equal, or the** tangent $PR$ bisects the **angle** $SPM$.

It is likewise evident that $SP$ and $MP$ make equal angles with $RP$ produced towards $t$, as in the figure of Art. 21, Cor.; and that if $PR$, or $PR$ produced, meet the axis in $T$, **the** angles at $P$ and $T$ in the triangle $SPT$, and therefore also the sides $SP$ and $ST$, will be equal to one another.

(ii) **Or we may proceed as** follows, taking EUCLID'S definition of a tangent.

Draw the straight line bisecting the angle $SPM$, and take any point upon it. The distance of any such point from $S$ is equal to its distance from $M$, and **therefore greater than** its distance from the directrix, **except when the point** coincides with $P$. Hence every point except $P$ on the bisector of the angle $SPM$ lies without **the curve, and the bisector of** $SPM$ **is therefore the tangent at** $P$.

*Corollary* **1.**

It **is evident** from the above that the tangent at $P$ bisects the angle $SRM$ between the directrix and the focal distance of the point $R$ in which it meets the directrix; and, in like manner, that the second tangent $RQ$ from $R$ bisects the supplementary angle $SRN$ between $SR$ and the directrix. Hence the two tangents to a parabola from any point on its directrix, or at the extremities of any focal chord, are at right angles; and conversely, the directrix of a parabola is the locus of the intersection of tangents at right angles.

### Corollary 2.

To draw tangents to a parabola from any external point $E$, with centre $E$ and radius $ES$ describe a circle cutting the directrix in $M$ and $N$, and let the diameters through those points meet the curve in $P$ and $Q$, which will be the points of contact of the tangents required; since, as readily appears, $EP$ bisects the angle $SPM$ and $EQ$ bisects the angle $SQN$.

### Corollary 3.

The subtangent at any point is double of the abscissa; since, in the next figure, $ST = SP = NX = AN + AS$, and therefore $AN = ST - AS = AT$, or $NT$ the subtangent is equal to $2AN$.

### PROPOSITION VII.

**26.** *The normal at any point of a parabola bisects the interior angle between the diameter and the focal distance of the point.*

If the normal at $P$ meet the axis in $G$, then, by Art. 19,

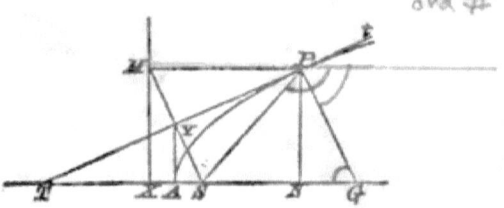

$SG = SP$; and therefore $PG$ makes equal angles with $SP$ and the axis, and bisects the angle which $SP$ makes with the diameter through $P$.

The same might have been deduced as a corollary from the preceding proposition.

*Corollary.*

If $AN$ be the abscissa of $P$, then since $SG = SP = NX$, therefore $NG = NX - SN = SX = 2SA$, or the **subnormal is equal to the semi-latus rectum**.

## PROPOSITION VIII.

**27.** *The tangent at the vertex of a parabola is the locus of the foot of the focal perpendicular upon the tangent at any point; and the focal perpendicular is a mean proportional to the focal distances of the vertex and of the point of contact of the variable tangent.*

(i) Let the diameter at any point $P$ of a parabola be produced to meet the directrix in $M$, and let the tangent at $A$ meet $SM$ in $Y$.

Then, $SY$ being evidently equal to $MY$, and $SP$ being equal to $PM$, and $PY$ common to the triangles $SPY$, $MPY$; therefore $PY$ is at right angles to $SM$, and it bisects the angle $SPM$, and is therefore the tangent at $P$.

Hence $Y$, which by construction lies on the tangent at $A$, is the foot of the focal perpendicular upon the tangent at $P$; and conversely, the locus of the foot of the focal perpendicular on the tangent at $P$ is the tangent at $A$.

This suggests an obvious method of drawing a second tangent to a parabola from a given point on the tangent at its vertex.

(ii) Again, since the two tangents from $Y$ to the parabola subtend equal angles at $S$, the right-angled triangles $SAY$, $SYP$ are similar, so that

$$SA : SY = SY : SP,$$

or
$$SY^2 = SA.SP.$$

*Corollary.*

Since (**Art. 24, Cor.**) any **two** intersecting tangents to a parabola are in the subduplicate ratio of the focal distances of their points of contact, **they** are in the same ratio as the focal perpendiculars upon them.

## PROPOSITION IX.

**28.** *The exterior angle between any two intersecting tangents to a parabola is equal to the angle which either of them subtends at the focus; and the inclination of either tangent to the axis is equal to that of the other to the focal distance of their common point.*

(i) Let the tangents at $P$ and $Q$ intersect in $R$, and meet the axis in $T$ and $U$; and let $O$ be a point in $AS$ produced.

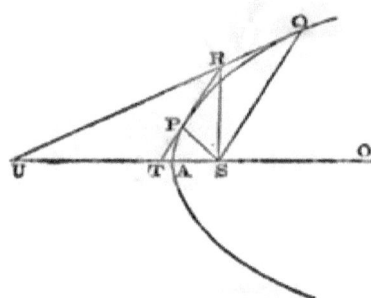

Then the exterior vertex angle $PSO$ of the isosceles triangle $PST$ being double of the interior base angle $STP$, and the angle $QSO$ in like manner being double of $SUQ$, therefore by subtraction, in the figure drawn,

$$\angle PSQ = 2 TRU.$$

Therefore, since the two tangents subtend equal angles at $S$, the angle subtended by either is equal to the exterior angle $TRU$ between them.

Hence $\angle TRU$ will be acute or obtuse according as the focus lies without or within the segment of the curve cut off by $PQ$. In either case it will be seen that the *acute angle* between the tangents is equal to half the angle which their chord of contact subtends at the focus.

(ii) Since the angle $TRU$ is equal to $PSR$, therefore

$$\angle SRU = PSR + PRS = SPT$$
$$= STP;$$

or the angles which $QR$ makes with $SR$ are equal to those which $TR$ makes with the axis.

Hence also, subtracting the angle $TRU$,
$$\angle SRP = STP - TRU = SUQ$$
$$= SQU.$$

*Corollary* 1.

If from any **point** $R$ on the tangent **at a fixed point** $P$ the second tangent $RQ$ be drawn, **the angle** $SRQ$ will be *constant*, since the equal angle $SPR$ is fixed. If one of the tangents be the tangent at the vertex, $SR$ will be the focal perpendicular upon the other.

*Corollary* 2.

The triangles $SPR$, $SRQ$ are similar, having their **angles at** $S$ equal, and likewise those opposite to $SP$ and $SR$ **respectively**. Hence $SR^2 = SP.SQ$, or the focal distance of the intersection of any two tangents to a parabola is a mean proportional to the focal distances of their points of contact; and each tangent is to the other as $SR$ to the focal distance of the point of contact of the latter.

*Corollary* 3.

If two fixed tangents be cut by any third in points **P** and $Q$, as in the next figure, the triangle $SPQ$ will have its angles constant, since, by Cor. 1, its angle at $P$ is constant, and likewise its angle at $Q$. Again, in the same figure, if the three tangents be fixed, and if any fourth cut them in points $L$, $M$, $N$, then, the angles of the triangles $SLN$, $SMN$, being constant by the former case, the ratio of $LN$ to $MN$ is constant. Conversely, the envelope **of a straight** line which is cut in a constant ratio **by three fixed straight** lines is a **parabola** touching the three fixed lines.

PROPOSITION X.

29. *The circumscribed circle of any triangle whose three sides touch a parabola passes through the focus.*

Let $PQR$ be any triangle whose three sides touch a parabola, and let $PR$ meet the axis in $T$. Then, by Art. 28,
$$\angle SRQ = STP = SPQ.$$

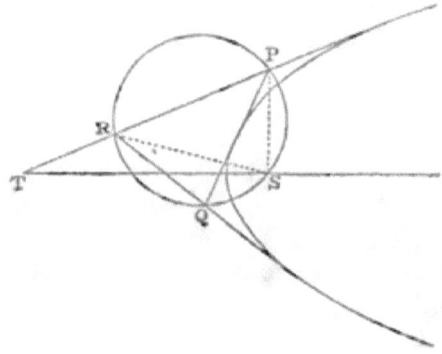

Therefore the points $S, P, R, Q$ are concyclic; that is to say, the focus $S$ lies on the circumscribed circle of the triangle $PQR$.

*Corollary* 1.

Let $p, q, r^*$ be the **points** of contact of the tangents $QR$, $RP, PQ$, and let $PQ$ meet the directrix in $D$, so that $\angle rSD$ is a right angle. Let the perpendicular drawn from $R$ to $PQ$ meet the directrix in $O$, and $SD$ in $N$. Then the angles $PQR$ and $QSr$ are equal by Prop. IX., and therefore their complements are equal, so that

$$\angle QRN = QSN,$$

or $N$ lies on the circle $QRS$, which also circumscribes the triangle $PQR$. Moreover, $PQ$ bisects the angle $ODN$ (Art. 25), and therefore also the line $ON$, to which it is at right angles. Hence $O$ is the orthocentre of the triangle $PQR$, or *if any parabola be inscribed in a triangle, **its directrix will pass through the orthocentre.***

*Corollary* 2.

If four **tangents to a** parabola **be** given, its *focus* is determined by the intersection of the circumscribed circles of any **two of** the triangles formed by the four tangents, and its *directrix* **is the straight** line joining the orthocentres of any two of them. Hence **it** appears that one parabola can in general be **described** touching four given straight lines.

---

* See the lithographed figure, No. 2.

## Corollary 3.

Since (Art. 13, Cor.) $RQ$ and $Pq$ subtend equal angles at $S$; and since, in the circle, $\angle SPq = SQR$; therefore the triangles $SPq$, $SQR$ are similar, so that

$$QR : Pq = SR : Sq.$$

Hence, and by Art. 28, Cor. 2,

$$QR : Pq = Rp : Rq = Qp : PR;$$

or, if two tangents to a parabola be cut by any third, their alternate segments will have the same ratio, and this ratio will be *constant* if the two tangents be fixed.

### SCHOLIUM.

The above proposition, with several deductions therefrom, is found in Section I. §§ 15—24 of I. H. LAMBERT's *Insigniores Orbitæ Cometarum Proprietates* (Augustæ Vindelicorum, 1761). The proposition itself, together with Cor. 1, may be applied to prove certain properties of the straight line and circle, as below.

(i) In any given triangle, and with any point on its circumscribed circle as focus, suppose a parabola to be inscribed. Then, since the sides of the triangle are tangents to the parabola, the feet of the three focal perpendiculars upon them must lie on the tangent at the vertex. Hence, if from any point on the circumscribed circle of a triangle perpendiculars be let fall upon its three sides, the feet of the three perpendiculars will be collinear.

(ii) Supposing a parabola to be described touching four given straight lines, its focus must lie on the circumscribed circle of the triangle formed by any three of the said lines. Hence the circumscribed circles of the four triangles formed by any four straight lines meet in a point.

(iii) The directrix of the parabola touching four given straight lines passes through the orthocentres of the four triangles formed by those lines. Hence the orthocentres of the four triangles formed by any four straight lines are collinear.

For the proof of Cor. 1 given above I am indebted to Mr. Rawdon Levett, of St. John's College, Cambridge. Another elementary proof, based upon the property that the feet of the focal perpendiculars on the three tangents are collinear, was given in No. 160, p. 63, of the *Lady's and Gentleman's Diary* (1863). The theorem in question, which is in reality a particular case of Brianchon's theorem (Salmon's *Conic Sections*, Art. 268), was propounded by J. STEINER in Crelle's *Journal für die reine und angewandte Mathematik*, vol. II. p. 191 (Berlin, 1827), and was demonstrated by him in Gergonne's *Annales de Mathématiques pures et appliquées*, vol. XIX. p. 59 (Paris, 1828), with the help of Pascal's theorem, as follows. If $S$ be any point on a conic, $PQR$ an inscribed triangle, and $PP'$, $QQ'$ chords through any point $O$; then in the hexagon $PP'SQQR$ the points $(P'S, QR)$ and $(SQ', RP)$ will lie on a straight line through $O$. If the conic be a *circle*, and $O$ the *orthocentre* of the triangle, the straight line through $O$ will evidently meet the continuations of the perpendiculars from $S$ to the sides of the triangle at distances from $S$ which are respectively double of those perpendiculars, and will therefore be the directrix of the parabola drawn with $S$ as focus to touch the sides. Steiner himself likewise applied his theorem as in § iii. (Crelle, II. 97; Gergonne, XIX. 59).

### PROPOSITION XI.

30. *The portion of any diameter intercepted by any tangent and the ordinate of its point of contact with respect to that diameter is bisected at the curve.*\*

Let the diameter at $P$ be met by the tangent at $Q$ in $T$, and by the ordinate of $Q$ in $V$; and let the tangent at $P$ meet that at $Q$ in $R$.

Complete the parallelogram $QRPO$ by drawing $PO$ parallel

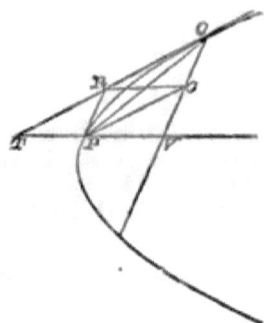

to $RQ$. Then the diagonal $RO$ bisects the diagonal $PQ$, which is also the chord of contact of the tangents $RP$, $RQ$. Therefore $RO$ is a diameter of the parabola, and hence, all diameters being parallel,
$$PV = RO = PT,$$
or $VT$ is bisected at $P$.

---

\* This is included in Art. 23, Cor. 3. See also Art. 18, Cor., and Art. 25, Cor. 3.

## THE PARABOLA.

### *Corollary* 1.

Any triangle whose base is parallel to the axis of a parabola has its remaining sides in the ratio of the parallel tangents; for supposing those sides parallel to the tangents in the figure, their ratio will be that of $RP$ to $RT$, whereof the latter is equal to $RQ$.

### *Corollary* 2.

If from any external point $R$ there be drawn a tangent meeting the curve in $P$, and a chord meeting the curve in $M$, $N$, and the diameter through $P$ in $V$, then by Cor. 1, and by Art. 16, Cor. 1, it is easily shewn that $RV^2 = RM.RN$.

### *Corollary* 3.

It may be deduced from the proposition that the intercepts upon any diameter made by any two tangents and the ordinates of their points of contact are equal; and hence, that the area between the two tangents and the diameter is equal to half the area between their chord of contact, the ordinates of its extremities, and the diameter; and hence, that the triangle made by any three tangents is equal to half the triangle formed by joining their points of contact.

## QUADRATURE.

### PROPOSITION XII.

32. *The area of the parabolic segment upon any chord as base is equal to once and* **one-third of a triangle** *having the same base and altitude.*[*]

Take $RR'$ as the base of the segment, and suppose it parallel to the tangent at $P$.

Let the diameters through $R$ and through an adjacent point $Q$ on the curve meet the tangent at $P$ in $M$ and $O$; and let the diameter through $P$ meet $RQ$ in $T$, and $RR'$ in $U$;

---

[*] This theorem, one of the great discoveries of ARCHIMEDES, was the first example of the exact quadrature by infinitesimals of a continuous curvilinear area. It forms the twenty-fourth and last proposition in his special treatise on the Quadrature of the Parabola. See the Oxford edition of his works, p. 33 (1792).

and let $PV$ be the abscissa of $Q$. Complete the parallelogram $UTLR$ by drawing $TL$ parallel to the base of the segment to meet $RM$ produced.

Let $Q$ coalesce with $R$, so that the chord $QR$ becomes a tangent, and $P$ becomes the middle point of $VT$, and therefore $PM$ bisects the parallelogram $QL$.

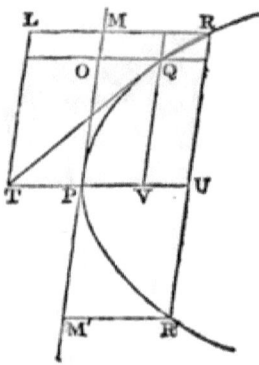

Hence, and by Euclid I. 43,

the parallelogram $QU = QL = 2QM$.

Through any number of points on the arc $PR$ draw parallels to $RR'$ and $PU$, so as to form with $PU$ two series of parallelograms, the one corresponding to $QU$ and the other to $QM$, and let the number of the points be increased and their successive distances diminished indefinitely.

Then, as above, the several parallelograms in the former series become double of those in the latter, and the sum of the former, which is ultimately the parabolic area $RPU$, becomes double that of the latter, or of the parabolic area $RPM$.

Hence the semi-segment $RPU$ is equal to two-thirds of the parallelogram $MU$, or to four-thirds of the triangle $RPU$; and the whole segment $RPR'$ is equal to four-thirds of the triangle $RPR'$, which has the same base and altitude.

*Corollary.*

Let the tangents at $R$, $R'$ meet in $T$. Then the area of the segment is equal to two-thirds of the triangle formed by these tangents and its base, or the portion of the triangle on the concave side of the arc is double of that on its convex side.

This might have been proved by connecting the points $R$, $R'$ by an infinity of consecutive chords, drawing the tangents at their extremities, and shewing, after the manner of Prop. XI., **Cor. 3**, that the area between the successive chords and $RR'$ is double of that between the corresponding tangents and the tangents at $R$, $R'$.

## *EXAMPLES.*

81. The radius of the circle through the vertex and the extremities of the latus rectum of a parabola is equal to five-eighths of the latus rectum.

82. A point on a parabola being given, if the focus also be given the envelope of the directrix will be a circle; or if the directrix be given the locus of the focus will be a circle.

83. If two parabolas have a common focus their common chord passes through the intersection of their directrices and bisects the angle between them.

84. The common chord of two parabolas which have a common directrix bisects the straight line joining their foci at right angles.

85. Deduce from Prop. I. that the ordinate of the middle point of a chord whose direction is given is of constant magnitude.

86. The perpendicular to a chord of a parabola from its middle point and the ordinate of that point intercept on the axis a length equal to the semi-latus rectum. Hence shew that the locus of the middle point of a focal chord, or of any chord which meets the axis in a fixed point, is another parabola.

87. Prove the following construction. Let $AN$ be the abscissa of any point $P$ on a parabola, and let $MP$ be equal and parallel thereto. Divide $NP$ into any number of equal parts and through the points of section draw parallels $p_1, p_2, p_3 \ldots$ to the axis, and divide $MP$ into the same number of equal parts in points 1, 2, 3.... Then will the lines $p_1, p_2, p_3 \ldots$ meet $A1$, $A2$, $A3\ldots$ respectively on the parabola.

88. If $PQ$ be a focal chord of a parabola, $SA.PQ = SP.SQ$.

89. If the ordinates or the focal distances of all points on a parabola be cut in a given ratio the locus of the points of section will in either case be a parabola.

90. Circles whose radii are in arithmetical progression touch a given straight line on the same side at a given point. If to each circle a tangent parallel to the given line be drawn it will cut the circle next larger in points lying on a parabola.

91. Find the locus of the centre of a circle which passes through a given point and touches a given straight line; or which touches a given circle and a given straight line.

92. If a parabola be made to roll upon an equal parabola, their vertices being initially coincident, the locus of the focus of the former will be the directrix of the latter.

93. Find the locus of a point which moves so that its shortest distance from a given circle is equal to its perpendicular distance from a given diameter of that circle.

94. The circle described on any focal chord of a parabola as diameter touches the directrix; and the circle on any focal radius touches the tangent at the vertex.

95. Given the focus, or the directrix, and two points of a parabola, shew how to construct the curve, and state the number of solutions in each case.

96. The diameters through the extremities of any focal chord of a parabola meet the chords joining them to the vertex upon the directrix and intercept upon it a length which subtends a right angle at the focus.

97. Two circles whose centres are on the axis of a parabola touch the parabola and one another. Prove that the difference of their radii is equal to the latus rectum.

98. Semicircles being described upon the segments of a focal chord, shew that the squares of their common tangents vary as the length of the chord.

99. The arms of any angle in a focal segment of a parabola meet the directrix at distances from the axis to which the semi-latus rectum is a mean proportional.

100. Shew how to place in a given parabola a focal chord of given length.

101. A parabola being given, find its axis, focus, vertex, and directrix.

102. If a chord be drawn to a parabola from the foot of its directrix, the rectangle contained by its segments will be equal to the rectangle contained by the segments of the parallel focal chord.

103. If $AQ$ be a chord drawn from the vertex $A$ of a parabola, and $QR$ be a perpendicular to it at its extremity $Q$ meeting the axis in $R$, then will $AR$ be equal to the focal chord parallel to $AQ$.

104. If $PQ$ be a focal chord of a parabola, and $R$ any point on the diameter through $Q$, then will $\dfrac{PR^2}{PQ}$ be equal to the focal chord parallel to $PR$.

105. Find the locus of the points which divide parallel chords of a parabola into segments containing a constant rectangle.

106. The latus rectum is a mean proportional to the ordinates of the extremities of any chord which passes through the focus or through the foot of the directrix; and the rectangle contained by the abscisses of the extremities of the chord is equal to the square of the focal distance of the vertex.

107. If a chord subtends a right angle at the vertex, shew that it passes through a fixed point on the axis, and that the latus rectum is a mean proportional to the ordinates, and likewise to the abscisses, of its extremities.

108. Shew that the absciss cut off by any chord from any diameter is a mean proportional to the abscisses of its extremities with respect to that diameter, and that the corresponding ordinates are proportionals.

109. The distances from the focus at which the straight lines joining the ends of a focal chord to the vertex meet the latus rectum are alternately equal to the ordinates of the ends of the chord.

110. In a given parabola inscribe, and about it circumscribe, a triangle whose sides shall be parallel to three given straight lines.

111. A chord of a parabola and the chord joining the two points on the curve at which it subtends right angles intercept on the axis a length equal the latus rectum.

112. Deduce from Ex. 54 that the intercept on the axis made by any polar and the ordinate of its pole is bisected at the vertex.

113. The intercepts upon any diameter by any two polars and the ordinates of their poles with respect to that diameter are equal.

114. On a chord through a fixed point $O$ a mean proportional $OM$ is taken to the segments of the chord. Shew that the locus of $M$ is a diameter.

115. A circle being described on a chord of a parabola which is parallel to a given line, shew that its centre is at a constant distance from the middle point of its opposite chord of intersection with the parabola.

116. If a circle cut a parabola in four points the ordinates of the points of section on one side of the axis will be together equal to the ordinate or ordinates of the point or points of section on the other side.

117. If three of the points of section coalesce their common ordinate will be equal to one-third of the ordinate of the fourth point; and the common chord of the circle and the parabola will be equal to four times their common tangent measured from its point of contact to the axis.

118. Three chords of a parabola drawn at right angles to a focal chord through its extremities and the focus are proportionals.

119. If from the vertex of a parabola a pair of chords be drawn at right angles, find the locus of the further vertex of the rectangle of which they are adjacent sides.

120. If a chord perpendicular to the axis be produced to meet the tangent at an extremity of the latus rectum, the rectangle contained by its segments will be equal to the square of its distance from the focus.

121. A chord of a parabola drawn from a given point on the curve is intersected by any ordinate of the diameter through that point and by the diameter through the extremity of the ordinate. Shew that the chord is a third proportional to its segments estimated from the given point to the ordinate and the diameter through its extremity respectively.

122. The ordinate of a point $Q$ on the curve being intersected by its diameter in $V$, by any other diameter in $R$, and by the straight line joining the vertices of those diameters in $R'$, shew that
$$QV^2 = VR \cdot VR'.$$

123. The straight lines joining any point on a parabola to the extremities of a given chord meet any diameter at distances from its extremity which have the same ratio as the segments into which it divides the chord.

124. If from the point of contact of any tangent straight lines be drawn to two points on the curve, each to intersect the diameter through the other point, the two points of intersection will lie on a parallel to the tangent.

125. If the diameter through any point $P$ of a parabola meet a given chord in $a$, and the tangents at its extremities in $b$, $c$, shew that $Pa^2 = Pb \cdot Pc$, and deduce the theorem of Ex. 73* for the case of the parabola.

126. Three fixed points and a variable point being taken on a parabola, shew that the chords joining the latter to two of the fixed points cut off abscisses from the diameter through the remaining fixed point which are in a constant ratio.

---

\* In the example referred to, for "is" read *varies as*.

127. Shew that the preceding theorem is a special case of Ex. 80.

128. If a parabola which bisects the sides of a triangle $ABC$ meet its sides again in $a$, $b$, $c$, then will $Aa$, $Bb$, $Cc$ be parallels.

129. Through a given point within a parabola draw a chord which shall be divided in a given ratio at that point.

130. Describe a parabola through four given points; or through three given points, and having its axis in a given direction; and shew that the latter is a particular case of the former.

131. Shew that a circle can be described touching any two diameters of a parabola and the focal radii to their extremities; and hence, that any two intersecting tangents to a parabola subtend equal angles at the focus.

132. The four points of intersection of two parabolas whose axes are at right angles lie on a circle, and the sums of the ordinates of their points of intersection on opposite sides of the axis of either are equal.

133. If $ACP$ be a sector of a circle of which $CA$ is a fixed radius, and if a circle be drawn to touch $CA$, $CP$ and the arc $AP$, the locus of its centre will be a parabola.

134. If a circle and a parabola touch at one point and intersect in two others, the diameters of the parabola at the latter points will meet the circle again on a parallel to the tangent at the former.

135. If a straight line be drawn from a fixed point on a circle to bisect any chord parallel to the diameter through that point, find the locus of its intersection with the diameter through an extremity of the variable chord.

136. If the two tangents from any point on the axis of a parabola be cut by any third tangent, their alternate segments will be equal.

137. If $OP$, $OQ$ be the tangents at $P$ and $Q$ to a parabola, and $Pp$, $Qq$ chords parallel thereto, the distances of $O$ from $pq$ and $PQ$ will be in the ratio of five to one.

138. If there be three tangents to a parabola, whereof one is parallel to the chord of contact of the other two, shew that the three **tangents contain an area equal to half** the area of the triangle **whose vertices are at their points of contact,** and apply this **result to prove Prop. XII.**

139. The locus of the vertex of a parabola which has a given focus and a given tangent is a circle.

140. If the tangents at $P$ and $Q$ intersect in $R$, the circle through $P$ touching $QR$ in $R$ passes through the focus.

141. The tangent at any point meets the directrix and the latus rectum at equal distances from the focus.

142. A chord of a parabola being drawn through a given point, determine when the rectangle contained by its segments will be a minimum.

143. Two equal parabolas have the same axis and directrix, and from a point on one of them two tangents are drawn to the other; shew that the perpendicular from that point to the chord of contact of the tangents is bisected by the axis.

144. If a leaf of a book be folded so that one corner moves along an opposite side the line of the crease will envelope a parabola.

145. The three straight lines drawn through the points of intersection of three tangents to a parabola at right angles to their respective focal distances meet in a point.

146. The centre of the circle through any two points on a parabola and the pole of the straight line joining them lies upon the focal chord at right angles to the focal distance of the said pole.

147. Any two parabolas which have a common focus and their axes in opposite directions intersect at right angles.

148. The portion of any tangent intercepted by the tangents at two fixed points subtends a constant angle at the focus. In what case will the subtended angle be a right angle?

149. If $QV$ be the ordinate of any point $Q$ on a parabola, and if the diameter bisecting $QV$ meet the curve in $P$, then will $VP$ meet the tangent parallel to $QV$ at a distance from its point of contact equal to $\frac{2}{3}QV$.

150. The tangent from the vertex of a parabola to the circle round $SPN$, where $PN$ is the principal ordinate of a point $P$ on the curve, is equal to $\frac{1}{2}PN$.

151. The focal vectors to the points of contact of a common tangent to a parabola and the circle on its latus rectum as diameter are equally inclined to the axis. Express the distance between the points of contact in terms of the latus rectum.

152. Describe an equilateral triangle about a given parabola; and shew that the focal distances of its vertices pass each through the opposite point of contact, and that the centre of gravity of the triangle must lie upon a certain fixed straight line perpendicular to the axis.

153. The segments of the sides of a regular polygon circumscribing a circle subtend equal angles at the centre. State an analogous property of the parabola.

154. Find the envelope of a straight line which cuts the sides $OA$, $OB$ of a given triangle $OAB$ in points $P$, $Q$ such that the rectangle $OP.OQ$ is equal to $AP.BQ$.

155. Find the envelope of the straight line connecting the feet of the perpendiculars let fall from any point of a parabola upon the axis and the tangent at the vertex.

156. If $PQ$ be a chord at right angles to the axis of a parabola, the perpendicular from $P$ to the tangent at $Q$ will

cut off from the diameter at $Q$ a length equal to the latus rectum.

157. A circle being drawn through the focus of a parabola to cut the curve in two points, compare the angles between the tangents to the parabola and the tangents to the circle at those points.

158. If the tangents to a parabola at $P$ and $Q$ meet in $O$, and if the diameter through $O$ meet $PQ$ in $V$, shew that
$OP.OQ = 2OS.OV.$

159. If the tangents at $P$, $Q$ intersect in $O$ and meet the tangent at $R$ in $P'$, $Q'$, then will $OR$ pass through the intersection of $PQ'$ and $P'Q$.

160. A parabola being inscribed in a triangle so as to bisect one of its sides, shew that the perpendiculars from the vertices of the triangle upon any tangent are in harmonical progression.

161. The vertex of a constant angle whose sides envelope a parabola traces a hyperbola having the same focus and directrix.

162. In Art. 29, if $PB$ and $QM$ be perpendiculars to $QR$ and the directrix, $O$ the point in which the perpendicular from $Q$ to $PR$ meets the directrix, and $PP'$ a diameter of the circle, shew that
$$SQ : QM = QB : PQ = QO : PP',$$
and deduce Steiner's theorem.

163. To two parabolas which have a common focus and axis, two tangents are drawn at right angles. Shew that the locus of their intersection is a straight line perpendicular to the axis; and examine the case in which the directrices of the two parabolas coincide.

164. Chords of a parabola being drawn to touch an equal parabola having the same vertex, their axes being in opposite directions, shew that the locus of the middle points of the chords is a parabola, whose linear dimensions are one-third of those of the original parabola.

165. Two parabolas have a common vertex, and their axes are in opposite directions. If the latus rectum of one of them be eight times that of the other, the intercept on any tangent to the former made by their common tangent and the axis will be bisected by the latter.

166. If the vertex of an angle of constant magnitude move on a fixed straight line, and one of its arms pass through a fixed point, the other will envelope a parabola of which the fixed point and line are the focus and a tangent.

167. If a focal chord meet any tangent at a given angle, determine the locus of their point of intersection.

168. If the tangents to a parabola at points $P$ and $Q$ intersect in $O$ and meet the tangent at any point $R$ in $P'$ and $Q'$, and if $OR$ meet $PQ$ in $Z$, then
$$PZ : QZ = P'R^2 : Q'R^2.$$

169. The locus of the foot of the focal perpendicular upon a normal chord of a parabola is a parabola.

170. If $PQ$ be a chord normal at $P$ and parallel to the focal chord $FF'$, then
$$PQ : FF' = SY : SA,$$
where $A$ is the vertex and $SY$ the focal perpendicular upon the tangent at $P$.

171. If from a given point on a parabola any two chords be drawn making equal angles with the normal at that point, the focal distances of their further extremities will contain a constant rectangle.

172. The intercept on any tangent made by the curve and the tangent at the further extremity of the normal at its point of contact is bisected by the directrix.

173. If $T$ be the pole of a chord $PQ$ normal at $P$, and $AN$ be the abscissa of $P$, shew that
$$PQ : PT = PN : AN.$$

174. The perpendicular to a normal to a parabola at the point in which the normal meets the axis envelopes an equal parabola, and the focal vector of the point at which the normal is drawn meets the envelope at the point in which the perpendicular touches it.

175. The normals at the ends of a focal chord intersect upon its diameter, and the locus of their intersection is a parabola.

176. The normal, terminated by the axis, is a mean proportional to the segments of the focal chord to which it is at right angles.

177. The squares of the normals at the extremities of a focal chord are together equal to the square of twice the normal perpendicular to the chord.

178. The normal at any point is equal to the ordinate which bisects the subnormal at that point.

179. The locus of the centre of the circle circumscribing the triangle $SYP$, where $SY$ is the focal perpendicular on the tangent at any point $P$, is a parabola.

180. All circles which have their centres on a parabola and touch the tangent at its vertex are cut orthogonally by a circle which touches the parabola at its vertex and whose diameter is equal to the latus rectum.

181. From a point on any double ordinate $QQ'$ a perpendicular is drawn to its polar to meet the polar in $M$ and the axis in $N$. Shew that $M$, $N$, $Q$, $Q'$, and the point in which the polar meets the axis are concyclic with the focus.

182. The continued products of the focal vectors to any three points on a parabola and of those to the poles of the chords joining the three points are equal.

183. If the tangents at $p, q, r$ intersect in $P, Q, R$, as in Art. 29, and if $O$ be the point in which the diameter through $r$ meets $pq$, shew that

$$pQ : QR = RP : Pq = Qr : rP = pO : qO.*$$

Shew also that

$$PQ.PR = Qr.Rq\,;\quad QR.QP = Rp.Pr\,;\quad RP.RQ = Pq.Qp\,;$$

and $\quad PQ.QR.RP = Pq.Qr.Rp = Pr.Qp.Rq.$

184. Prove that in general two parabolas† and any number of central conics can be drawn through four given points; and that no two parabolas or other conics can intersect in more than four points; and that no two parabolas can touch one another in more than one point.

185. If one triangle can be inscribed in a given circle (or ellipse) so that its three sides touch a given parabola, shew that any number of triangles can be so inscribed, and that the locus of their centroids is a straight line.

186. Any number of parabolas being described with the same vertex and axis, the polars with respect to them of all points on a fixed ordinate to the axis will meet in a point.

187. If a polygon be described about a parabola the continued products of the abscissæ of its vertices and of its points of contact respectively will be equal.

188. If $T$ be the point of concourse of the tangents to a parabola at $P$ and $Q$, and if $p, q$ be the points in which any third tangent intersects them, then

$$\frac{Tp}{TP} + \frac{Tq}{TQ} = 1.$$

189. If from any point on the chord of contact of any two tangents to a parabola parallels to them be drawn each to intersect the other tangent, the points of intersection will lie

---

\* This is proved by APOLLONIUS in Lib. III., Prop. 41, of his Conics.
† Two chords of a parabola being given, it may be deduced from Art. 30, Cor. 2, that there are two possible directions of its axis.

on the tangent at the extremity of the diameter through the assumed point.

190. If $P$ and $Q$ be any two points on a parabola, and if $PM$, $QN$ be the principal ordinates of $P$, $Q$, and $AL$ the principal abscissa of the pole of $PQ$, show that $PM.QN = 4AS.AL$.*
Shew also that if $O'$ be the pole of any chord drawn through any point $O$, and $O'V'$, $OV$ be the ordinates of $O'$ and $O$ with respect to the diameter at any point $P$ on the curve, then

$$OV.O'V' = 2SP(PV + PV').$$

191. If two parabolas be described each touching two sides of an equilateral triangle at the points in which it meets the third side, prove that they have a common focus and that the tangent to either of them at their point of intersection is parallel to the axis of the other.

192. If two parabolas be described each touching two sides of any triangle at the points in which it meets the third side, determine the area common to the two curves; and if three parabolas be so described, determine the area common to the three.

193. Any two tangents to a parabola intercept on two fixed tangents lengths which are in a constant ratio.

194. If $P$ and $Q$ be fixed points on a parabola, and $RR'$ any double ordinate of a given diameter, then will $RP$ and $R'Q$ meet that diameter at distances from the curve which will be in an invariable ratio.

195. The projections of any two tangents upon the directrix by lines radiating from the vertex are equal.

196. A triangle is revolving round its vertex in one plane; prove that at any instant the directions of motion of all the points of its base are tangents to a parabola.

---

* This follows with the help of Examples 108 and 112, whereof the former may be deduced from Art. 30, Cor. 2, or from Prop. IV.

EXAMPLES.

197. If three parabolas be inscribed in a given triangle, when will the area of the triangle formed by joining their foci be a maximum?

198. The area of the parabolic sector cut off by any two focal radii is equal to half the area bounded by the arc of the segment, the diameters through its extremities, and the directrix.

199. The difference of the ordinates of two points on a parabola being equal to $QD$, shew that the chord joining them will cut off a segment whose area is equal to $\dfrac{QD^3}{24AS}$. What is the envelope of a straight line which cuts off an area of given magnitude from a given parabola?

200. If the foci of four parabolas whereof each touches the straight lines joining the foci of the other three lie on a circle, the tangents at the vertices of the four parabolas will meet in a point.

# CHAPTER IV.

## CENTRAL CONICS.

In this chapter we shall deal with the common properties of the Central Conics, and in the next chapter with certain properties, viz. those of its asymptotes, which are peculiar to the Hyperbola.

The *Abscissæ* or *Abscisses* of any point with respect to any diameter of a central **conic are the segments of that** diameter **made by the** ordinate **of the point; and the** *Central Abscissa* **is the distance of the foot of the ordinate from the centre of the conic.**

## THE ORDINATE.

### PROPOSITION I.

**33.** *The square of the principal ordinate of any point on a central conic varies as the rectangle contained by its abscissæ.*

Let the straight lines connecting the vertices $A$, $A'$ of a central conic with any point $P$ on the curve meet the directrix in $Z$ and $Z'$;* and let $PN$ be the ordinate of $P$, and $X$ the point in which the directrix meets the axis.

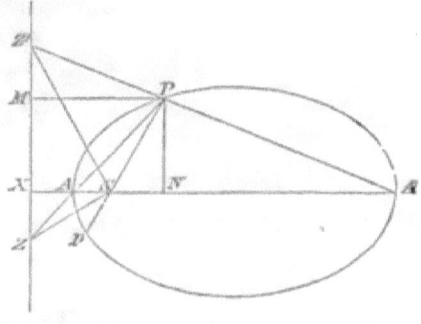

---

* For the hyperbola, use the figure on p. 80, supplying the lines $PAZ$, $PZ'A'$.

Then (Art. 12) the intercept $ZZ'$ subtends a right angle at $S$, and therefore $ZX \cdot Z'X$ has the *constant* magnitude $SX^2$.

And since $PN : AN = ZX : AX$,
and $PN : A'N = Z'X : A'X$;
therefore $PN^2 : AN \cdot A'N = ZX \cdot Z'X : AX \cdot A'X$
$= SX^2 : AX \cdot A'X$,

which is an invariable ratio.

Let $N$ be taken at the centre $C$ of the conic, and let $PN$, in virtue solely of the above proportion, and without reference to the form of the curve, become equal to $CB$, so that

$$CB^2 : CA^2 = SX^2 : AX \cdot A'X.$$

Then $PN^2 : AN \cdot A'N = PN^2 : CA^2 \sim CN^2 = CB^2 : CA^2$.

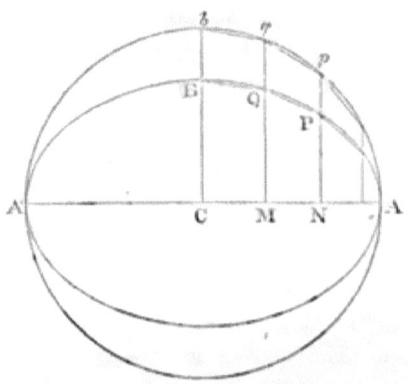

In the ellipse it is evident that $CB$ is equal to half the conjugate axis. In the hyperbola the conjugate axis does not meet the curve; nevertheless, for the sake of uniformity of expression, we shall define $CB$ as the half of its length,* and the point $B$ and a corresponding point $B'$ equidistant from the axis as its *extremities*.

### Corollary 1.

If the ordinate of $P$ be divided in the ratio of the transverse to the conjugate axis at the point $p$, then $pN^2 = AN \cdot A'N$.

---

\* The conjugate axis of any central conic is occasionally called its *Minor Axis*, although not necessarily less than the transverse **axis**, unless the curve be an ellipse.

Hence, when $P$ lies on an ellipse, the locus of $p$ is the circle described upon its major axis as diameter, and when $P$ lies on a hyperbola, the locus of $p$ is the hyperbola whose transverse axis is $AA'$, and whose conjugate axis is equal to $AA'$.

### Corollary 2.

If $Pn$ be the ordinate of $P$ with respect to the conjugate axis, it follows from the proposition that

$$PN^2 : CA^2 \sim Pn^2 = CB^2 : CA^2,$$
and
$$Pn^2 : CB^2 \sim Cn^2 = CA^2 : CB'^2.$$

Hence also it may be shewn, after the manner of Cor. 1, that the locus of the point which divides $Pn$ in the ratio of the conjugate to the transverse axis is either the circle on $BB'$ as diameter, or the hyperbola whose transverse axis is $BB'$, and whose conjugate axis is equal to $BB'$. Some of the uses of this corollary and the preceding will be pointed out in the chapter on Orthogonal Projection.

### Corollary 3.

From Cor. 1 it appears, conversely, that if the ordinates of any number of points lying on the circle upon $AA'$ as diameter be cut in any given ratio of minority $CB : CA$, the points of section will lie on an ellipse whose transverse axis is $AA'$, and whose conjugate axis is equal to $2CB$. The smaller $CB$ in comparison with $CA$, the less nearly circular is the ellipse; and ultimately, when $CB$ vanishes, the ellipse becomes coincident with its major axis $AA'$. In like manner, the "complement"\* of $AA'$ is the limit to which the hyperbola described upon it as transverse axis tends when its conjugate axis is indefinitely diminished.

### PROPOSITION II.

**34.** *The square of the ordinate of any point on a central conic with respect to any diameter is in a constant ratio to the rectangle contained by its abscissæ on that diameter.*

---

\* It may sometimes be convenient to speak of the remainder of an unlimited straight line from which any part has been taken away as the *Complement* of that part.

For by Art. 16, if $QQ'$ be any chord parallel to a fixed straight line, and $PP'$ be a fixed diameter meeting the chord in $V$, then the ratio $QV.VQ' : PV.VP'$ is constant. It follows as a special case that, if $QQ'$ be a double ordinate of the diameter $PP'$, then $QV^2 : PV.VP'$ is a constant ratio.

It is evident in the case of the ellipse that this result is equivalent to

$$QV^2 : PV.VP' = QV^2 : CP^2 - CV^2 = CD^2 : CP^2,$$

where $CD$ is the semi-diameter parallel to the ordinate $QV$.

In the case of the hyperbola, supposing that $CP$ meets and $CD$ does not meet the curve, we might define the length of the semi-diameter $CD$ and the position of its *extremity* $D$ by the condition that $CD^2$ must be to $CP^2$ in the above-mentioned constant ratio, viz. that of the focal chords parallel to $CD$ and $CP$, so that

$$QV^2 : PV.VP' = QV^2 : CV^2 - CP^2 = CD^2 : CP^2,$$

and therefore $\quad QV^2 + CD^2 : CD^2 = CV^2 : CP^2;$

but we shall at present merely remark that such a definition would be in accordance with the conventions usually adopted *

## THE SECOND FOCUS AND DIRECTRIX.

### PROPOSITION III.

35. *Every central conic has a second focus* **and** *directrix; and* **the sum of the focal distances** *of any point on the curve in* **the** *case of the ellipse, or* **the** *difference of the same in the case* **of the** *hyperbola, is constant* **and equal** *to the transverse axis.*

The existence of a second focus and directrix has been proved in Art. 14, Cor. 3; but it may also be deduced from the relation

$$PN^2 : CA^2 \sim Pn^2 = CB^2 : CA^2,$$

---

* Upon this subject, see Scholium **C.**

where $PN$ and $Pn$ are the ordinates of any point $P$ on the curve with respect to its transverse and conjugate axes.

For since the ordinates with respect to either axis of the

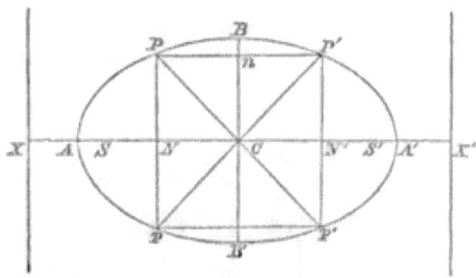

points in which a parallel to that axis meets the curve are equal, and since, from the above proportion, the corresponding ordinates with respect to the other axis must consequently be equal, it is evident that points on a central conic may be determined in pairs as $P$, $P'$ or $p$, $p'$ on opposite sides of and equidistant from the one axis, and likewise in pairs as $P$, $p$ or $P'$, $p'$ on opposite sides of and equidistant from the other axis.

The curve is therefore divided symmetrically by its conjugate axis as well as by its transverse axis, and it has a second focus $S'$ equidistant with $S$ from the centre, and a corresponding directrix meeting the axis at a point $X'$ whose distance from the centre is equal to $CX$.

Hence, if $P$ be any point on the curve,

$$S'P : NX' = S'A' : A'X' = SA : AX = SP : NX.$$

Therefore $\quad SP \pm S'P : NX \pm NX' = SA : AX.$

**Hence,** in the ellipse, since $NX + NX'$ or $XX'$ is constant, therefore $SP + S'P$ is equal to a constant length, viz. to $SA + S'A$ or $AA'$; that is to say, the sum of the focal distances of any point of the curve is equal to the major axis.

In the hyperbola, in like manner, $NX \sim NX'$ is constant, and therefore $SP \sim S'P$ is constant and equal to $SA \sim S'A$ or

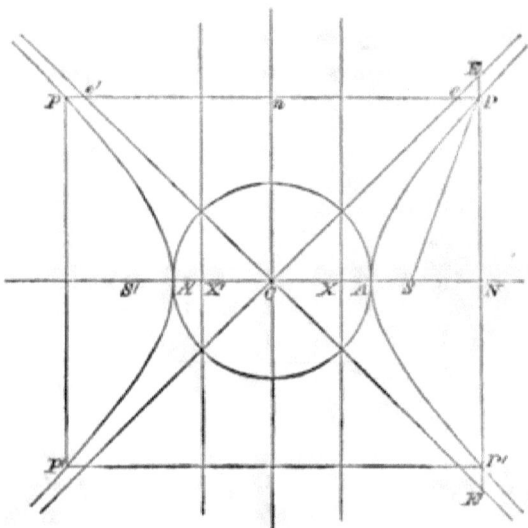

$AA'$; that is to say, the difference of the focal distances of any point on the curve is equal to the transverse axis.

*Corollary* 1.

Since $SA : AX = SA' : A'X = \frac{1}{2}(SA \pm SA') : \frac{1}{2}(AX \pm A'X)$; and since the latter ratios are equal respectively to $CA : CX$ and $CS : CA$, or *vice versa*; therefore

$$CS : CA = CA : CX = SA : AX,$$

and 
$$CS \cdot CX = CA^2.$$

*Corollary* 2.

It may now be shewn, as in the next proposition, that

$$CB^2 = CS^2 \sim CA^2 = AS \cdot A'S.$$

Hence, in the hyperbola, $CS^2 = CA^2 + CB^2$, or $CS$ is equal to the distance $AB$ between two adjacent extremities of the axes; and in the ellipse, $CS^2 + CB^2 = CA^2$, or $SB$ is equal to the semi-axis major. In the case of the ellipse it follows more directly that, $SB = S'B = CA$; and hence, conversely, that

$$CB^2 = CA^2 - CS^2 = AS \cdot A'S.$$

*Corollary* 3.

Hence $CS \cdot SX = CS^2 \sim CS \cdot CX = CS^2 \sim CA^2 = CB^2$,
and $SX : CX = CS \cdot SX : CS \cdot CX = CB^2 : CA^2$.

## THE LATUS RECTUM.

### PROPOSITION IV.

36. *The latus rectum is a third proportional to the transverse and conjugate axes.*

Since $CS : CA = SA : AX = SA' : A'X$;
therefore $CS + CA : CA = SA + AX : AX = SX : AX$,
and $CS \sim CA : CA = SA' \sim A'X : A'X = SX : A'X$.

Hence, and by Art. 33,

$$CS^2 \sim CA^2 : CA^2 = SX^2 : AX \cdot A'X = CB^2 : CA^2,$$

and therefore

$CS^2 \sim CA^2$, or $AS \cdot A'S$, is equal to $CB^2$.*

Hence, by Prop. I., if $SL$ be either of the ordinates corresponding to $AS$ and $A'S$ as abscissæ,

$$SL^2 : AS \cdot A'S = SL^2 : CB^2 = CB^2 : CA^2,$$

or the semi-latus rectum $SL$ is a third proportional to $CA$ and $CB$, and therefore the whole latus rectum is a third proportional to $AA'$ and $BB'$.

*Corollary.*

If $FF'$ be any focal chord, $CD$ the parallel radius, and $LL'$ the latus rectum, then by the proposition, and by Art. 34,

$$FF' : LL' = CD^2 : CB^2 = CD^2 : \tfrac{1}{2} LL' \cdot CA,$$

or $FF' \cdot CA = 2 CD^2$;

---

* The axis being a focal chord, it follows, from Art. 15, Cor., that $SL \cdot CA$ is equal to $AS \cdot A'S$, which in the ellipse may be shewn to be equal to $CB^2$. In the case of the hyperbola, the length $CB$ may then be *defined* as a mean proportional to $CA$ and $SL$. APOLLONIUS (Lib. III. 45) defined the points which we call the foci of central conics as certain points dividing the axis into segments whose rectangle is equal to $SL \cdot CA$; but he nowhere mentions the focus of the parabola.

that is to say, any focal chord is a third proportional to the transverse axis and the diameter parallel to the chord.

### SCHOLIUM A.

THE LATUS RECTUM of the axis, according to APOLLONIUS, was a certain straight line drawn at right angles to it from the vertex, equal to the double focal ordinate but defined without reference to the focus. The axis being regarded as Πλαγία, the *transverse* side of the "figure," the straight line drawn as above was taken as its 'Ορθία, or *erect* side; the term "figure" being used to denote the rectangle contained by those lines.

If $AL$ be the latus rectum, and if the ordinate $PN$ in Prop. I. meet $A'L$ in $Q$, then

$$PN^2 : AN.A'N = AL : AA' = QN : A'N = QN.AN : AN.A'N.$$

It follows that the square of the ordinate $PN$ is greater than the rectangle $AL.AN$ contained by the latus rectum and the abscissa in the case of the hyperbola, and less than the same rectangle in the case of the ellipse; and hence the name HYPERBOLA, which signifies *excess*, and the name ELLIPSE, which signifies *deficiency*. The PARABOLA was so called from the *equality* of the square of the ordinate of any point upon it to the rectangle contained by its abscissa and the latus rectum. The names of the three conics have indeed been otherwise explained, but the interpretations of them given above are in accordance with the manner in which Apollonius introduces them. See Halley's edition of his works, Lib. I., props., XI., XII., XIII., pp 31—37 (Oxon., 1657). Moreover, it is reported by Proclus in his Commentaries on the first book of Euclid, at the commencement of the forty-fourth proposition, upon the authority of "the Familiars of Eudemus," that the terms parabola, hyperbola, and ellipse had been used by the Pythagoreans to express the equality or inequality of areas, and were subsequently transferred to the conic curves for the reason given above. The passage is quoted in the original Greek on page 13 of E. F. August's *Zur Kenntniss der geometrischen Methode der Alten in besonderer Beziehung auf die Platonische Stelle im Meno 22d* (Berlin, 1843), and it may be seen in English in Thomas Taylor's translation of the Commentaries of Proclus, Vol. II. p. 198 (London, 1789). The whole work in the original Greek was printed at the end of *editio princeps* of Euclid's Elements (ed. Simon Grynæus, BASIL., 1533); and it has also been edited separately by Godfr. Friedlein (Teubner, LEIPZIG, 1873).

More generally, the *Latus Rectum* of any diameter was a length measured from its extremity upon the tangent thereat, equal in the case of the parabola to the parallel focal chord (although defined without reference to the focus), and in other cases a third proportional to the said diameter and its conjugate. Some later writers, as Mydorge, used the term *Parameter* for Latus Rectum in all cases, that of the axis being distinguished as *Recta Parameter.* Hamilton gave the alternative, "Latus Rectum sive Parameter istius diametri, &c."; but these expressions are now seldom used in geometrical treatises otherwise than as defined above on **pages 1 and 44.**

## THE TANGENT.

### PROPOSITION V.

37. *The tangent at any point of a central conic makes equal angles with the two focal distances of that point.*

Let the tangent at any point $P$ to an ellipse, or a hyperbola, whose foci are $S$ and $S'$, meet the directrices in $R$ and $R'$;

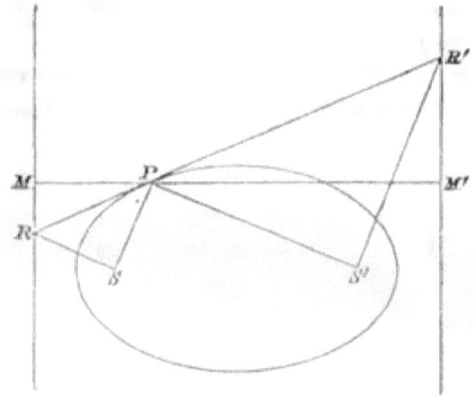

and let a parallel through $P$ to the axis meet the directrices in $M$ and $M'$.

Then since
$$SP : SP' = PM : PM' = PR : PR',$$
and since $PR$ and $PR'$ subtend right angles at $S$ and $S'$ respectively, therefore the triangles $SPR$, $S'PR'$ are similar, having their angles at $P$ equal; that is to say, the tangent at $P$ makes equal angles with $SP$ and $S'P$.

In the ellipse, the tangent lies without the angle $SPS'$ and bisects it *externally*.

In the hyperbola (fig. p. 80), the tangent must cut the axis between $A$ and $A'$ (since otherwise it could not lie wholly without the curve), and therefore between $S$ and $S'$; it must therefore bisect the angle $SPS'$ *internally*.*

### Corollary.

If $P$ be one of the four points of intersection of an ellipse and a hyperbola which have the same foci $S$ and $S'$, their tangents at $P$, being the two bisectors of the angle $SPS'$, will be at right angles to one another. Hence *confocal conics intersect at right angles.*

### PROPOSITION VI.

**38.** *The projections of the foci upon the tangent at any point of a central conic lie on its auxiliary circle; and the semi-axis conjugate is a mean proportional to the distances of the foci from their respective projections.*

Let $S$ and $H$ be the foci, $Y$ and $Z$ their respective projections upon the tangent at any point $P$. Then will $Y$ and $Z$

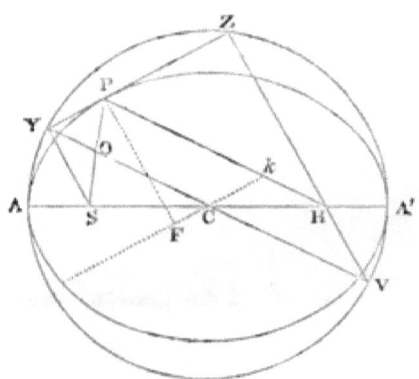

lie on the circle described upon the axis $AA'$ as diameter, and the rectangle $SY.HZ$ will be equal to $CB^2$.

(i) For let a parallel to $HP$ through $Y$ meet $SP$ in $O$. Then the tangent makes equal angles with $SP$ and the parallel

---

\* If, in accordance with the principle of the note on p. 22, the distance of any point $P$ on a hyperbola from $S'$ be estimated *within* the curve, in which case it will be the "complement" of $S'P$, we may say of the hyperbola, as of the ellipse, that the tangent bisects the angle between the focal distances *externally*.

to $HP$, viz. the angles $OPY$, $OYP$, and the complements of these angles, viz. $OSY$, $OYS$ are equal, therefore
$$OP = OY = OS.$$

Hence, the parallel to $HP$ bisects $SH$, viz. at the centre $C$

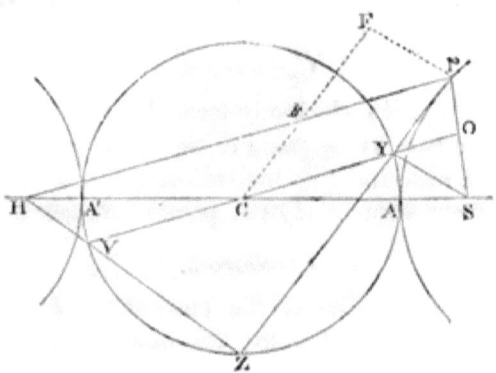

of the conic, and
$$CY = CO \pm OY = \tfrac{1}{2}HP \pm \tfrac{1}{2}SP = CA,^*$$
the upper signs being taken in the case of the ellipse, and the lower in the case of the hyperbola.

Hence, the auxiliary circle is the locus of the foot of the perpendicular from either focus to the tangent at any point of the conic; and, conversely, the straight lines drawn from any point on the auxiliary circle at right angles to the two focal distances of that point are tangents to the conic.

(ii) Let $ZH$ meet the circle again in $V$; then, the angle at $Z$ being a right angle, $VY$ passes through the centre $C$.

Hence, evidently $SY$, is equal to $HV$; and therefore,
$$SY.HZ = HV.HZ = HA.HA' = CB^2,$$
or $CB$ is a mean proportional to the focal perpendiculars upon any tangent.†

---

\* This result might also have been arrived at by supposing $SY$, $HP$ to meet in a point $S'$, and shewing that $CY = \tfrac{1}{2}HS' = CA$.

† In the figure on p. 88, if $T$ be the point of concourse of *any* two tangents, then $SY : HZ' = SY' : HZ$. It follows that angle $STY = HTZ'$; or, conversely, from Prop. XVIII. it may be deduced that $SY . HZ = CB^2$.

### Corollary 1.

This proposition enables us to draw a tangent to the conic which shall be parallel to any given straight line, viz. by drawing $SY$ (to a point $Y$ on the auxiliary circle) at right angles to the given straight line, and drawing $YP$ at right angles to $SY$.

### Corollary 2.

The points $V, Z$ in which any focal chord meets the auxiliary circle lie upon parallel tangents to the conic, and the semi-axis conjugate is a mean proportional to the perpendiculars $HV, HZ$ from either focus $H$ upon any two parallel tangents.

### Corollary 3.

The diameter parallel to the tangent at $P$ intercepts on either focal distance $HP$ a length $Pk$ equal to $CY$ or $CA$.

### PROPOSITION VII.

39. *The distance from the centre at which any tangent meets a given diameter varies inversely as the central abscissa of its point of contact with respect to that diameter.*

Let the tangent at any point $Q$ meet a given diameter in $T$, and let $CV$ be the abscissa of $Q$ with respect to that diameter; then will the rectangle $CV.CT$ be of constant magnitude.

(i) Let $CT$ meet the curve in $P$, and let the tangents at $P$ and $Q$ intersect in $R$; complete the parallelogram $QRPO$.

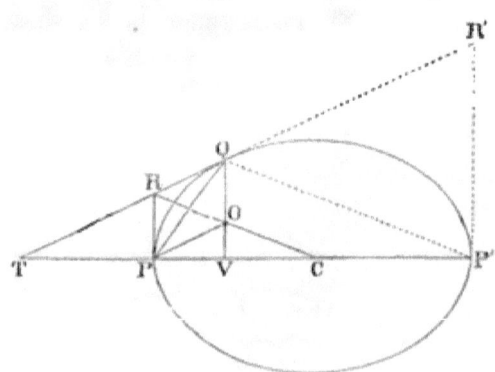

Then since, in the parallelogram, $RO$ bisects $PQ$, and since $PQ$ is the chord of contact of the tangents from $R$, therefore $RO$ passes through the centre $C$.

Hence, by parallels,
$$CV : CP = CO : CR = CP : CT,$$
or $CV.CT$ has the constant magnitude $CP^2$, and $CT$ varies inversely as the abscissa $CV$.

(ii) Next, let $t$ be the point of concourse of $QT$ with any diameter of a hyperbola *which does not meet the curve;* and let the ordinate $QV$ in the former case of the proposition be supposed an ordinate of the conjugate diameter, and therefore parallel to $Ct$.

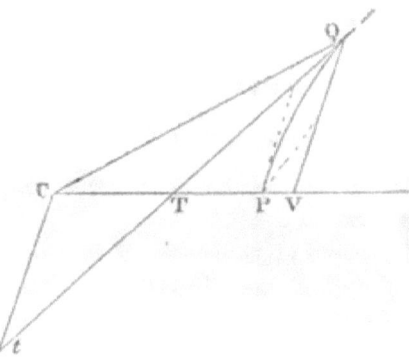

Then since $Ct : CT = QV : VT$,
therefore $QV.Ct : CV.CT = QV^2 : CV.VT$.

And since, by the former case, $CV.CT$ is equal to $CP^2$, therefore
$$QV.Ct : CP^2 = QV^2 : CV^2 - CP^2;$$
and therefore (Art. 34) $QV.Ct$ is equal to the square of the "semi-diameter" $CD$ parallel to $QV$; that is to say, $Ct$ varies inversely as $QV$, or as the abscissa of $Q$ with respect to the diameter on which $Ct$ is estimated.

*Corollary* 1.

The relation $CV.CT = CP^2$ implies that any diameter $PP'$ which meets the conic is divided harmonically (Art. 18, Cor.)

at $V$ and $T$. This likewise involves the equality of the rectangle $TC.TV$ to $TP.TP'$, which may also be deduced independently from the evident parallelism of $P'Q$ and $CR$, thus

$$TP : TV = TR : TQ = TC : TP'.$$

Or again, supposing the tangents at $Q$ and $P'$ to meet in $R'$, we might have inferred from Art. 16, Cor. 1, that

$$RP : RQ = R'P' : R'Q,$$

and thence that $TP$, $TV$, $TP'$ are in harmonic progression.

### Corollary 2.

If $CV$ and $CT$ be estimated on the transverse axis, their product will be equal to $CA^2$; or if on the conjugate axis, it will be equal to $CB^2$.

## THE DIRECTOR CIRCLE.

### PROPOSITION VIII.

40. *The locus of the vertex of a right angle whose sides envelope a central conic is a circle.*

Let $T$ be the point of concourse of a pair of tangents at right angles; $Y$ and $Z$ the projections of the foci upon one of them; $Y'$ and $Z'$ their projections upon the other.

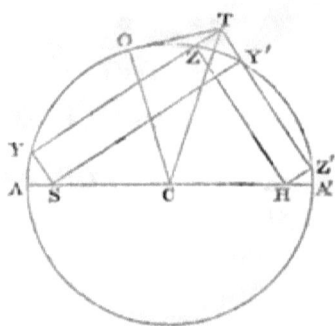

(i) Draw the auxiliary circle through $Y$, $Z$, $Y'$, $Z'$; and first, if $T$ lie without it (that is to say, in the case of the ellipse), let $TO$ be drawn touching it in $O$.

Then
$$TO^2 = TY'.TZ' = SY.HZ = CB^2,$$
and therefore $CT^2 = CO^2 + CB^2 = CA^2 + CB^2,$
or the locus of $T$ is a circle concentric with the conic, and whose diameter is equal to the diagonal of the rectangle contained by the axes.

(ii) Next let $T$ lie within the circle, as may happen in the case of the hyperbola. Then it may be shewn in like manner that
$$CA^2 - CT^2 = TY.TZ = CB^2,$$
or
$$CT^2 = CA^2 - CB^2.$$

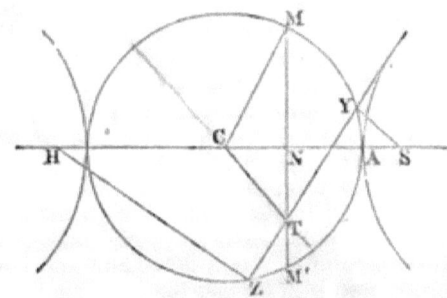

Hence, provided that the transverse be greater than the conjugate axis, the locus of $T$ will be a circle, the square of whose radius is equal to $CA^2 - CB^2$; but if the conjugate axis be the greater, the locus will be imaginary, or an "obtuse" hyperbola *can have no real tangents at right angles.**

## Corollary 1.

In like manner it may be shewn that, if the sides of a right angle envelope two *confocal* conics whose semi-axes are $CA$, $CB$ and $Ca$, $C\beta$ respectively, its vertex will lie on a concentric circle the square of whose radius is equal to $CA^2 \pm C\beta^2$, or $Ca^2 \mp CB^2$. When $C\beta$ vanishes, one arm of the right angle passes through a focus, and its vertex $Y$ (Art. 38) lies on the

---

* In this case it will be seen that the theorem is applicable to the Conjugate Hyperbola. In the limiting case in which the axes are equal the locus reduces to a point, the only tangents at right angles being the **Asymptotes**.

auxiliary circle. In the proposition itself, if $CB$ be supposed to vanish, the director circle becomes the circle on $SH$ as diameter.

*Corollary* 2.

A conic and its director circle are so related that any rectangle circumscribing the former is inscribed in the latter.

SCHOLIUM B.

The term DIRECTOR CIRCLE has been used of late years to denote the locus of intersection of tangents at right angles to a conic, with reference to the circumstance that when the conic degenerates into a parabola this locus becomes coincident with the directrix (Art. 25). The analogous term "Director Sphere" was introduced by Professor Townsend in the *Quarterly Journal of Pure and Applied Mathematics*, vol. VIII. p. 10 (1867). DE LA HIRE proved in his *Sectiones Conicæ* (PARIS, 1685), Lib. VIII. props. 27, 28, that the tangents to a conic from any point on a concentric circle whose radius is equal to $\sqrt{(CA^2 \pm CB^2)}$ meet the circle again at the ends of a diameter, and therefore contain a right angle. He also gave the equivalent of Cor. 2.

In some treatises the term Director Circle is used to denote the circle described about the further focus $H$ with radius equal to the transverse axis, which possesses a property analogous to that of the directrix of the parabola; for if $Y$ be any point on the circumference of the circle, and if $HY$ meet the conic in $P$, then $SP = PY$, or the focal distance of $P$ is equal to its normal distance from the circle. This circle affords a construction, analogous to that in Art. 25, Cor. 2, for drawing tangents to a central conic from any external point. Nevertheless it scarcely deserves a distinctive name; whereas the "director circle," according to the former definition, is of considerable importance in the higher geometry of conics. The analogy of the circle in question to the directrix of the parabola was pointed out by BOSCOVICH, (*Sectionum Conicarum Elementa*, § 102), but he did not give it a name.

## POLAR PROPERTY.

PROPOSITION IX.

41. *The tangents at the extremities of any chord drawn through a given point intersect on a fixed straight line parallel to the ordinates of the diameter through that point.*

(i) Through a given point $O$ within or without the conic draw any chord, and let the tangents at its extremities meet

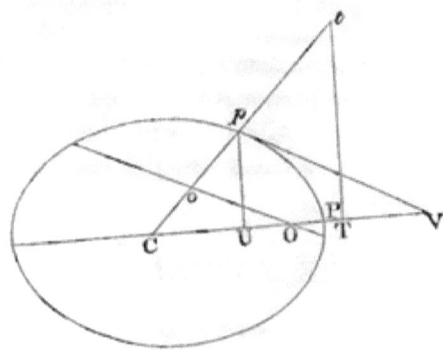

in $t$. Draw $Ct$ bisecting the chord in $o$, and meeting the curve in $p$, and let $pU$ and $tT$ be ordinates of the diameter through $O$.

Let $CO$ meet the curve in $P$, and let it meet the tangent at $p$, which is parallel to $oO$, in $V$.

Then $CO : CV = Co : Cp$,

and $CU : CT = Cp : Ct$;

and therefore, the lengths $Co$, $Cp$, $Ct$ being continued proportionals (Art. 39),

$$CO.CT = CU.CV = CP^2.$$

Hence $T$ is a fixed point, and $tT$, which was ordinately applied to the diameter $CP$, is a fixed straight line.

Conversely, the chord of contact of the tangents drawn to a conic from any point $t$ on a fixed ordinate passes through a fixed point $O$ situated on the diameter of that ordinate.

(ii) In the case of the hyperbola, either or both of the diameters $Co$, $CO$ may not meet the curve.

If $Co$ only do not meet the curve, let $Pu$ be the ordinate of $P$ with respect to it, and let the tangent at $P$ meet it in $v$; then, $Co.Ct$ being equal to $Cu.Cv$ (Art. 39, § ii),

$$CO : CP = Co : Cu = Cv : Ct = CP : CT,$$

as in the first case. But if $CO$ only do not meet the curve, the first proof is applicable as far as $CO.CT = CU.CV$; and it is known from Art. 39, § ii. that the latter rectangle is of constant magnitude.

In the case in which neither $Co$ nor $CO$ meets the curve, if $Cp$ be taken a mean proportional to $Co$ and $Ct$, and if $pV$ be drawn parallel to $oO$,* it will be seen that $CU.CV$ is still constant, and $tT$ will be a fixed straight line as in the cases previously considered.

## THE NORMAL.

### PROPOSITION X.

**42.** *The normal at any point of a central conic bisects the angle between the two focal distances of that point.*

If $P$ be any point on a conic whose foci are $S$ and $H$, it follows as a corollary from **Prop. v.** that the normal at $P$ bisects the angle $SPH$ in the case of the ellipse, and its supplement in the case of the hyperbola. The same may also be deduced from Art. 10 (where $S$ may be either focus) as follows.

If the normal meet the axis in $G$, then

$$SG : SP = CS : CA = HG : HP;$$

and therefore $PG$ bisects the angle $SPH$ internally or externally, according as $G$ lies in $SH$, as in the case of the ellipse, or in the "complement" of $SH$, as in the case of the

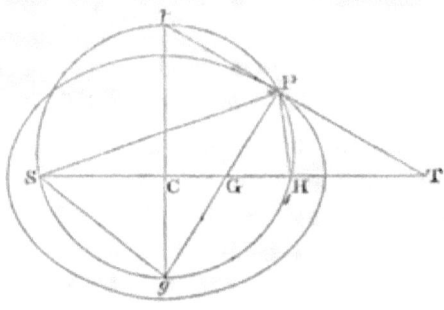

---

* The line thus drawn is the tangent at $p$ to the *Conjugate Hyperbola*, which will be defined in the next chapter. For another proof of the proposition, see Art. 17. The proof in Art. 41, § i. applies, with obvious modifications, to the Parabola.

hyperbola. In either case the normal bisects the angle between those portions of the focal distances which fall *within the curve*.

If the circle round $SPH$ meet the conjugate axis in $g$ and $t$, it is evident that $Pg$ and $Pt$ will be the two bisectors of the angle $SPH$; that is to say, they will be the tangent and the normal at $P$. This suggests an obvious method of drawing the two tangents or normals to a conic from any point on its conjugate axis.

## Corollary 1.

The tangent and normal to a conic whose foci are $S$ and $H$ divide the straight line $SH$ harmonically, and $CS$ is a mean proportional to the lengths $CG$, $CT$ which they intercept on the axis.

## Corollary 2.

It is likewise evident by similar triangles that
$$PG.Pg = PT.Pt = SP.HP;$$
and it will be shewn that each of these rectangles is equal to the square of the semi-diameter parallel to the tangent at $P$.

### PROPOSITION XI.

43. *At any point of a central conic the* normal, *terminated by either* axis, *varies inversely as the central* perpendicular *upon the tangent, and directly as the* radius parallel *to the* tangent.

Let the tangent and the normal at $P$ meet the transverse

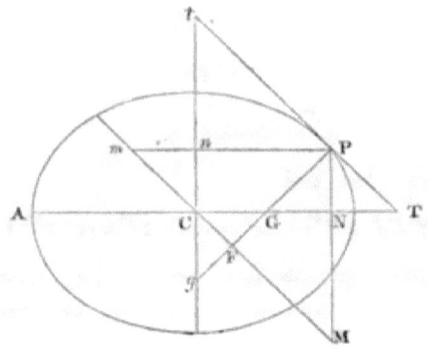

axis in $T$ and $G$, and the conjugate axis in $t$ and $g$, respectively; draw perpendiculars $PN$, $Pn$ to those axes, and let them meet the diameter parallel to the tangent in $M$ and $m$; and let the normal meet that diameter in $F$.

(i) Then (Art. 39, Cor. 2), the angles at $N$ and $F$ being right angles,
$$PG.PF = PN.PM = Cn.Ct = CB^2;$$
and, in like manner, the angles at $n$ and $F$ being right angles,
$$Pg.PF = Pn.Pm = CN.CT = CA^2;$$
that is to say, $PG$ and $Pg$ vary inversely as $PF$, which is equal

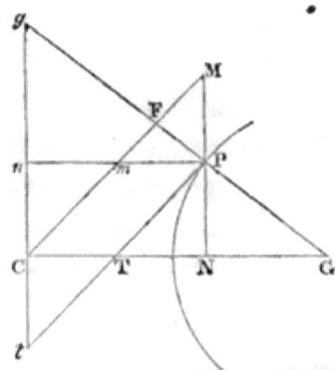

to the central perpendicular upon the tangent at $P$.

(ii) It will be proved in the section on conjugate diameters that
$$PG : CD = CD : Pg = CB : CA,$$
where $CD$ is the semi-diameter parallel to the tangent at $P$.

### Corollary.

It is hence evident that
$$NG : CN = NG : Pn = PG : Pg = CB^2 : CA^2,$$
or *the subnormal varies as the abscissa*. In like manner it may be shown that the subnormal $ng$ on the conjugate axis varies as the abscissa $Cn$.

## CONJUGATE DIAMETERS.

### PROPOSITION XII.

44. *Supplemental chords are parallel to conjugate diameters.*

Let $OP'$, $OP$ be any two supplemental chords, and $CQ$, $CR$ the diameters to which they are parallel; then will $CQ$, $CR$ be

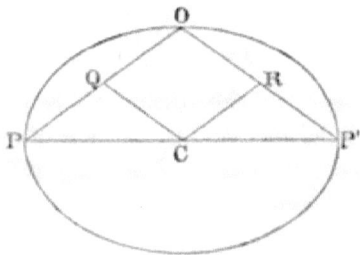

conjugate. For it is evident that these diameters bisect $OP$ and $OP'$ respectively; that is to say, each of them bisects a chord parallel to the other.

This enables us to determine the relation between the directions of any two conjugate diameters, for, in Art. 33,

$$PN^2 : AN.NA' = CB^2 : CA^2,$$

where the supplemental chords $AP$, $A'P$ may be supposed parallel to any assumed pair of conjugate diameters. Hence, if the ratio of $PN$ to $AN$ (or the direction of one of the diameters) be given, the ratio of $PN$ to $A'N$ (or the direction of the conjugate diameter) is determined.[*]

### Corollary 1.

It readily follows that if $P$ and $D$ be points on two diameters of a central conic whose ordinates $PN$ and $DR$ (as in Art. 45) are so related that

$$PN : CR = DR : CN = CB : CA,$$

---

[*] An equivalent result may be deduced from Art. 14, Cor 4 and Art. 35, Cor. 3. If $\theta$ and $\phi$ be the inclinations of two conjugate diameters to the axis, $\tan\theta \tan\phi = \mp \dfrac{b^2}{a^2}$ (where $a$, $b$ are the semi-axes), the negative sign being taken for the ellipse and the positive for the hyperbola.

then will $CP$ and $CD$ be conjugate, provided that they lie in adjacent **quadrants in the** case of the ellipse, **or in the** same quadrant or **in opposite** quadrants in the case of the hyperbola. If the **ordinates of** $P$ and $D$, in the case of the ellipse, be produced to **meet the auxiliary circle in** $p$ and $d$, as in the next figure, *the angle $pCd$ will a right angle*, as will **be noticed more particularly** in the chapter on Orthogonal **Projection.**

### Corollary 2.

In the hyperbola it is evident that of every two supplemental chords one must lie **wholly** within and the other **wholly without the curve;** and hence that *one and one only of every two conjugate diameters meets the curve.*

### Corollary 3.

To draw **a pair of** conjugate diameters inclined at a given angle, let **a segment of a circle containing the** given angle, and described on **any diameter** $PP'$ **as base,** meet the conic again in $O$; then will the diameters **parallel to** $OP$ and $OP'$ be inclined at the given angle.

### PROPOSITION XIII.

**45.** *The sum of the squares of any two conjugate diameters is constant in the ellipse, and the difference of the same is constant in the hyperbola.*

(i) If $CP$, $CD$ be any two radii of an ellipse, and $CN$, $CR$ the central abscisses of their extremities $P$ and $D$ respectively, then, by Art. 33,

$$PN^2 : CA^2 - CN^2 = DR^2 : CA^2 - CR^2 = CB^2 : CA^2.$$

Let $CP$ and $CD$ be supposed to lie in adjacent quadrants, and let

$$CN^2 + CR^2 = CA^2;$$

then the above proportions reduce to

$$PN : CR = DR : CN = CB : CA,*$$

---

* The same proportions will hold when the ordinates are oblique, if $CA$ and $CB$ be the lengths of the corresponding semi-diameters.

and therefore (Art. 44, Cor. 1) $CP$, $CD$ are conjugate, and conversely.

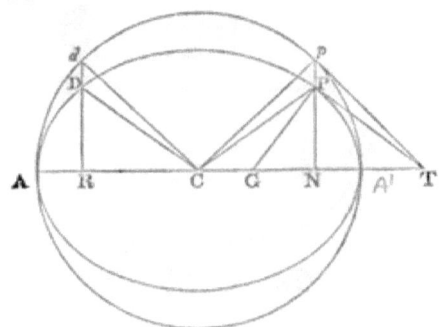

Hence also $\quad PN^2 + DR'^2 = CB^2$,

and $\quad CP^2 + CD^2 = CN^2 + PN^2 + CR'^2 + DR'^2 = CA^2 + CB^2$;

or the sum of the squares of **any two conjugate semi-diameters** of an ellipse is constant and **equal to the sum of the squares of the semi-axes**.

(ii) In the case of the hyperbola, let $P$ be a point on the curve, and $D$ a point on the hyperbola whose transverse and *conjugate* axes are $BB'$ and $AA'$ respectively; then, by Art. 33,

$$PN^2 : CN^2 - CA^2 = DR^2 : CR^2 + CA^2 = CB^2 : CA^2.$$

Let $CP$ and $CD$ be supposed to lie in the same quadrant or in vertically opposite **quadrants**, and let $CD$ be regarded as *terminated* at the point $D$.*

Let $\quad CN^2 - CR^2 = CA^2$;

then the above proportions reduce to

$$PN : CR = DR : CN = CB : CA,$$

---

\* The radius $CD$ which does not meet the curve (Art. 44, Cor. 2) is here regarded as terminated by the Conjugate Hyperbola. For another proof of the proposition, in which the length of $CD$ is defined as suggested in Art. 34, the reader is referred to the next chapter; and for a third proof, to the chapter on Orthogonal Projection. It will be seen that the above conventions with regard to the length of $CD$ are consistent with one another, but the true definition of the lengths of diameters which do not meet the curve is that given in Scholium C.

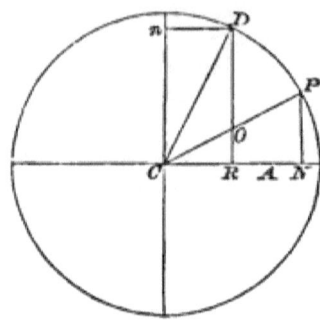

and therefore (Art. 44, Cor. 1) $CP$, $CD$ are conjugate, and conversely.

Hence also $DR^2 - PN^2 = CB^2$,

and $CP^2 - CD^2 = CN^2 + PN^2 - (CR^2 + DR^2) = CA^2 - CB^2$;

or the difference of the squares of any two conjugate "semi-diameters" of a hyperbola is constant and equal to the difference of the squares of the semi-axes.

### Corollary 1.

Let the normal at $P$, which is at right angles to $CD$, meet the transverse and conjugate axes in $G$ and $g$ respectively; then it may be shewn by similar triangles that

$$PG : CD = PN : CR = CB : CA,$$

and $$Pg : CD = CN : DR = CA : CB.$$

Hence $CD$ is a mean proportional to $PG$ and $Pg$.

### Corollary 2.

If $P$ be any point on a conic whose foci are $S$ and $H$, then since $C$ is the middle point of $SH$,

$$2CP^2 + 2CS^2 = SP^2 + HP^2 = 4CA^2 \mp 2SP \cdot HP.$$

Hence $CP^2 \pm SP \cdot HP = CA^2 + CA^2 - CS^2 = CA^2 \pm CB^2$,

and therefore, by the proposition, $SP \cdot HP$ is equal to $CD^2$.

### PROPOSITION XIV.

**46.** *The area of any parallelogram whose sides are equal and parallel to two conjugate diameters of a central conic is equal to the area of the rectangle contained by the axes.*

(i) Let a parallelogram be formed by **drawing parallels to** two conjugate diameters $PP'$, $DD'$ of a central conic through

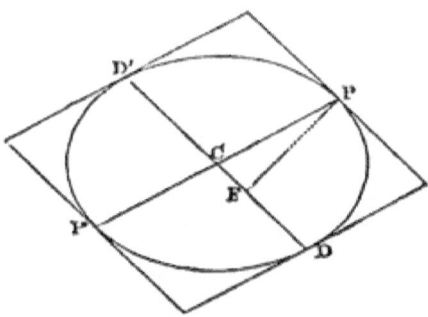

their extremities, **and let the normal at** $P$ **meet** $DD'$ **in** $F$, and let it meet the axis in $G$.

Then since (Art. 45, **Cor. 1**),

$$PG : CD = CB : CA,$$

therefore $\quad PF.PG : PF.CD = CB^2 : CA.CB$

But, by Art. 43, the antecedents of this proportion are equal.

Therefore $\quad PF.CD = CA.CB,$

**and the area of the whole** parallelogram is equal to $4PF.CD$, **that is, to** $2CA.2CB$, **or to the rectangle** contained by the axes.

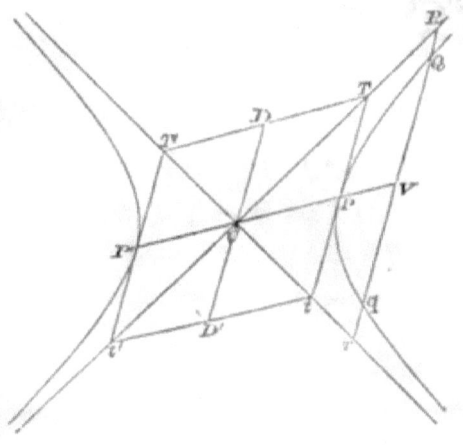

In the case of the ellipse the parallelogram described as above completely envelopes the curve; but in the case of the hyperbola two of its sides only touch the curve (Art. 44, Cor. 2).

(ii) The proposition might also have been proved by shewing that, with the construction of Art. 45,

$$\triangle PCD = PDRN - PCN \mp DCR = \tfrac{1}{2}(CN.DR \pm CR.PN),$$

which last expression reduces to $\tfrac{1}{2} CA.CB$.

## PROPOSITION XV.

**47.** *The intercept upon any* **tangent** *by any two conjugate diameters is divided at the curve into segments to which the parallel radius is a mean* **proportional**.

Let the tangent at $P$ meet any two conjugate diameters in $T$ and $t$, and let $CD$ be the radius parallel to the tangent.

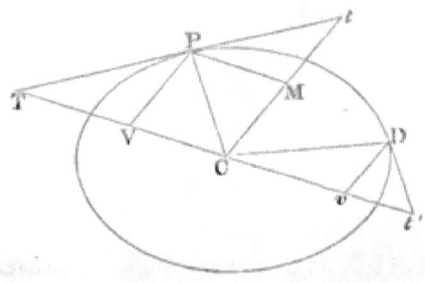

Let $PV$, $Dv$ be ordinates of the diameter $CT$, and $PM$ an ordinate of $Ct$; and let the tangent at $D$ meet the former diameter in $t'$.

Then, by similar triangles, the tangent at $D$ being parallel to $CP$,

$$PT : CT = CD : Ct',$$

and $$Pt : CV = Pt : PM = CD : Cv.$$

Therefore* $$PT.Pt : CV.CT = CD^2 : Cv.Ct';$$

and the consequents in this proportion being equal by Art. 39, therefore $$PT.Pt = CD^2,$$

---

* More briefly, the condition that $CT$, $Ct$ should be conjugate (note, p. 96) gives at once the relation $PT.Pt : CP^2 = CD^2 : CP^2$.

or $CD$ is a mean proportional to $PT$ and $Pt$; and when the point $P$ is given the rectangle $Pt.PT$ is constant.

*Corollary.*

If the tangent at any point $Q$ meet any two parallel tangents in $R$ and $R'$, then will the radius parallel to the tangent be a mean proportional to $QR$ and $QR'$. For, in the figure of Art. 39, it is evident that $CR$ and $CR'$ are parallel to the supplemental chords $P'Q$, $PQ$, and are therefore conjugate diameters. Moreover, if $RC$ meet $R'P'$ in $r$, the semi-diameter parallel to $RP$ will be a mean proportional to $P'r$, $P'R'$, and therefore to $PR$, $P'R'$.

### SCHOLIUM C.

Although the conjugate axis of a hyperbola and its other diameters which do not meet the curve are commonly regarded as terminated by the conjugate hyperbola, this convention is by no means accurate, but the true account of the matter is that given below.

Given the relation (Art. 33) between the coordinates of any point on the hyperbola

$$PN^2 : CN^2 - CA^2 = CB^2 : CA^2,$$

the true length of the semi-axis conjugate is found, by making $CN$ vanish, to be $\sqrt{(-1)} CB$. Let this be denoted by $C\beta$, so that

$$PN^2 : CA^2 - CN^2 = -CB^2 : CA^2 = C\beta^2 : CA,$$

which shews that the hyperbola may be regarded as *an ellipse whose minor axis is a certain imaginary quantity*. In like manner the true length of the semi-diameter conjugate to $CP$, in the second case of Art. 34, is $\sqrt{(-1)} CD$; and if this be denoted by $C\delta$, we may write

$$QV^2 : CP^2 - CV^2 = C\delta^2 : CP^2,$$

as in the case of the ellipse. Now, treating $\beta$ and $\delta$ as if they were real points on the curve and supposing $\rho$ to be the projection of $\delta$ upon the axis, we have, precisely as in Art. 45, § i,

$$CN^2 + C\rho^2 = CA^2; \quad PN^2 + D\rho^2 = C\beta^2; \quad CP^2 + C\delta^2 = CA^2 + C\beta^2;$$

which will be seen to be equivalent to the results of Art. 45, § ii,

$$CN^2 - CR^2 = CA^2; \quad PN^2 - DR^2 = -CB^2; \quad CP^2 - CD^2 = CA^2 - CB^2.$$

And so in other cases (cf. Art. 40) we may pass at once to properties of the ellipse, in so far as they involve $CB^2$ and $CD^2$, by writing in place thereof $C\beta^2$ and $C\delta^2$, that is to say, *by changing the signs of $CB^2$ and $CD^2$*.

Next consider the hyperbola as a particular form of ellipse whose determining ratio has become one of majority. When this ratio has increased up to unity the further focus and vertex are at infinity; but as soon as it exceeds unity they at once come back from infinity to $H$ and $A'$ on the opposite of $A$. The true length of the axis of the hyperbola is therefore $A \infty A'$, the complement of $AA'$, and the distance between its foci is $S \infty H$, the complement of $SH$. In like manner the distance of any point $P$ on the curve from the further focus $H$ is to be regarded as $H \infty P$, the complement of $HP$; and thus the fundamental bifocal property assumes the indeterminate form $SP + H \infty P = A \infty A'$. In order to pass to the determinate form of the bifocal property of the hyperbola from the property of the ellipse, we may remark that a point moving in a straight line from the further focus $H$ of the former to a point $P$ on $S$-branch of the curve may be supposed to move either (1) *within the curve*, from $H$ to $\infty$ and from $\infty$ to $P$, always in the same direction, or (2) along the finite length $HP$ *in the opposite direction*; and we may therefore regard $HP$, drawn towards the convexity of the $S$-branch, as essentially *negative*, and the finite axis $AA'$ as likewise negative. The ellipse property, $SP + HP = AA'$, thus becomes, in the case of the hyperbola, $SP + (-HP) = (-AA')$. So in Art. 15, Cor., in the case in which the focal chord meets the nearer branch of a hyperbola in $Q$ and its further branch in $P$, we must write $\dfrac{1}{(-SP)} + \dfrac{1}{SQ} = \dfrac{2}{L}$. In accordance with the same principle the normal to a hyperbola at $P$ bisects the *interior* angle between $PS$ and $P \infty H$; and if two tangents $TO$, $TO'$ be drawn from $T$ to the same branch of a hyperbola, the angle $STO$ will be equal to the angle between $TO'$ and $T \infty H$.

Combining the results of the two preceding paragraphs, we infer from the property $SY \cdot HZ = CB^2$ in the ellipse, that $SY \cdot (-HZ) = (-CB^2)$ in the hyperbola; and from $SP \cdot HP = CD^2$ in the former, that $SP \cdot (-HP) = (-CD^2)$ in the latter. On the same principle, Prop. I. assumes the form

$$PN^2 : AN \cdot (-A'N) = (-CB^2) : CA^2$$

in the case of the hyperbola. In Art. 36, Cor., if the focal chord $FF'$ be positive $CD^2$ will be negative, and *vice versa*. From the result $PF \cdot CD = CA \cdot CB$, obtained in evaluating the conjugate circumscribing parallelogram of the ellipse, we deduce in the case of the hyperbola that $(-PF) \cdot \sqrt{(-1)} \, CD = (-CA) \cdot \sqrt{(-1)} \, CB$; and the final result is independent of the factor $-\sqrt{(-1)}$.

## THE BIFOCAL DEFINITION.*

### PROPOSITION XVI.

48. *The tangent to a bifocal conic makes equal angles with the focal distances of its point of contact.*

---

\* The theorems in this section have for the most part been already proved in other ways; but they are here derived from the bifocal property, $SP \overset{+}{\underset{-}{}} HP = AA'$,

(i) Let $TPQt$ be a chord which meets the conic in the adjacent points $P$ and $Q$, and let $SM$ be taken equal to $SP$

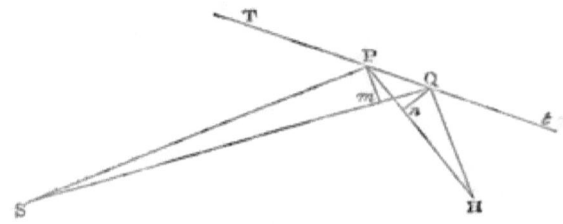

upon $SQ$, and $Hn$ equal to $HQ$ upon $HP$, so that

$$Pn = HP \sim HQ = SQ \sim SP = Qm.$$

Now in the isosceles triangles $SmP$, $HnQ$, as the angles at $S$ and $H$ are diminished indefinitely, each of the remaining angles becomes very approximately a right angle; and hence ultimately, when $Q$ coalesces with $P$, the triangles $PmQ$, $PnQ$ become *right-angled* at $m$ and $n$.

And since these triangles have the common hypothenuse $PQ$, and it has been shewn that $Pn$, $Qm$ are equal, therefore the angle $SQT$ of the one triangle is equal to the angle $HPt$ of the other. That is to say, the point $Q$ being supposed to have coalesced with $P$, the angle $SPT$ is equal to $HPt$, or the tangent at $P$ makes equal angles with the focal distances of that point.

(ii) Or, conversely, taking EUCLID's definition of a tangent, we may proceed as follows.

In the case of the ellipse, $P$ being any point on the curve, and $S$, $S'$ the foci, in $SP$ produced take $Ps$ equal to $PS'$, draw the bisector of the angle $S'Ps$, and take any point $Y$ upon it.

Then since $S'Y$ is evidently equal to $sY$,

$$SY + S'Y = SY + sY > Ss > SP + S'P;$$

that is to say, the sum of the focal distances of any point other than $P$ upon the bisector of the angle $SPs$ is greater than the

---

by which the ellipse and hyperbola are sometimes *defined*. From this property it is evident that an ellipse may be traced with the point of a pencil moved along a string, whereof the ends are fixed, so as to keep it stretched.

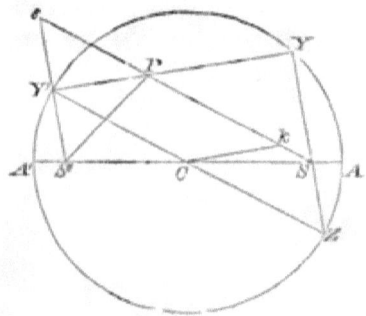

transverse axis; every such point therefore lies without the curve, and the bisector of the angle $S'Ps$ is the tangent at $P$. The proof applies *mutatis mutandis* to the hyperbola.

### Corollary.

The normal at $P$ bisects the angle between the focal distances of $P$, *estimated within the curve;* and hence it easily follows that $G$ being the point in which it meets the axis

$$SG : SP = HG : HP = CS : CA.$$

### PROPOSITION XVII.

49. *The two tangents to a bifocal conic from any external point subtend equal or supplementary angles at either focus.*

This may be deduced from the LEMMA that *a circle can be inscribed in any quadrilateral which is such that the sum or difference of two of its sides is equal to the sum or difference of its other two sides.*

If four straight lines touch a circle at points $a$, $b$, $c$, $d$, and if they form by their intersections the quadrilaterals $POQR$,

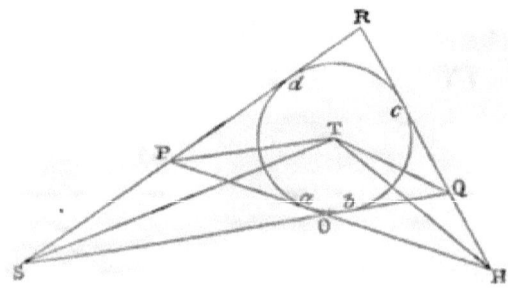

$SPHQ$, $SOHR$, as in the diagram, it readily follows, from the equality of the two tangents (as $Oa$, $Ob$) from any external point to a circle, that

$$OP + QR = OQ + PR; \quad SP + HP = SQ + HQ;$$
and
$$SO - HO = SR - HR;$$

and, conversely, if one of these relations be given, a circle can be drawn touching the four straight lines.

The LEMMA being assumed, if $P$, $Q$ be points on a conic whose foci are $S$ and $H$, then since

$$SP \pm HP = SQ \pm HQ,$$

a circle can be described touching $SP$, $SQ$, $HP$, $HQ$, and having its centre $T$ at the point of concourse of the external or internal bisectors of the angles $SPH$, $SQH$; that is to say, at the point of concourse of the tangents at $P$ and $Q$.*

And since $SP$ and $SQ$ touch a circle whose centre is $T$, therefore $ST$ bisects the angle between them; and in like manner $HT$ bisects the angle between $HP$ and $HQ$. That is to say, the tangents $TP$, $TQ$ to the conic subtend equal (or supplementary) angles at either focus.

*Corollary.*

One or other of the angles between any two tangents to a conic is equal to half the sum or difference of the angles which their chord of contact subtends at the foci.† Taking the case of a pair of tangents $TP$, $TQ$ to an ellipse whose chord of contact (as in the next figure) does not pass between the foci, we may shew, by equating the angles of the quadrilaterals $TPSQ$, $TPHQ$ to eight right angles, that the exterior angle between the tangents is an arithmetic mean to $PSQ$ and $PHQ$.

---

* If with $S$ and $H$ as foci an ellipse be drawn through $P$, $Q$ and a hyperbola through $O$, $R$, their common diameter through $T$ will bisect $PQ$ in the one curve and $OR$ in the other. Hence the middle points of the three diagonals of any quadrilateral in which a circle can be inscribed lie upon one diameter of the circle.

† This theorem, with its analogue for the parabola (Art. 28), was proved by BOSCOVICH (*Sectionum Conicarum Elementa*, §184). In a slightly different form it will be noticed again in Scholium D.

## PROPOSITION XVIII.

50. *The two tangents to a bifocal conic from any external point are equally inclined to its two focal distances, each to each.*

Let $TP$, $TQ$ be the tangents at $P$, $Q$ to a conic whose foci are $S$ and $H$; then will the angles $STQ$, $HTP$ be *equal* to one another, unless $P$ and $Q$ lie on the same branch of a hyperbola, in which case they will be *supplementary*.

(i) In the case of the ellipse, produce $HP$ to any point $V$, and let $SP$, $HQ$ intersect in $O$.

Then since the tangent at $P$ bisects the angle $SPV$, and

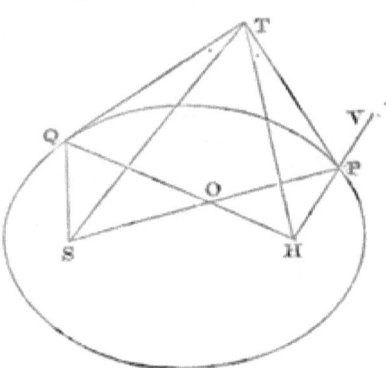

$HT$ bisects the angle $PHQ$, therefore

$$\angle HTP = TPV - THP = \tfrac{1}{2}SPV - \tfrac{1}{2}PHQ = \tfrac{1}{2}POH;$$

and in like manner it may be shewn that the angle $STQ$ is equal to $\tfrac{1}{2}QOS$ or $\tfrac{1}{2}POH$.

Hence $TP$ and $TQ$ make equal angles with the focal distances of $T$.*

(ii) If the tangents be drawn to opposite branches of a hyperbola, they will still make equal angles with $ST$ and $HT$.

(iii) If both tangents be drawn to the $S$-branch of a hyperbola, and if $HT$ be *produced* to a point $H'$, it will be

---

\* In Art. 37 the points $S$, $P$, $M$, $R$ and $S'$, $P'$, $M'$, $R'$ are concyclic, as will likewise be the case (Art. 9, Cor. 1) if $RR'$ be *any straight line*, and $P$ its polar. Hence, if $X$, $X'$ be the feet of the directrices, $SPR = SMX = S'M'X' = SPR'$. It follows by the proposition that *the intercepts on any chord made by the curve and its directrices subtend equal (or supplementary) angles at the pole of the chord.*

CENTRAL CONICS. 107

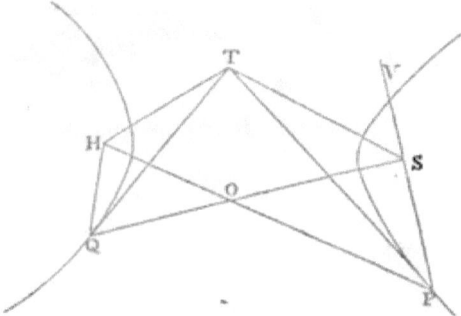

seen that they make equal angles with $ST$ and $H'T$ respectively, which is in accordance with the principles of Scholium C.

### Corollary 1.

The bisectors of the angles between any two tangents to a bifocal conic are the tangent and normal to the two confocals through their point of concourse.

### Corollary 2.

*The triangle whereof the base is equal to the transverse axis of a conic and its remaining sides to the focal distances of any external point has its vertex angle equal to the angle between the tangents to the conic from that point and its remaining angles to the angles which either tangent subtends at the foci.* For in the first figure, if $SP$ produced to a point $H'$ be equal to the transverse axis, it is evident that the triangle $TH'P$ is identically equal to $THP$, and hence that the triangle $STH'$ is of the specified linear and and angular dimensions.

#### SCHOLIUM D.

The theorem of Cor. 2 may be expressed as follows. If the straight lines in either diagram be regarded as a framework jointed at their ends, and if $S$ and $H$ be drawn apart (or brought together) until the distance between them is equal to $AA'$, then will $SPH$ and $SQH$ become straight lines, and the angles at the joints (except $SPH$ and $SQH$) will be equal to the angles at the same points in the original figure, the inner and outer angles at $T$ being *interchanged*. The angle $STH$ in the deformed figure is in general equal to the angle between the tangents, but in the third case of the proposition it becomes *supplementary* thereto. The following are some applications of this theorem.

(i) If the angle between the tangents be *a right angle*, then $ST^2 + HT^2 = AA'^2$, whence the property of the director circle readily follows. In any case, the angle $\theta$ between the two tangents from any given point $T$ satisfies the relation

$$ST^2 + HT^2 - 2ST \cdot HT \cos\theta = AA'^2.$$

(ii) Since the **exterior vertex** angle of the triangle is equal to the sum of its angles at $S$ and $H$, therefore the supplement of the angle between any two tangents to an ellipse is equal to *the sum of the angles which either tangent subtends at the foci*. In the hyperbola, as may be proved in like manner, the *difference* of the angles which either tangent subtends at the foci will be equal or supplementary to the angle between the tangents, according as they are drawn to opposite branches of the curve or to the same branch. In the one case, $P$ and $Q$ in the deformed figure will lie on opposite sides of $SH$, and in the other case on the same side of it.

(iii) It may be shewn from the deformed figure that

$$TP^2 : TQ^2 = SP \cdot HP : SQ \cdot HQ.$$

Hence, having deduced (Art. 45, Cor. 2) from the bifocal definition that

$$SP \cdot HP + CP^2 = CA^2 + CB^2 = SQ \cdot HQ + CQ^2,$$

we infer, taking the case in which $CP$, $CQ$ are *conjugate*, and therefore equal and parallel to the tangents at $Q$ and $P$, that

$$SP \cdot HP = CQ^2; \quad SQ \cdot HQ = CD^2; \quad CP^2 + CQ^2 = CA^2 + CB^2;$$

and hence, that *in all* cases the tangents are as the parallel radii.

## PROPOSITION XIX.

51. *At any point of a bifocal* conic the projection of the normal terminated by *the conjugate axis upon the distance of the point from either focus is equal to the semi-axis* transverse.

Let the normal at $P$ meet the conjugate axis in $g$, and draw $gk$ and $gl$ perpendicular to $SP$ and $HP$ respectively. Then since the normal bisects the angle $kPl$, therefore

$$Pk = Pl, \text{ and } gk = gl.$$

And since also the hypothenuses of the right-angled triangles $gkS$, $glH$ are equal, therefore the side $Sk$ of the one is equal to the side $Hl$ of the other.

Hence, in the first figure,
$$SP - Pk = Pl - HP = Pk - HP;$$

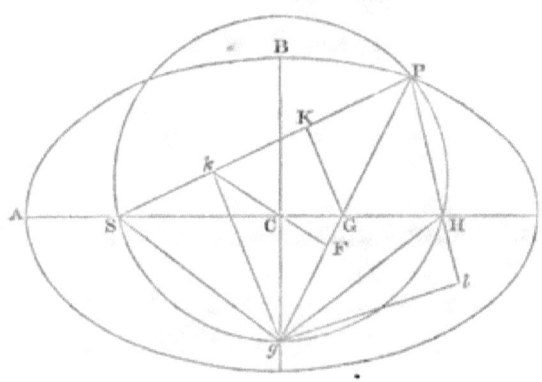

and in the second figure,
$$SP + Pk = HP - Pl = HP - Pk.$$

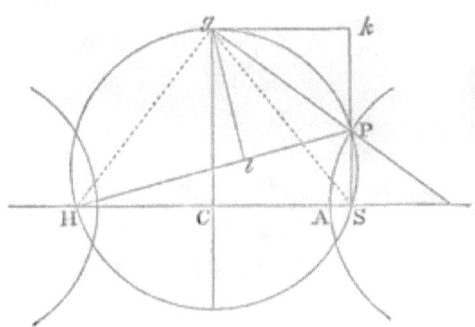

Therefore $\quad Pk = \tfrac{1}{2}(SP \pm HP) = CA.$

*Corollary 1.*

If $GK$ be drawn, as in Art. 11, it follows from this proposition, together with Art. 52, Cor. 2, that $PK \cdot CA = CB^2$.

*Corollary 2.*

By the converse of Art. 38, Cor. 3, the diameter $Ck$ is parallel to the tangent at $P$, or perpendicular to the normal;

if therefore it meet the normal in $F$,
$$Pg . PF = Pk^2 = CA^2,$$
and
$$PG . PF = PK . Pk = CB^2.$$

## PROPOSITION XX.

**52.** *The distance of any point on a bifocal conic from either focus varies as its distance from a corresponding fixed directrix perpendicular to the axis.*

Bisect the angle between the focal distances of any point $P$ of the locus, estimated *within* the curve, and let the bisecting

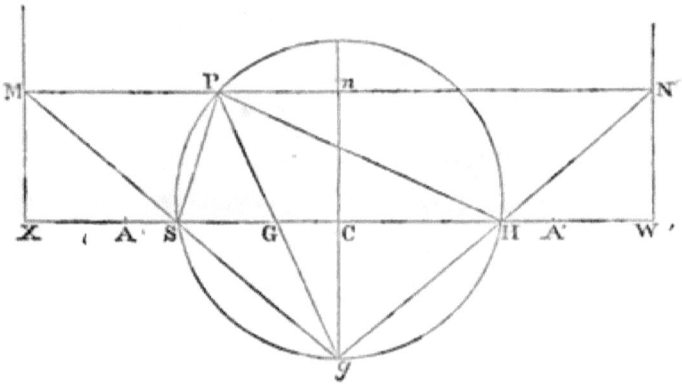

line meet the axes in $G$ and $g$. Let a parallel to the transverse axis through $P$ meet $gS$ in $M$ and $gH$ in $N$; and let perpendiculars be drawn to the axis through those points, meeting

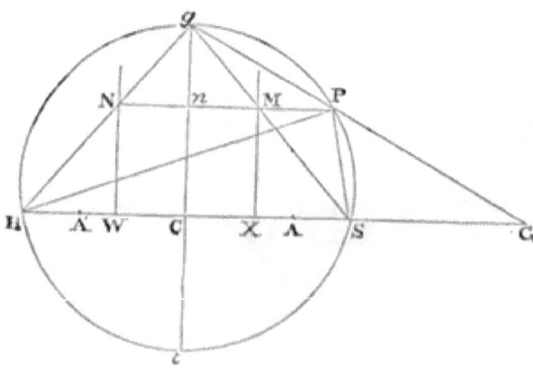

it in $X$ and $W$. We shall shew that these perpendiculars are *directrices*.

Since $\angle SMP = gSH = gHS = GPS$, the points $g$, $S$, $P$, $H$ being evidently concyclic; therefore, in the similar triangles $SPM$, $SGP$,

$$SP : PM = SG : SP = SG \pm GH : SP \pm PH = CS : CA;$$

and therefore $SP$ is in *a constant ratio* to $PM$.

And since $SH : MN = SG : PM = CS^2 : CA^2$, therefore $MN$, or $2CX$, is constant, and $MX$ is *a fixed straight line*.

The curve may therefore be described with $S$ and $MX$, or, in like manner, with $H$ and $NW$, as FOCUS and DIRECTRIX.

### Corollary 1.

The line $PG$ in the above construction being evidently the normal at $P$, it follows that *the focal distances of the point in which the normal at any point of the curve meets the conjugate axis pass through the feet of the perpendiculars from that point to the directrices.* Hence an obvious construction for the normals from $g$ with the help of the ruler only.

### Corollary 2.

Since $Gg : Pg = SG : PM = CS^2 : CA^2$, therefore $PG : Pg = CA^2 \sim CS^2 : CA^2 = CB^2 : CA^2$.

#### SCHOLIUM E.

The name Focus has reference to the *optical* property in relation to the conic of the points to which it is now commonly applied, viz. that rays proceeding from one of them and reflected at the curve would converge to or diverge from the other (Art. 37), or become parallel to the axis (Art. 25).

APOLLONIUS, who introduces the foci somewhat late in his treatise, proves their properties in the following order (Lib. III. props. 45—52). Starting with the property, $AS \cdot A'S = CB^2$, he shews that the intercept on the tangent at any point $P$ by the tangents at the vertices subtends right angles at the foci; that the tangent at $P$ and either of the fixed tangents make equal angles with the focal vectors to their point of concourse (a special case of Art. 50); that a pair of the focal vectors to its points of concourse with the two fixed tangents intersect on the normal at its point of contact; that it makes equal angles with the focal distances of that point; that the axis subtends a right angle at the foot of the focal perpendicular $SY$ upon the tangent; that the diameter parallel to $HP$ meets the tangent upon the circumference of the circle on $AA'$, viz. in the same point $Y$; and lastly, that $SP \mp HP = AA'$.

The property of the focus and directrix by which we defined the general conic is given in the Mathematical Collections of PAPPUS, Lib. VII. prop. 238 (vol. II. p. 1012, *ed.* Hultsch, BEROL., 1877).

NEWTON, to whom some later writers were indebted for their acquaintance with the property, mentions it in the *Principia*, at the end of Lib. I., Sect. IV., in connexion with his construction (a modification of De la Hire's Lib. VIII. 25, to which he refers) for an orbit whereof a focus and three points are given, viz. by determining a point of the directrix upon the chord joining each pair of given points on the curve.

## EXAMPLES.

201. If $AB$ and $CD$ be equal portions of two straight bars, and if they be connected by hinges with two equal bars $AD$, $BC$ in such a manner that initially $AB$ and $CD$ form opposite sides of a rectangle and $AD$ and $BC$ its diagonals; then (1) if a *side* of the rectangle be fixed, the remaining parts of the framework being moved about in any way in a plane, the intersection of the cross bars will trace an ellipse; or (2) if a *diagonal* be fixed, the continuations of $AB$ and $CD$ will cross one another upon the arc of a hyperbola.

202. The sides $AD$, $DC$ of a rectangle $ABCD$ are divided into the same number of equal parts, and straight lines are drawn from $B$ and $A$ respectively to the points of section. Shew that the corresponding lines of the two series meet on an ellipse whose axes are equal to the sides of the rectangle.

203. A parallelogram $ABCD$ has its diagonal $AC$ at right angles to the side $AB$. If $CD$ be divided into any number of equal parts and straight lines be drawn from $A$ to the points of section, and if $AC$ be divided into the same number of equal parts and straight lines be drawn from $B$ to the points of section, then will the corresponding lines in the two series meet on a hyperbola.

204. The straight lines joining the vertices of a conic to opposite ends of any segment of the directrix which subtends a right angle at the focus intersect on the curve. Hence show how to trace an ellipse or hyperbola, and prove that the ellipse is a closed curve giving wholly between the perpendiculars to the axis at its vertices.

205. The locus of the point at which the distances of either vertex from a focus and directrix subtend equal angles is the auxiliary circle. Hence show that the ellipse cuts the ordinates of its auxiliary circle in a constant ratio.*

206. If $AM$ and $A'M$ be taken on the axis of a conic equal to the focal distances of any point $P$ on the curve, show that

$$CP^2 = CB^2 + CM^2;$$

and deduce that the square of the ordinate of $P$ varies as the rectangle contained by its abscisses.

207. If a circle be drawn through the vertices of a conic and any point on the curve, find the locus of the second point in which it meets the ordinate of the former.

208. If two ellipses whose major axes are equal have a common focus, they will intersect in two points only; and their common chord will be at right angles to the straight line joining their centres.

209. Given a chord of a parabola and the direction of its axis, the locus of the focus is a hyperbola whose foci are at the extremities of the chord.

210. The straight lines from either focus of a conic to the ends of a diameter make equal angles with the tangents thereat.

211. A circle can be drawn through the foci and the intersections of any tangent with the tangents at the vertices.

212. The intercept on any tangent by the tangents at the extremities of a focal chord subtends a right angle at the focus.

---

* If $p$ be a point on the locus, then $Sp : PX = SA : AX = SA' : A'X$, and therefore $A'p$ is the outer bisector of the angle $SPX$, and is at right angles to $Ap$. Hence $pN^2 - PN^2 = Sp^2 - SP^2 = e^2(PX^2 - NX^2) = e^2 \cdot PN^2$, and $PN^2$ varies as $pN^2$, or as $AN \cdot A'N$.

213. The major axis is the maximum chord of an ellipse, and the minor axis is its least diameter. When is the angle subtended at the curve by the straight line joining the foci a maximum?

214. The focal vectors to any two points on a conic meet in two other points lying on a confocal conic, and the tangents at the two pairs of points cointersect. Examine the case in which the focal vectors are drawn to opposite extremities of a diameter.

215. Find the locus of the centre of a circle which touches two fixed circles. Also, if two circles, the sum or difference of whose radii is constant, be described about fixed points, find the locus of the centre of a circle of given magnitude which touches both of them.

216. If $PQ$ be a chord of an ellipse, and if the ordinates of its extremities to either axis be produced to meet the corresponding auxiliary circle in $p$ and $q$, then will $PQ$ and $pq$ meet on that axis; and if the tangent at $p$ meet the same axis in $T$, then will $TP$ touch the ellipse at $P$, and the abscissa of $P$ will be a third proportional to $CT$ and $CA$.

217. If any tangent to an ellipse meet the axes in $T$ and $t$, the tangents from those points to the major and minor auxiliary circles respectively will be parallel two and two, and their four points of contact will lie on two diameters of the ellipse.

218. The perpendiculars to the axes from the points in which a common diameter meets the two auxiliary circles of an ellipse intersect two and two on the curve. Hence shew how to construct an ellipse with the help of two fixed concentric circles.

219. If two points on a straight line move along the arms of a right angle, any other point on the line will trace an ellipse whose semi-axes are equal to the segments of the line between that point and the former two.*

---

\* The moving line may be supposed parallel to the "common diameter" in Ex. 218. This theorem explains the construction of the *Elliptic Compasses*.

220. The ordinates of the points in which any tangent to an ellipse or hyperbola meets the curve the tangents at its vertices and the conjugate axis are proportionals, and the product of the extremes or the means is equal to the square of the semi-axis conjugate.

221. If any ordinate to either axis be bisected in $O$, and $AO$, $A'O$ meet the tangents at $A'$ and $A$ in $T'$ and $T$ respectively, then will $TT'$ be the tangent at the extremity of the ordinate; and the straight line joining the intersections of $ST$, $HT'$ and $ST'$, $HT$ will be the normal at that point.

222. If two ellipses with equal axes be placed vertex to vertex, and one of them then roll upon the other, each of its foci will describe a circle about a focus of the latter.

223. Given a central conic, shew how to find its centre, axes, foci, and directrices.

224. Given a focus and two points of an ellipse, the locus of the other focus will be a hyperbola. If instead of one of the two points the length of the axis be given, determine the loci of the centre and of the second focus.

225. Given one focus of a conic inscribed in a triangle, shew how to determine the other focus. If an ellipse inscribed in a triangle have one focus at the orthocentre, its other focus will be at the centre of the circumscribed circle, and its auxiliary circle will be the nine-point circle of the triangle. Examine the case in which a focus of the inscribed conic is at the centre of the inscribed circle of the triangle.

226. The angular points and the sides of a triangle being taken as the centres and directrices of three ellipses which have a common focus at the orthocentre, shew that the sum of the squares of their major axes is double of the sum of the squares of the sides of the triangle; their minor axes are equal to one another, the square of each being equal to the sum of the squares of the three latera recta; the sum of the squares of the eccentricities is equal to two; the intercepts made by the ellipses upon the sides of the triangle are conjugate diameters;

the perpendiculars of the triangle are the common chords of the ellipses, and their six poles lie on three focal chords parallel to the sides of the triangle.

227. If $CP$ be any radius of a conic, and if a parallel to it be drawn from the vertex $A$ to meet the curve in $Q$ and the conjugate axis in $R$, shew that $AQ.AR$ is equal to $2CP^2$.

228. A straight line equal to the radius of a circle slides with one end on a fixed diameter and the other end on the convexity of the circumference. Shew that any intermediate point on the line traces an arc of an ellipse.

229. From a fixed point $O$ a straight line $OP$ is drawn to a given circle. Find the envelope of a straight line drawn through $P$ at a constant inclination to $OP$.

230. The straight line joining the foci of a conic subtends at the pole of any chord an angle equal to half the sum or difference of the angles which it subtends at the extremities of the chord.

231. If $CR$ be the projection of any radius $CD$ upon the axis of the conic, and $OL$ the ordinate of the middle point $O$ of any chord* parallel to $CD$, prove that

$$OL.DR : CL.CR = CB^2 : CA^2.$$

Hence shew that the diameters of conics are straight lines, and obtain the relation between the inclinations of any two conjugate diameters to the axis.

232. The ellipse has a pair of equal conjugate diameters, which coincide in direction with the diagonals of the rectangle formed by the tangents at the ends of its axes, and which are equal to the sides of a square whose diagonals are equal to those of the said rectangle.

What is the corresponding property of the hyperbola?†

---

\* Art. 33 gives $PN^2 \sim QM^2$ in terms of $CN^2 \sim CM^2$; whence the required result readily follows.

† It has two pairs of (infinite) conjugate diameters which are in a ratio of equality, and each pair coincide in direction with one of the equi-conjugate diameters of the ellipse which has the same axes.

233. The common diameters of equal similar and concentric ellipses are at right angles to one another.

234. Find the loci of the centres of the four circles which touch the axis of a conic and the two focal vectors to any point on the curve.

235. A chord of a conic which subtends a right angle at the vertex passes through a fixed point on the axis.

236. If a hyperbola touch the sides of a quadrilateral inscribed in a circle, and if one focus lie on the circle, the other will also lie on the circle.

237. If three circles be described on the transverse axis and the two focal distances of any point of a conic as diameters, determine their radical centre.

238. If the tangent at a point $P$ whose ordinate to either axis is $PN$ meet the corresponding auxiliary circle in $Y$ and $Z$, shew that $C$, $N$, $Y$, $Z$ lie on a circle, and that $PN$ bisects the angle $YCZ$.

239. If an ellipse and a hyperbola have the same axes, the director circle of the one will pass through the foci of the other.

240. The diagonals of any rectangle circumscribing a conic are conjugate diameters.*

241. The diagonals of any parallelogram circumscribing a conic are conjugate diameters, and the sides of any inscribed parallelogram are parallel to conjugate diameters.

242. The sum or difference of the reciprocals of any two focal chords at right angles, or of the squares of any two diameters at right angles, is constant.

243. The locus of the centre of an ellipse which slides between two straight lines at right angles is a circle.

244. The circle described upon the straight line joining the foci of a conic meets the conjugate axis in two points such that

---

* This appears from De la Hire's original proof of the property of the director circle, in which he assumed the theorems of Arts. 39, 45.

the sum of the squares of the perpendiculars therefrom to any tangent is equal to half the square upon the transverse axis.

245. Determine the positions of a chord of an ellipse which subtends a right angle at each of the foci; and also the locus of the pole of a chord which subtends supplementary angles at the foci.

246. The opposite sides of a quadrilateral described about an ellipse subtend supplementary angles at either focus.

247. The angle which a diameter of an ellipse subtends at an extremity of the major axis is supplementary to that which its conjugate subtends at an extremity of the minor axis.

248. If a focal chord of a conic be drawn to meet at a given angle any tangent, or any chord subtending a constant angle at one of the foci, the locus of the point of intersection will be a circle.

249. To what does the theorem that confocal conics intersect at right angles reduce when the two foci coalesce?

250. The circle described about any point on the axis of a hyperbola so as to cut its auxiliary circle orthogonally meets the ordinate through that point upon the circumference of an equilateral hyperbola.

251. The pole of any straight line with respect to a central conic may be found by joining the points in which it meets the directrices to the nearer foci, and drawing perpendiculars through the latter to the joining lines.

252. The straight lines joining any point to the intersections of its polar with the directrices touch a confocal conic.

253. In Art. 4, if the centre of the circle be taken midway between the vertices of the conic, shew that the directrix will be the polar of the focus with respect to the circle. Hence shew that every chord of an ellipse or hyperbola which passes through its centre is bisected at that point, and that the curve is consequently symmetrical with respect to its conjugate axis.

254. The locus of the vertex of a triangle whose base and the ratio of whose sides are given is a circle, whereof a diameter is determined by dividing the base of the triangle externally and internally in the given ratio. Hence shew that a straight line parallel to the axis of a conic meets the curve in general in two points, and that the two points are equidistant from a fixed straight line parallel to the directrix.*

255. Apply the same method to determine the points in which any assumed straight line meets the conic;† and also to shew that the diameters of conics are straight lines.

256. Prove the following construction for drawing tangents to a conic whose foci are $S$ and $H$ from a given external point $O$. About $O$ with radius $OS$ describe a circle, and about $H$ with radius equal to the transverse axis describe a circle; then will the required points of contact lie upon the straight lines drawn from $H$ to the points in which the two circles intersect. Prove also that the two tangents as thus determined subtend equal or supplementary angles at either focus.

257. If the diagonals of a quadrilateral circumscribing an ellipse meet at its centre the quadrilateral must be a parallelogram.

258. If a principal ordinate meet an ellipse in $P$ and its auxiliary circle in $Q$, the distance of the former point from either focus will be equal to the perpendicular from that focus to the tangent at $Q$.

259. The locus of the middle point of a focal chord of a conic is a similar conic. In what other cases will the locus of the middle point of a chord of a conic be a similar conic?

---

* The two points are determined as follows. Let the parallel meet the directrix in $Q$, and let $Z$ and $Z'$ divide $SQ$ in a ratio equal to the eccentricity. Describe the circle on $ZZ'$, and let it cut the parallel in $P$ and $P'$. The projection of the centre of this circle upon the axis of the conic evidently lies midway between the projections of $Z$ and $Z'$ upon the same, that is to say, midway between the vertices of the conic.

† If the assumed line meet the directrix in $Q$ and make an angle $\alpha$ with the axis, divide $SQ$ in the ratio $e \cos \alpha$, and upon the intercept between the two points thus determined describe a circle cutting the assumed line in $P$ and $P'$, which will be the points required.

260. If tangents be drawn to a conic at the extremities of a pair of conjugate radii, the focal vectors to their point of concourse will meet those diameters in four points lying on a circle.

261. If $SY$ be a focal perpendicular upon the tangent at $P$ and $CD$ the radius parallel to the tangent,
$$SY^2 : CB^2 = SP^2 : CD^2 = SP : 2CA \pm SP.$$

262. The feet of the perpendiculars from one focus of a conic to a pair of tangents are on a straight line at right angles to the distance of their point of intersection from the other focus.

263. Find the greatest or least value of the sum of the squares of the focal perpendiculars on any tangent.

264. The normal at any point of an ellipse is a harmonic mean to the focal perpendiculars upon the tangent at that point.

265. If one focus of a conic which touches the sides of a triangle be at its centroid, the distances of the other focus from its sides will be as the lengths of those sides.

266. If the normal at $P$ meet either axis in $G$, shew that any circle through those points will intercept on the focal distances of $P$ chords whose sum or difference has one or other of two constant values.

267. The diameters parallel to the tangent and normal at $P$ intercept on $SP$ a length equal to $HP$; and the latter diameter meets $SP$ on the circumference of a circle.

268. The circle described upon the central abscissa of the foot of the normal at any point is cut orthogonally by a circle described about that point and equal to the minor auxiliary circle.

269. The intercepts on the focal vectors to the points of contact of a conic with any circle which touches it in two points have one or other of two constant values.

270. Any tangent and its normal meet either axis in points $T$ and $G$ such that $CG \cdot CT = CS^2$. In what cases does the normal meet the conjugate axis without within or upon the curve? If a circle touch an ellipse in two points and also touch its directrices, its centre will be at an end of the minor axis.

271. The intercept made by the directrices upon any normal chord of a conic subtends at the pole of the chord an angle equal to half the sum or difference of the angles which the distance between the foci subtends at the extremities of the chord.*

272. If two chords be drawn from any point of a conic equally inclined to the normal at that point, the tangents at their further extremities will intersect upon the normal.

273. Supplemental chords of a conic which are equally inclined to the curve at their common point have their poles upon the director circle, and their sum or difference is equal to the diameter of the same.†

274. The normals at the extremities of any two conjugate radii meet on the diameter which is at right angles to the chord joining those points; and any two normals at right angles to one another intersect on the diameter which bisects the chord joining the points at which the normals are drawn.

275. If on the normal at $P$ a length $PQ$ be taken equal to the semi-diameter conjugate to $CP$, the locus of $Q$ will be a circle of radius $CA \pm CB$.

276. If normals be drawn to an ellipse at the ends of any chord parallel to one of its equal conjugate diameters, the locus of their intersection will be a line perpendicular to the conjugate diameter.

---

\* If the normal at $P$ meet the $S$-directrix in $R$, and $O$ be the pole of the normal, the circle on $OR$ will pass through $S$ and $P$.

† See Wolstenholme's *Book of Mathematical Problems*, No. 493 (London and Cambridge, 1867). If $PQ$, $PQ'$ be two such chords they will evidently touch a confocal, and the parallel chords will also touch the same. Hence the tangent at $Q$ will make equal angles with $PQ$ and a parallel to $PQ'$, and will be at right angles to the tangent at $P$. Otherwise thus: the tangent and normal at $P$ divide $QQ'$ harmonically, and the normal is an ordinate of $QQ'$ and parallel to the tangent at $Q$.

277. With the construction of Art. 52, shew that
$$Sg : Pg = PG : SM = Pg : gM = CS : CA.$$
Shew also that the loci of the middle points of $PG$ and $Pg$ are conics.

278. If $N$ be the projection of any point $P$ on a conic upon the tangent at a given point $O$, find the relation between the lengths of $ON$ and $PN$.*

279. If $PQ$ be a chord of a conic which subtends a right angle at a given point $O$ on the curve, and $MN$ be the projection of the chord upon the tangent at $O$, shew that
$$\frac{PM}{OP^2} \pm \frac{QN}{OQ^2} = \text{a constant},$$
and that $PQ$ passes through a fixed point on the normal at $O$.†

280. If $PG$ be the normal at any point $P$ on an ellipse, $Q$ the point in which the ordinate of $P$ meets the auxiliary circle, and $R$ the point in which $CQ$ meets the ellipse, then will $QG$ be parallel to the normal at $R$. Moreover, if any two chords of an ellipse be at right angles, its diameters conjugate to the corresponding chords of its auxiliary circle will be conjugate diameters of a certain concentric ellipse.

281. If $N$ be the projection of any point in the plane of a conic upon its transverse or conjugate axis, and $T$ the point in which its polar meets the same, shew that $CN.CT$ is equal to $CA^2$ in the former case and to $CB^2$ in the latter.‡

282. If $P, P'$ be any two points whereof the one lies on the polar of the other, and $N, N'$ be their projections on the transverse axis, then
$$\frac{CN.CN'}{CA^2} \pm \frac{PN.P'N'}{CB^2} = 1.$$

---

\* Let the normal at $O$ meet the curve again in $H$, and let $NP$ meet the tangent at $H$ in $M$; then $ON$ varies as a mean proportional to $PN$ and $PM$ (Ex. 73, and note p. 65), and also as $PN + PM \sim OH$.

† The intercept on the normal varies as the diameter conjugate to $CO$.

‡ Let the diameter through the assumed point $O$ meet the directrix in $V$, and let its polar meet the directrix in $R$, which will be the orthocentre (Art. 14) of the triangle $STV$. Then, $TV$ being parallel to $SO$ (Art. 17), $CN : CX = CO : CV = CS : CT$.

**283.** If $P$ and $Q$ be any two points on a conic, $M$ and $N$ their projections on either axis, and $T$ the point in which $PQ$ meets the same, shew that

$$CT(PM - QN) = PM.CN - QN.CM.^*$$

**284.** Shew also that if $CL$ be the central abscissa upon either axis of the point of concourse of the normals at $P$ and $Q$,† then $CL.CT'$ varies as $CM.CN$.

**285.** If the normals at four points of a conic cointersect, and if the three pairs of chords joining those points meet either axis in $T, T'$; $U, U'$; $V, V'$; shew that

$$CT.CT' = CU.CU' = CV.CV'.$$

**286.** If the normals at four points of an ellipse or hyperbola cointersect, each pair of chords joining the four points will be parallel to a pair of conjugate diameters of the hyperbola or ellipse which has the same axes, and will meet either axis in points which are conjugates in an involution determined by the latter curve.‡

**287.** The sum or difference of the squares of the perpendiculars from the extremities $P$ and $D$ of any two conjugate semi-diameters of a conic upon a fixed diameter of the same is constant; and if $CN$ and $CR$ be the abscissæ of $P$ and $D$ upon that diameter, and $2CP'$ be its length and $2CD'$ the length of its conjugate, then

$$CN^2 \pm CR^2 = CP'^2; \quad PN^2 \pm DR^2 = CD'^2;$$

and
$$PN : CR = DR : CN = CD' : CP'.$$

---

* Equate the areas $(CPT - CQT)$ and $(CPM + PMNQ - QCN)$.

† If $L$ be the projection upon either axis of any point $O$ on the normal at $P$, and $G$ be the point in which the normal meets that axis, then $OL : PM = CG \sim CL : MG$; and $CG$ and $MG$ vary as $CM$.

‡ The pairs of points in Ex. 285 determine an involution whose centre is $C$, and to which the ends of the axis in question likewise belong. If these points be on the transverse axis, $CT.CT' = \&c. = CA.CA' = -CA^2$; and if on the conjugate axis, $CT.CT'$ is equal to $-CB^2$ in the case of the ellipse, and to $+CB^2$ in the case of the hyperbola.

288. The vertices of the conjugate circumscribing parallelogram of a conic lie on a similar conic, and their polars envelope another similar conic.

289. The inscribed parallelogram whose diagonals are at right angles envelopes a circle, the reciprocal of the square of whose radius is equal to $\dfrac{1}{CA^2} \pm \dfrac{1}{CB^2}$.

290. If the polar of any point on an ellipse with respect to its minor auxiliary circle meet the major and minor axes in $H$ and $K$, shew that
$$\frac{CB^2}{CH^2} + \frac{CA^2}{CK^2} = \frac{CA^2}{CB^2}.$$

291. Supplemental chords being drawn to a conic from its vertices, the perpendiculars to them at their common point make an intercept equal to the latus rectum upon the axis.

292. If an ordinate of any diameter meet the curve in $P$, the diameter in $M$, and any two supplemental chords drawn from its extremities in $Q$ and $R$, shew that $PM$ is a mean proportional to $QM$ and $RM$, and that $QR$ is bisected by the tangent at the intersection of the chords.

293. If two conjugate semi-diameters $CP$, $CD$, or their prolongations, make an intercept $P'D'$ upon a line which is parallel to $PD$ and meets the conic in $Q$, shew that
$$QP'^2 \pm QD'^2 = PD^2.$$

294. From extremities of two conjugate diameters of an ellipse a pair of parallels are drawn to any tangent; if any diameter meet these parallels in $P$ and $Q$ and the tangent in $R$, shew that
$$CP^2 + CQ^2 = CR^2.$$

295. If $PP'$ and $DD'$ be conjugate diameters of a hyperbola and $Q$ any point on the curve, then will $QP^2 + QP'^2$ exceed $QD^2 + QD'^2$ by a constant quantity.

296. Find the axes of a given conic by means of Art. 44, Cor. 3.

297. Given two conjugate diameters of an ellipse, shew that the locus of the centre of the circle through their common point and the points in which a tangent parallel to one of them meets any two conjugate diameters is a straight line perpendicular to the other.

298. Given two conjugate semi-diameters $CP$, $CD$ of an ellipse or hyperbola, prove the following construction for its axes. In the direction $CP$ or in the opposite direction take $PQ$ a third proportional to $CP$ and $CD$; draw the tangent at $P$, and through $C$, $Q$ draw a circle having its centre upon it; join $C$ with the points in which the circle cuts the tangent. The directions of the axes being thus determined, shew how to find their extremities.

299. Given two conjugate diameters $CP$, $CD$ of an ellipse, with centre $C$ and radius $CP$ describe a circle, and let $KK'$ be its diameter at right angles to $CP$; then will the axes of the ellipse be equal to $KD \pm K'D$, and parallel to the bisectors of the angle $KDK'$.*

300. Shew also that, if $DN$ be taken equal to $CP$ and be placed so as to cut it at right angles in a point $L$, and if $AB$ be that diameter of the circle round $CLN$ which passes through $D$, then will $CA$, $CB$ be the directions and $DB$, $DA$ the lengths of the semi-axes of the ellipse.

301. Any point of or in the same straight line with a rod which slides between two fixed straight lines describes an arc of an ellipse.†

---

\* See the *Oxford, Cambridge, and Dublin Messenger of Mathematics*, vol. III. pp. 151, 227 (1866); and the *Messenger of Mathematics* (New Series), vol. V. p. 122 (1876).

† Of two conjugate radii of an ellipse let $CP$ be the shorter and $CD$ the longer; draw a perpendicular $DE$ to $CP$, and in the prolongation of or within $ED$ take $DQ$ equal to $CP$. Then if a straight line $KM$ equal to $EQ$ slides between $CP$ and $CQ$, the point which divides it into segments $KO$ and $OM$ equal to $CP$ and $DE$ will be a point on the curve, viz. one whose abscissa on $CD$ is terminated by the perpendicular from $M$ to $CP$. As in the special case of Ex. 219, $KO \cdot OM = CA \cdot CB$. See Leslie's *Geometrical Analysis, and Geometry of Curve Lines*, p. 257 (Edinburgh, 1821).

302. The chords joining the extremities of two diameters of a conic and of their conjugates respectively are either parallel or conjugate in direction. If a series of chords pass through a fixed point, the chords of the corresponding conjugate arcs have the same property; and the diameters through the two fixed points are at right angles.

303. With the orthocentre of a triangle as centre two ellipses are described, the one touching its sides and the other passing through its angular points; prove that these ellipses are similar, and that their homologous axes are at right angles.

304. The perpendiculars from opposite foci of a conic upon two conjugate diameters intersect on a concentric conic passing through the foci.

305. If a chord $AP$ drawn from the vertex $A$ be divided in such a manner that $AQ : PQ = CA^2 : CB^2$, shew that the perpendicular from $Q$ to the line joining $Q$ to the foot of the ordinate of $P$ divides the transverse axis in the same ratio.

306. From the foot of the ordinate of any point $P$ on a conic a parallel is drawn to $AP$ to meet the diameter through $P$ in $Q$; shew that $AQ$ is parallel to the tangent at $P$. Shew also that the bisectors of the angles $ASP$ and $AHP$ intersect on the tangent at $P$.

307. If two conics whose transverse axes are equal be inscribed in the same parallelogram, their foci will be at the corners of an equiangular parallelogram.

308. Any one of a series of conterminous circular arcs may be trisected by drawing a pair of hyperbolas whose determining ratio is equal to two, and whose centres and vertices trisect the chord of the arc. How does it appear from this construction that the problem, to trisect a given angle, admits of three solutions?

309. If any two conics have a common focus, one pair of their common chords cointersect with the corresponding directrices, and the other pairs subtend equal or supplementary angles at that focus.

310. A diameter meets the conic in $P$, its auxiliary circle in $Q$, and the tangent at either vertex in $T$. Prove that when the diameter through $T$ coalesces with the axis $PT$ and $QT$ are in the duplicate ratio of the axes.

311. The perpendicular drawn through any point of a conic to one of its focal distances and terminated by the conjugate diameter varies inversely as the principal ordinate of the said point, and the perpendiculars from the vertices upon the tangent at any point meet its focal distances upon fixed circles.

312. A parabola of given linear dimensions being drawn to touch any two conjugate diameters of a conic symmetrically, find the locus of its focus.

313. The tangent at $P$ meets any two conjugate diameters in $T$, $t$, and $TS$, $tH$ meet in $Q$; prove that the triangles $SPT$, $HPt$, $TQt$ are similar, and also that the area of the triangle $CPT$ varies inversely as $CPt$.

314. The two points of a conic at which a given chord subtends the greatest and least angles are at the extremities of a diameter equal to that which bisects the chord.*

315. If an ellipse touch a given ellipse at adjacent extremities $A$, $B$ of its axes and also pass through its centre, the tangent at the latter point will be parallel to $AB$.

316. With the normal and tangent at any point of a conic as axes a conic is described touching an axis of the former at its middle point; shew that the foci of the conic so drawn lie on fixed circles, whose diameters are equal to the sum and difference of the axes of the given conic.

317. Two fixed points being taken in given parallel lines, a straight line revolves about each point and meets the opposite parallel. If the envelope of the line joining the points of

---

* The chord must subtend *equal* angles at either point and a consecutive point on the curve. The two points therefore lie on segments of circles described upon the chord so as to touch the conic.

concourse be a conic touching the parallels at the fixed points, determine the locus of the point in which the revolving lines intersect.

318. A chord of a circle which subtends a right angle at a fixed point envelopes a conic whose foci are the fixed point and the centre of the circle.

319. A straight line being drawn through a fixed point $S$ to meet a given pair of parallels in $Y$ and $Z$, shew that the envelope of the circle on $YZ$ as diameter is a conic, of which the parallels are directrices and $S$ is a focus.

320. On the axis of a hyperbola whose determining ratio is equal to two a point $D$ is taken at a distance from the focus $S$ equal to the distance of $S$ from the further vertex $A'$, and $A'P$ is drawn through any point $P$ on the curve to meet the latus rectum in $K$. Prove that $DK$ and $SP$ intersect on a certain fixed circle.

321. The parallelograms whose diagonals are any two diameters of a conic and their conjugates respectively are of equal area.

322. If tangents $TP$ and $TQ$ be drawn to an ellipse whose foci are $S$ and $H$, and $CP'$ and $CQ'$ be the parallel semi-diameters,
$$TP.TQ + CP'.CQ' = TS.TH.*$$

323. Find the locus of a point such that the tangents therefrom to a central conic contain with the semi-diameters to their points of contact an area of constant magnitude; and the locus of a point such that the product of its focal distances varies as the product of the tangents.

324. The distance between any point and any point on its polar is cut harmonically by the tangents at the extremities of any chord through either point.†

---

\* We have to shew that the triangle $STH'$ (Art. 50, Cor. 2) is equal to $PTQ + P'CQ'$; which follows from Ex. 321, taking into account that $PCQ = \frac{1}{2}(PSQ + PHQ)$.

† In the tractate, *De Linearum Geometricarum Proprietatibus Generalibus*, forming the Appendix to *A Treatise on Algebra, &c.*, by COLIN MACLAURIN, late Professor of

325. If $O$ and $P$ be any two points in the plane of a conic whose centre is $C$, the perpendiculars from $O$ and $C$ to the polar of $P$ are to one another as the perpendiculars from $P$ and $C$ to the polar of $O$.

326. If two conics be concentric and similarly situated, the pole of any tangent to the one with respect to the other will trace a concentric conic; and if the two conics be also similar the third will be similar to both. If the axes of the two be identical, the pole of any tangent to the one with respect to the other will lie on the former.

327. Find a point which has the same polar, and a line which has the same pole, with respect to three conics, whereof one has double contact with the other two.

328. An ellipse has double contact with each of two circles, whereof one lies within the other. Shew that its chords of contact with them meet in a fixed point on the line joining their centres; the locus of its centre is a circle passing through their centres; its eccentricity is constant; and the locus of its foci is a circle concentric with the outer given circle.

329. Any diameter of an ellipse varies inversely as the perpendicular focal chord of its auxiliary circle.

330. If a parallelogram circumscribing a conic have two of its angular points on the directrices, the other two will lie on the auxiliary circle.

331. If two parallelograms be constructed, the one by joining the ends of two parallel focal chords of a conic, and the other by drawing tangents to it at those points; the area of the

---

Mathematics in the University of Edinburgh (LONDON, 1779), it is shewn (Sect. I. §§ 9—11. Cf. Salmon's *Higher Plane Curves*, Art. 60), that if a straight line revolving about a fixed point $P$ meet a curve of the $n^{th}$ order in $n$ points, and the tangents at those points meet any assumed straight line through $P$ in $K$, $L$, $M$, &c., then will $\frac{1}{PK} + \frac{1}{PL} + \frac{1}{PM} + $ &c. the constant; and if the assumed line through $P$ meet the curve, viz. in the $n$ points $A$, $B$, $C$, &c., this constant will be equal to $\frac{1}{PA} + \frac{1}{PB} + \frac{1}{PC}$ + &c. In the particular case of Ex. 324, it is evident that the point $P$, its polar, and the tangents at the extremities of any chord through $P$ divide any straight line from $P$ harmonically.

one will vary directly, and that of the other inversely, as the projection of one of the focal chords upon the conjugate axis.

332. If $SY$, $HZ$ be the focal perpendiculars upon the tangent at $P$ to an ellipse, and $SY''$, $HZ'$ perpendiculars upon the tangents from $P$ to a confocal ellipse, then will the rectangle $YY'.ZZ'$ be equal to the difference of the squares of their major (or minor) semi-axes.

333. Determine the condition that the intercept on any tangent to a bifocal conic by two fixed tangents may subtend equal angles at the foci.

334. The tangents to a conic from any point on a circle through its foci meet the circle again in two points such that the second tangents therefrom intersect upon the circle.

335. Given an ellipse and a circle through its foci; prove that their common tangents touch the circle in points lying upon the tangent to the ellipse at an extremity of its conjugate axis.

336. If the tangent to a conic at a given point be met by any two parallel tangents, the focal distances of the points of concourse will meet on a fixed circle, whose centre will be on the normal at the given point.

337. The product of the tangents to a conic from any point is to the product of its focal distances as the distances of the point from the centres of the chord of contact and of the conic respectively. If the tangents from any point to a conic be in a constant ratio to the parallel diameters, determine the locus of the point.

338. Given an ellipse and one of its "cercles directeurs,"[*] shew that an infinity of triangles can be described about the one and inscribed in the other, and that all will have the same orthocentre.

339. An ellipse may be described by means of an endless string passing round two fixed points. If one focus be taken

---

[*] See the second paragraph of Scholium B (p. 90) and compare Ex. 225.

anywhere on a given straight line whilst the other remains constantly fixed, the envelope of all the ellipses described with a given string will consist of two arcs of parabolas.

340. Shew that the line joining any point outside a conic to its centre, and the radical axis of a pair of circles through the point, the one passing through the points of contact of the tangents from the point, and the other passing through the foci, are equally inclined to the focal distances of the point.

341. If a normal to a conic meet the curve again in $Q$ and the directrices in $R$, $R'$; and if $O$ be the pole of the chord and $S$, $S'$ the foci; prove that $SR$, $OR'$ and $S'R'$, $OR$ intersect on the normal at $Q$. In the case of the parabola, any normal chord produced to meet the directrix subtends a right angle at the pole of the chord; and the polar of the middle point of the chord meets the focal vector to its point of concourse with the directrix upon the normal at its further extremity.

342. At any point $P$ on the auxiliary circle of an ellipse a tangent is drawn meeting the axis in $T$, and $PA$, $PA'$ are drawn to the vertices meeting the ellipse again in $D$ and $E$; prove that the chord $DE$ passes through $T$.

343. The polar of any point $O$ with respect to a conic and the perpendicular to it from $O$ meet either axis in points $T$ and $G$ such that

$$CG.CT = CS^2.*$$

344. If a point be taken anywhere on a fixed perpendicular to either axis of a conic, the perpendicular from it to its polar will pass through a fixed point on that axis.

345. If perpendiculars $SY$, $HZ$, $CM$, $PN$ be drawn to the polar of any point $P$, and if $PN$ meet the axis in $G$, shew that $SY.HZ = CM.NG$; $CM.PG = CB^2$; and that the normal at a point on the curve which has the same central abscissa as $P$ is a mean proportional to $NG$ and $PG$.

---

* Comparing Ex. 281 (note), $CG : CS = CO : CV = CS : CT$.

346. The poles of a given straight line with respect to a series of confocal conics lie upon a second straight line perpendicular to the former. Hence shew that if a chord of one conic touch another conic having the same foci, the tangents at its extremities will meet on the normal at its point of contact, and conversely the foot of the perpendicular from their intersection to the chord will be its point of contact.

347. If a triangle inscribed in a conic envelope a confocal, its points of contact will lie severally on three of the four circles which touch the sides of the triangle.

348. If a chord of an ellipse be drawn touching a confocal ellipse, the tangents at its extremities meet the diameter parallel to the chord on the circumference of a fixed circle, and the intercept on the chord by the diameter parallel to either tangent is of constant length;* the chord varies as the parallel focal chord of the outer ellipse, and conversely a chord which so varies envelopes a confocal; the projection upon the chord of the normal (terminated by either axis) at an extremity thereof is of constant length; if any circle touch a given ellipse in two points, the chords which can be drawn to the circle from either point of contact so as to touch a fixed confocal are of constant length, and conversely the envelope of a chord of constant length drawn to the circle from either point of contact is a confocal ellipse. Examine the case in which the minor axis of the inner ellipse is evanescent.

349. If the arms of a right angle envelope two confocal ellipses the line joining the points of contact will envelope a third ellipse confocal with the former two; and if two parallel positions of each arm be taken, the perimeter of the parallelogram formed by joining the points of contact will be constant,

---

* If $a$, $b$ and $a'$, $b'$ be the semi-axes of the outer and inner ellipses and $\lambda^2 = a^2 - a'^2 = b^2 - b'^2$, the intercept on the chord is $\dfrac{ab}{\lambda}$; the projections upon it of the normals are $\dfrac{\lambda b}{a}$ and $\dfrac{\lambda a}{b}$; and its length in terms of the parallel focal chord $f$ is $\dfrac{\lambda f}{b}$.
See the *Oxford, Cambridge, and Dublin Messenger of Mathematics*, vol. IV. pp. 11–22 (1868), in which article several of the examples following are likewise solved.

and each pair of its adjacent sides will make equal angles with the tangent at their point of concourse.

350. An ellipse which has double contact with each of two fixed confocals has a fixed director circle; and an ellipse which has double contact with one of two fixed confocal ellipses, and has its foci at the ends of any diameter of the other, has a fixed auxiliary circle.

351. Four tangents being drawn to a conic, if one pair of their points of intersection lie on a confocal each of the remaining two pairs will lie on a confocal. If $TP$ and $TQ$ be a pair of tangents to a conic, and tangents be drawn from $P$ and $Q$ to a confocal and intersect in points $S$ and $H$, then $TP$ and $TQ$ subtend equal or supplementary angles at $S$ and $H$; the four tangents from $P$ and $Q$ touch one and the same circle; and
$$SP \pm HP = SQ \pm HQ.\text{*}$$

352. Given two confocal ellipses, shew that the latus rectum of any ellipse which has its foci on the inner fixed ellipse and touches the outer is of constant length.†

353. The locus of the centre of a conic which has four-point contact with a given conic at a given point is a straight line through the centre of the fixed conic.

354. Prove Graves' Theorem, that the sum of the tangents from any point on an ellipse to a fixed confocal ellipse exceeds the intercepted arc of the latter by a constant quantity.‡ Prove also that the difference of the tangents to an ellipse from any point on a confocal hyperbola is equal to the difference of the segments into which the intercepted arc of the ellipse is divided by the hyperbola.

---

\* See the article referred to in the note on Ex. 348.

† With the notation used above, its length is $\dfrac{2\lambda^3}{ab}$.

‡ See Salmon's *Conic Sections*, Art. 399. The theorem may also be deduced from Ex. 351 as in the article referred to in the note. Adding the perimeter of the inner ellipse, we see that the outer confocal may be described with the help of a loop of string placed round the inner curve, a construction which becomes equivalent to that of the note on Art. 48 when the inner ellipse reduces (by the evanescence of its minor axis) to the line joining its foci.

355. If three ellipses be described as in Ex. 352, having their six foci at three points on the inner confocal, the sum of the areas of their minor auxiliary circles will be constant. Moreover, if any number of ellipses be described with the same number of points on a given ellipse as foci (every such point being a focus of two of them), and if the several ellipses touch as many fixed confocals, the areas of their minor auxiliary circles will be connected by a linear relation.*

356. If every vertex but one of a polygon circumscribing a conic trace a confocal conic, its remaining vertex will likewise trace a confocal, and the perimeter of the polygon will be constant.

357. If two points trace an ellipse (in the same direction) with velocities which are always as the focal chords parallel to the tangents at those points, the tangents will intersect on a fixed confocal ellipse, and their angular velocities about their points of contact will be as the central perpendiculars upon them.

358. If a parallelogram can be inscribed in an ellipse whose semi-axes are $A$ and $B$ so as to envelope a coaxal ellipse whose semi-axes are $a$ and $b$,
$$A^2 - a^2 : a^2 = b^2 : B^2 - b^2.$$

359. If a single quadrilateral can be described about one of two given conics and inscribed in the other, any number of quadrilaterals can be so described, and they will have one diagonal in common.†

360. If the normal at any point $P$ to an ellipse meet the two perpendicular tangents to a confocal ellipse in $K$ and $L$, shew that $PK.PL$ is constant and equal to the difference of the squares of their major or minor semi-axes.

---

* If $\beta$ be the minor semi-axis of one of the variable ellipses and $\phi$ the arc joining its foci, then $\beta^2 = c\, (\phi + c')$, where $c$ and $c'$ are certain constants (Exx. 352, 354).

† The quadrilateral and its circumscribing conic can be projected into a rectangle and a circle, which latter must be the director circle of the projection of the inscribed conic. See also Ex. 382.

361. If from the intersections of any two parallel tangents to an ellipse with the tangent at a fixed point four tangents be drawn to a given confocal ellipse, the four intersections of the latter will lie on a certain circle having its centre on the **normal** at the fixed point; and the radius of the circle and the intercept made by its centre upon the normal will **vary as the perpendicular diameter of the outer conic.**\*

362. If a tangent to a conic (or other curve) cuts off a constant area from another, it will be bisected at its point of contact, and conversely.†

363. A central conic which passes through four given points has a pair of conjugate diameters parallel to the axes of the two parabolas which can be drawn through the same **four points.**‡

364. Give a construction for finding a point $P$ such that if straight lines $PQ, PR, PS, PT$ be drawn from it to meet four given straight lines $AB, CD, AC, BD$ at given angles, the rectangle $PQ.PR$ may be in a specified ratio to $PS.PT$. Hence shew how to draw the tangent at any given point on the locus of $P$; and determine a pair of conjugate diameters of the same.§

365. If a parallelogram $ASPQ$ has its opposite vertices $A$ and $P$ on a conic, and its sides $AQ, AS$ meet the curve in

---

\* The four tangents in any assumed position will intersect on a circle (Art. 50). Any other quadrilateral inscribed in the same circle so as to envelope the inner ellipse will have the intersections of its opposite sides at points $P$ and $Q$ on the fixed tangent (Ex. 359); and it may be shewn conversely that the second tangents from $P$ and $Q$ to the outer ellipse are parallel. Making one of the parallel tangents coincide with the tangent at the fixed point, we see that the centre $O$ of the circle must lie on the normal. Let $M$ now be the intersection of the diagonals of the quadrilateral, $N$ the fixed point, $CD$ the semi-diameter conjugate to $CN$, and $\rho$ the radius of the circle; then (Ex. 348) $NO = \frac{ab}{\lambda^2} CD$; $MN = \frac{\lambda^2}{ab} CD$; $\rho^2 = \left(\frac{a^2b^2}{\lambda^4} - 1\right) CD^2$.

See also *Mathematical Questions, &c. from the* EDUCATIONAL TIMES, vol. XIII. p. 31.

† See SALMON's *Conic Sections*, Art. 396.

‡ Let $TP$, $TQ$ be tangents to an ellipse, and $OAB$, $OCD$ chords parallel to $TP$, $TQ$. Determine a diameter of each of the two parabolas through $A$, $B$, $C$, $D$ (Ex. 184); then it is easily seen that $PQ$ and the diameter through $T$ in the ellipse are parallel to the diameters of the parabolas.

§ See NEWTON's *Principia*, Lib. I. sect. v. lemma 19. The next ten examples are mostly solved in the same Section, which will repay a careful study. See also Book II. of Leslie's *Geometry of Curve Lines*.

$B$ and $C$; the straight lines **joining** any point on the curve to $B$ and $C$ will meet $PS$ and $PQ$ in points $T$ and $R$ such that $PR$ varies as $PT$, and conversely. Hence shew **how to** draw a tangent to the conic at any point; and shew how to draw a conic through five **given points**, and prove that one conic only can be so drawn.

366. If two straight **lines $BM$, $CM$** turn about **fixed poles** $B$ and $C$ so that their intersection $M$ moves along a **fixed straight** line or directrix, and if $BD$ and $CD$ be drawn at given angles to $BM$ and $CM$ respectively, the point of concourse $D$ of the second pair of lines will trace a conic passing through the poles $B$, $C$, and conversely.*

367. By the foregoing construction (or otherwise) determine any number of points on a conic passing through five given points.†

368. Describe a conic passing through four **points, three** points, two points, **or one** point, and likewise **touching one, two,** three, **or four straight** lines respectively. Examine the cases in which two or more of the given points or lines coalesce.

---

\* For distinctness of conception let the points $B$, $D$, $A$, $P$ in the following solution be supposed to lie (in the order specified) on one branch of a hyperbola and $C$ on the other branch, as in NEWTON's figure (lemma 21). Now when the moving point $M$ has an assumed position $N$ on the *directrix* let $P$ be the corresponding fixed point on the locus of $D$. Draw $BDT$, $CDR$ through any other position of $D$, and make the angle $BPT$ equal to $BNM$ and $CPR$ equal to $CNM$; then it may be shewn that $PT : NM = PB : NB$, and $PR : NM = PC : NC$. Hence $PT$ varies as $PR$, and therefore by the preceding lemma (Ex. 365)—as NEWTON abruptly concludes—the locus of $D$ is a conic through the points $B$, $C$, $P$. The last step (see Le Sueur and Jacquier's edition of the *Principia*) is explained as follows: when $NM$ becomes infinite, let $D$ assume the position $A$; then it may be proved that $BA$ is parallel to $PT$ and $CA$ to $PR$. Let $PT$, $CA$ meet in $S$ and $PR$, $BA$ in $Q$. Then $ASPQ$ is the parallelogram of Ex. 365, and $PT$ has been shewn to vary as $PR$.

† Still using the same figure, let $A$, $B$, $C$, $P$, $D$ be the given points. Take $ABC$ and $ACB$ as the given angles which are to rotate about $B$ and $C$ as poles. The other two points $P$ and $D$ enable us to determine two points $M$ and $N$ on the directrix, and the whole curve can then be described. Or again, if $A, B, C, D, E$ be five points on a conic, let $AC$, $BE$ meet in $F$, and draw from $D$ a parallel to $CA$ to meet $BE$ in $G$; then to determine the point $H$ in which the parallel meets the curve again, we have $DG.GH : BG.GE = AF.FC : BF.FE$. The diameter bisecting the parallels can now be drawn, and in like manner a second diameter can be determined. Hence the centre is known, &c.

369. If $AFP$, $BQG$ be the tangents to a conic at the ends of a diameter $AB$, and $FG$ and $PQ$ be tangents to the same and intersect in $O$, shew that
$$AF : BQ = FP : GQ = FO : OG,$$
and that $PG$, $FQ$, $AB$ cointersect.

370. **If in a parallelogram $LMIK$ any conic** be inscribed touching **the** sides $ML$, $IK$, $KL$, $IM$ in $A$, $B$, $C$, $D$, any fifth tangent **to the conic will meet those sides in** points $F$, $Q$, $H$, $E$ such **that**
$$ME : MI = BK : KQ,$$
and
$$KH : KL = AM : MF.$$

371. If the inscribed conic in **the preceding example be fixed**,
$$KQ \cdot ME = KH \cdot MF = \text{a constant}.$$
Moreover, **if a sixth tangent be drawn to meet** $DE$ in $e$ and $IQ$ in $q$, then
$$KQ : Me = Kq : ME = Qq : Ee.$$

372. Hence shew that the diagonals of the quadrilateral $EqQe$ are bisected by one and the same diameter of the conic, and that the locus of the centre of a conic inscribed in a given quadrilateral **is a straight** line bisecting its three diagonals.

373. If $IB$, $ID$ **be the tangents at given points** $B$, $D$ of a conic, and $EQ$ the intercept made by them on any other tangent to the same, shew that $IE \cdot IQ$ **varies as** $ED \cdot QB$.* Hence, if from two fixed points in a **given pair of straight lines any other two** lines be drawn, each **to meet the** opposite **fixed line, shew that if the** straight line joining the points of concourse envelopes a **conic touching** the fixed lines at the fixed points, the locus of the intersection of the variable lines will be a conic satisfying the same condition, and conversely. Examine the case in which the fixed lines are parallel.

---

* By making the sixth tangent in Ex. 371 coincide successively with $IE$ and $IB$ we deduce that the two rectangles are as $MI$ to $MD$. When $M$ is at infinity they become equal (Ex. 183).

374. Given five tangents to a conic, determine its five points of contact with them;\* and given five points on a conic, determine the five tangents thereat.

375. Given the centre of a conic and a self-conjugate triangle, shew how to determine the diameters conjugate to its sides and to describe the curve. Hence (or otherwise) shew how to describe a conic touching five given straight lines.

376. Given a tangent to an ellipse, its point of contact, and the director circle, shew how to construct the ellipse.

377. Two ellipses have a common focus and equal major axes, and one of them revolves about this focus in its own plane whilst the other remains fixed: prove that their chord of intersection envelopes an ellipse confocal with the fixed ellipse.

378. The condition that a straight line which makes intercepts $CB$ and $CD$ on two fixed straight lines should envelope a conic touching the fixed lines is of the form

$$a \cdot CB \cdot CD + b \cdot CB + c \cdot CD + d = 0,$$

where the ratios of $a$, $b$, $c$, $d$ are constant. Determine the points of contact of the envelope with the fixed lines; and explain the result when the intercepts are connected by a relation of the form

$$CB \cdot CD = \text{a constant}.$$

---

\* If $ABCD$ be the pentagon formed by the five tangents, the straight line joining $D$ to $(AC, BE)$ passes through the point of contact of $AB$, as appears most simply by supposing two sides of the enveloping hexagon in Brianchon's theorem to coalesce. When five points are given, the tangents thereat may be drawn and number of points on the curve may be found with the help of Pascal's hexagon. See Salmon's *Conic Sections*, Art. 269.

† Call the straight line bisecting the three diagonals of a quadrilateral its DIAMETER. The diameters of any two of the quadrilaterals formed by the five tangents determine the centre of the conic, and any one of the quadrilaterals gives a self-conjugate triangle. For another solution, in which the five points of contact are first found, see LIB. I. Sect. V. of the *Principia* (prop. 27, prob. 19); and see Ex. 374 (note), and Besant's *Conic Sections treated Geometrically*, Art. 229 (1875). It is evident that the diameters of the five quadrilaterals formed by five straight lines meet in one point, viz. the centre of the conic touching the five lines.

379. In order that the envelope in Ex. 378 may be a parabola the ratio of $a$ to the other constants must vanish.* Hence shew that the polars of a fixed point with respect to a series of confocal conics, and likewise the normals appertaining to the tangents drawn to them from that point, envelope a parabola touching the axes of the confocals.

380. If $OP$ and $OQ$ be the tangents from a fixed point $O$ to any conic which has two given points for foci, each of the corresponding normals is the polar of $O$ with respect to a conic having the same foci; and the circle about $OPQ$ passes through a second fixed point $F$, such that $CF$ and $CO$ lie on opposite sides of the transverse axis and make equal angles therewith, and $CF.CO = CS^2$.

381. A tangent being drawn from an extremity of one axis of an ellipse to a coaxal ellipse, find the length of its intercept on the other axis and the ordinate of its point of contact to either axis.

382. Deduce from Ex. 356 that, if a single $n$-gon can be described about a given conic and inscribed in a given confocal, any number of $n$-gons can be so described.

383. If a triangle can be circumscribed to two confocal ellipses, the straight lines joining the extremities of the axes of the outer must pass through the intersections of the tangents at the extremities of the axes of the inner ellipse.†

384. If $PQR$ be a triangle circumscribed to a pair of confocal ellipses and $P'$ be the point of contact of $QR$, shew that the confocal hyperbola through $P$ passes through $P'$ and the

---

* The general condition of Ex. 378 is implied in Ex. 373, and the condition that the envelope may be a parabola is inferred from Ex. 183. In what follows, supposing $h$, $k$ to be the coordinates of the fixed point, we see from Ex. 343 (or Ex. 270) that, if the enveloping line make intercepts $CL$ and $CM$ on the transverse and conjugate axes, $h.CL - k.CM = CS^2$; and consequently that the envelope is a parabola which makes intercepts $\frac{CS^2}{h}$ and $-\frac{CS^2}{k}$ on the axes.

† The proof may be simplified by considering the special case in which a side of the triangle is parallel to an axis of the ellipses. The semi-axes $a$, $b$ and $a'$, $b'$ of the outer and inner ellipses are connected by the relation $\frac{a'}{a} + \frac{b'}{b} = 1$.

point diametrically opposite thereto, and **that** if the outer ellipse be regarded as traced by means of a loop $PP'P$ passed round the inner, the loop will be bisected at $P'$.

385. The area **of** an ellipse is a **mean proportional to the** areas of its auxiliary circles.

386. A quadrilateral can be circumscribed to two **confocal ellipses** if the common difference of the areas of their major **and their minor** auxiliary circles be equal to the area of the inner ellipse; **the** locus of the pole of any chord of the outer ellipse **which touches** the inner is a circle whose diameter is equal to **the sum of the axes** of the latter; the **tangents to** the inner from any **point on the outer** ellipse are **as the parallel** focal chords of the latter; **the chord** joining the **ends of a pair of** semi-axes of the outer touches the inner ellipse and is **divided at its point** of contact into segments equal to the semi-axes **of the latter.**

387. Prove **Fagnani's** theorem, **that a quadrant of an ellipse can be** divided **into segments which differ by the difference** of its semi-axes, **the greater segment being that which is terminated by the minor axis.**\*

388. If $C$ be the common centre of an ellipse and a circle of equal area, $P$ **the point in** which the circle meets a quadrant $AQPB$ of the ellipse, and $CQ$ be equal to radius conjugate to $CP$; **shew that** the middle point of the quadrantal arc $AB$ lies within the arc $PQ$.

389. If a hexagon can be circumscribed **to two confocal** ellipses, and $AP$, $BQ$ be the tangents **to a quadrant** $A'B'$ of **the** inner from the extremities of the **semi-axes** $CA$, $CB$ of the outer ellipse, and $F$ be Fagnani's **point of** division of the quadrantal arc $AB$, shew that

$$\text{arc } B'Q - \text{arc } A'P = \text{arc } BF - \text{arc } AF = CA - CB.†$$

---

\* The point **of contact last** mentioned (Ex. 386) divides the inner ellipse in the manner specified. For another geometrical proof see Salmon's *Conic Sections*, Art. 400.

† If $\lambda^2 = CA^2 - CA'^2 = CB^2 - CB'^2$, and if the tangent at $B'$ meet the outer ellipse in $O$, it may be shewn that $\dfrac{1}{\lambda} = \dfrac{1}{CA} + \dfrac{1}{CB}$; $B'O = CA - \lambda$; $OP = \lambda$; $AO = CB$,&c.

390. Any circle through the focus $S$ and the further vertex $A'$ of a hyperbola whose eccentricity is *two* meets the curve in three points $P$, $Q$, $R$ which determine an equilateral triangle,\* and conversely the circumscribing circle of any equilateral triangle inscribed in a hyperbola whose eccentricity is two passes through a focus and the further vertex; the focal vectors $SP$, $SQ$, $SR$ meet the curve in three other points which likewise determine an equilateral triangle; if $P$ be any point on the $S$-branch of the curve the angle $A'SP$ is double of the angle $SA'P$, and if $Q$ be any point on the opposite branch the supplement of $A'SQ$ is double of the supplement of $SA'Q$; any chord through $S$ subtends a right angle at $A'$; the equilateral triangle $PQR$ envelopes a fixed parabola having $S$ and the $S$-directrix for focus and directrix; the tangents to the hyperbola at $P$, $Q$, $R$ form a triangle $P'Q'R'$ inscribed in a fixed hyperbola of eccentricity four; the tangents to the latter at $P'$, $Q'$, $R'$ form a triangle inscribed in a fixed hyperbola of eccentricity eight, and so on continually.

---

\* This hyperbola—whose directrix bisects $SA'$—is one of the TRISECTORS (Ex. **308**) of any circular arc whereof $A'S$ is the chord; and the meaning of the remarkable property that $PQR$ is an equilateral triangle is that *the problem of bisecting a given angle a admits of the three distinct solutions* $\frac{1}{2}a$, $\frac{1}{2}(2\pi \pm a)$. Since the solution must in any case be threefold, it is evident *a priori* that it cannot be effected by means of a straight line and circle, which can intersect in two points only. All this is fully pointed out by BOSCOVICH in his *Sectionum Conicarum Elementa*, §§ 274–279. NEWTON shewed (*Arithmetica Universalis*, prob. 36) that the locus of the vertex of a triangle on a given base and having one base angle differing from twice the other by a constant angle is a cubic curve, which reduces to the hyperbola in question when the constant angle vanishes; and he remarked that ($P$ being a point on the $S$-branch) the angle at $A'$ in the triangle $A'SP$ is equal to ONE THIRD of the exterior angle at $P$.

NOTE.

The undermentioned Examples and others are solved wholly or in part in vols. I–XXIX of *Mathematical Questions with their Solutions from the* EDUCATIONAL TIMES (London, 1864—78):

Ex. **79** (vol. XXII.); **174, 222, 226 (I.)**; 324 (XXII.); 328 (II.); 331, 332 (III.); 334 (XII.); **336 (XIII.); 338 (XXI.)**; 339, 340 (XXII.); 341 (XXVI.); 347, 348 (IV.); 361 (XIII.).

# CHAPTER V.

## THE ASYMPTOTES.

**53.** The *Asymptotes* of a hyperbola are two diameters equally inclined to the axis and such that, if $E$ be any point on either of them and $CN$ its central abscissa, then

$$EN : CN = CB : CA;$$

in other words, the asymptotes are the diagonals of a certain rectangle which is determined by the two axes of the hyperbola.

If any two conjugate diameters meet $EN$ and its prolongation in $L$ and $M$, it follows from Art. 44, and the above relation that

$$NL.NM : CN^2 = CB^2 : CA^2 = EN^2 : CN^2;$$

and hence that in the limiting case in which the diameter $CL$ coalesces with an asymptote $CE$ its conjugate $CM$ coalesces with the same, or *an asymptote may be regarded as a diameter conjugate to itself*.

Two hyperbolas are said to be conjugate when the transverse axis of each is coincident with the conjugate axis of the other; thus, the transverse and conjugate axis of a hyperbola being $AA'$ and $BB'$, those of the *Conjugate Hyperbola* will be $BB'$ and $AA'$. It is evident that a pair of conjugate hyperbolas have the same asymptotes but lie on opposite sides thereof.

**54.** *Limiting positions of Tangents.*

The asymptotes are so called because, being produced, they continually approximate to the curve (Art. 56) but without actually meeting it until produced infinitely. We shall shew that such lines may be regarded as *tangents whose points of contact are at infinity.*

The tangent at any point $P$ meets the axis in a point $T$ such that $CT$ varies inversely as $CN$ the abscissa of $P$ (Art. 39) and therefore vanishes when $CN$ is infinite. To determine the position of a pair of tangents which pass through the centre of the hyperbola, draw the tangents from $S$ to the Auxiliary

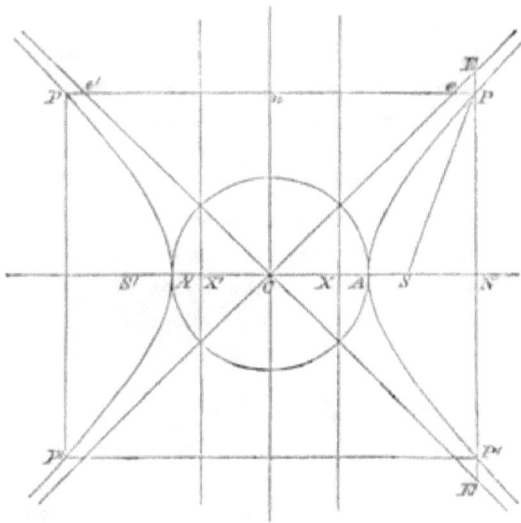

Circle (Art. 6), and draw the diameters through the points in which they meet the $S$-directrix. The points in question will lie on the circle, since the diameter $AA'$ of the circle is divided harmonically at $S$ and $X$, and the directrix is therefore the polar of $S$ with respect to the circle. The tangents to a hyperbola from its centre are therefore those diameters which pass through *the intersections of the directrices with the Auxiliary Circle*.

It is easily seen that the said diameters possess the property

$$EN : CN = CB : CA,$$

and are therefore identical with the asymptotes.

It is likewise evident that they possess the property,

$$CE : CN = CS : CA = \text{the eccentricity};$$

and hence that all hyperbolas which have the same (or parallel) asymptotes and lie on the same sides thereof are SIMILAR conics; and *the asymptotes* themselves taken together are the limiting form to which the curve tends when its axes are diminished indefinitely, and they *may be regarded as constituting a similar hyperbola.*\*

The hyperbola **may be called** *Acute* or *Obtuse* according as the interior **angle between its asymptotes is less** or greater than a right **angle; that** is to say, according **as its conjugate axis is less or greater** than its transverse **axis.** In the intermediate case, when they are equal, it is called *Rectangular*.

It is easily seen that any two tangents to the same branch **of the** hyperbola intersect within the interior angle between **the** asymptotes and themselves contain a greater angle; and likewise that any two tangents to opposite branches contain an **angle less** than **the** exterior **angle** between the asymptotes; **and hence that** *an obtuse hyperbola* **can have no** *real tangents at right angles.*

55. *A construction for the Normal.*

If $P$ be a point on the curve whose **ordinate to** either **axis** meets the nearer asymptote **in** $E$, and if the normal at $P$ meet the axis in $G$, then **(Art. 43, Cor.) in the case of the transverse axis,**

$$CN.NG : CN^2 = CB^2 : CA^2 = EN^2 : CN^2;$$

and therefore $CN.NG$ is equal to $EN^2$ and the angle $CEG$ is a right angle, **as may** be shewn likewise for the case of the conjugate axis. Hence **the** following construction for the normal at a given point $P$:

*Let the ordinate of $P$ to either axis meet the nearer asymptote in $E$, and through $E$ draw a perpendicular to $CE$ to* **meet the** *same axis in $G$ ; then will $PG$ be the* **normal at $P$.**

When $CN$ is infinite the normal itself **coincides with** $EG$ and is perpendicular to the asymptote.

---

\* Notice in Art. 38 that when $SY$ *touches the circle* its diameter through $Y$ should be **the tangent to the** hyperbola; **and** also that in this case $SY = CB = HZ$. Moreover (Art. **14, Cor. 1)** the diameter conjugate to $C\infty$ must be parallel to the tangent at $\infty$, **and must therefore** coalesce with $C\infty$ itself. The HYPERBOLA is distinguished as the conic **which** has a pair of tangents whose points of contact are at infinity and whose chord of contact is therefore the *Straight Line* **at** *Infinity* (Art. 17, Cor. 2); and the PARABOLA is distinguished as the conic to which the line at infinity is a tangent, since (Art. 27) $SY^2 = SA.SP = SA.ST$, **and** therefore when $SP$ becomes infinite the tangent $TY$ is removed wholly to infinity.

# THE ASYMPTOTES. 145

## PROPOSITION I.

56. *If a parallel to either axis of a hyperbola be **drawn** through any point on the curve to meet the asymptotes, the product of its segments between the point and the asymptotes will be equal to the square of the semi-axis **to** which **it** is parallel.*

First let a principal double ordinate $PP'$ be produced to

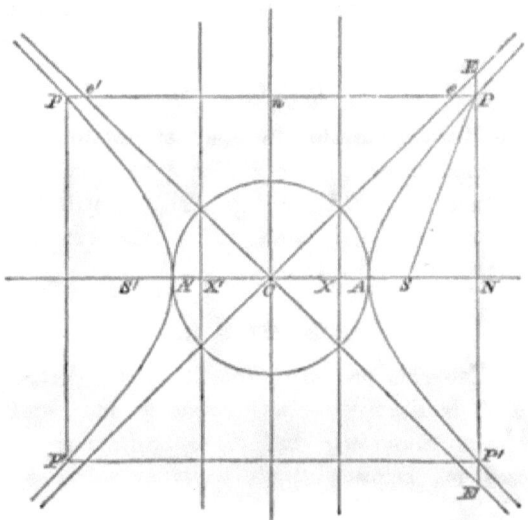

meet the asymptotes in $E$ and $E'$; then will $PE.PE'$ or $P'E.P'E'$ be equal to $CB^2$.

For by Art. 33 and by a property of the asymptotes, if $PP'$ meet the transverse axis in $N$,

$$PN^2 + CB^2 : CN^2 = CB^2 : CA^2 = EN^2 : CN^2,$$

or
$$PN^2 + CB^2 = EN^2;$$

and therefore

$$PE.PE' = P'E.P'E' = EN^2 \sim PN^2 = CB^2.$$

In like manner it **may** be shewn that $CA$ is a mean proportional to the segments $Pe$ and $Pe'$ of a straight line drawn through $P$ parallel to the transverse axis to meet the asymptotes.

Hence it appears that the distance of $P$ (or $P'$) from the nearer asymptote varies inversely **as** its distance from the

L

other, and when the latter distance is increased indefinitely the former is diminished indefinitely; the curve therefore as it branches out continually approximates to its asymptotes, but without actually meeting them at any finite distance from the centre.

It is easy to shew by a *reductio ad absurdum* that no diameter other than $CE$ or $CE'$ have the above property, or in other words, that the hyperbola cannot have more than two asymptotes.

### Corollary.

If $PO$ be drawn parallel to one asymptote $E'C$ to meet the other asymptote $EC$, then will $PO$ vary as $PE$ and $CO$ as $PE'$, and therefore $PO.CO$ (fig. Art. 60) will be constant; and it may be shewn by taking $P$ at the vertex that it is equal to $\frac{1}{4}CS^2$ or $\frac{1}{4}(CA^2 + CB^2)$.

### PROPOSITION II.

**57.** *The intercepts on any tangent to a hyperbola between the curve and its asymptotes are equal to one another and to the parallel semi-diameter; and the opposite intercepts on any chord between the curve and its asymptotes are equal to one another.*\*

(i) Let the tangent at $P$, supposed parallel to the semi-diameter $CD$, meet any two conjugate diameters in $L$ and $L'$; then by Art. 47

$$PL.PL' = CD^2.$$

Hence in the case in which $L'$ coalesces with $L$ and $CL$ therefore becomes an asymptote, $PL^2$ is equal to $CD^2$; and in like manner, if the same tangent meet the other asymptote in $M$, $PM^2$ is equal to $CD^2$.

Therefore $\qquad PL = PM = CD.$

---

\* The hyperbola and its *Asymptotes* being similar conics (Art. 54), the above is a special case of Ex. 50. The latter follows at once from Art. 14 since, when the *direction* of $CX$ and the *magnitude* of $CS : CX$ are given, if the direction of $CV$ be assumed that of $SV$ (which is perpendicular to the ordinates of $CV$) is determined. It is evident that a pair of conjugate hyperbolas also make equal intercepts $QQ'$, $qq'$ on any chord.

# THE ASYMPTOTES.

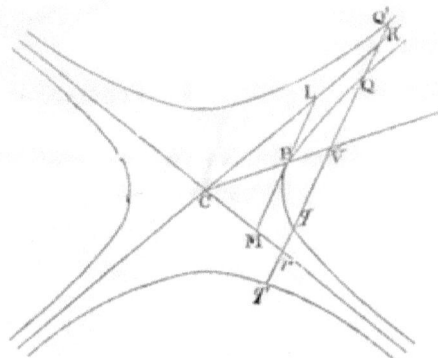

(ii) Next let $Qq$ be any chord of the hyperbola meeting the asymptotes in $R$, $r$, and let $LPM$ be one of the tangents to which it is parallel.

Then the diameter $CP$ bisects the chord $Qq$, and from above it is evident that it likewise bisects $Rr$; whence it follows that

$$QR = qr, \text{ and } Qr = qR.$$

### PROPOSITION III.

**58.** *The product of the segments into which any chord of the asymptotes is divided by either of the points in which it meets the curve is equal to the square of the parallel radius.*

(i) Using the same construction as in Prop. II., let $V$ be the middle point of the chord $Qq$. Then by Art. 34 and by parallels (first taking the case in which $Q$ and $q$ lie on the same branch of the curve),

$$QV^2 + CD^2 : CV^2 = CD^2 : CP^2 = PL^2 : CP^2$$
$$= RV^2 : CV^2,$$

or $$QV^2 + CD^2 = RV^2.*$$

Hence $$RQ \cdot Qr = Rq \cdot qr = RV^2 - QV^2 = CD^2,$$

or $CD$ is a mean proportional to $RQ$ and $Qr$, and to $Rq$ and $qr$.

The above proof may be adapted to the case in which $Q$, $q$

---

* Conversely, if this relation be assumed the point $R$ must always lie on one of two straight lines which continually approach the curve, that is to say, on one of the asymptotes.

lie on opposite branches of the hyperbola by writing $-CP^2$ for $CD^2$ and $-CD^2$ for $CP^2$.

(ii) These results may also be deduced as follows from Prop. I.

From **any point $Q$ on** the curve draw $QRR'$ **in any** given direction to meet the asymptotes, and draw $QEE'$ parallel to

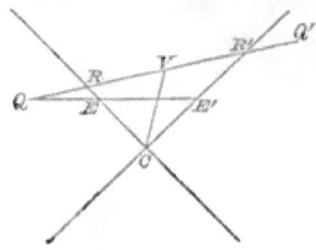

the transverse or the conjugate axis to meet the same. Then $QR$ varies as $QE$ and $QR'$ as $QE'$, and therefore, $QE.QE'$ being constant, $QR.QR'$ is likewise constant.

Supposing $RR'$ to become a diameter or a tangent, according as its direction cuts both branches of the curve or one only, we see that $QR.QR'$ is equal to the square of the parallel semi-diameter or of the intercept on the tangent between the curve and either asymptote.

### Corollary 1.

If the tangent at $P$ meet the asymptotes in $L$, $M$ (fig. Art. 60), and if $O$ be the middle point of $CL$, and $OP$ be therefore parallel to $CM$, then (Prop. I. Cor.)

$$CL.CM = 2CO.2PO = CS^2;$$

and therefore the area $LCM$ is constant, that is to say, *the area of the triangle bounded by the asymptotes and any tangent is of constant area*, and it is equal to $CA.CB$. It is otherwise evident that the triangle in question is one-fourth of the conjugate circumscribing parallelogram (Art. 46).

### Corollary 2.

Moreover, if $PK$ be drawn perpendicular to $CL$,

$$CP^2 \sim PL^2 = CK^2 \sim LK^2 = 2CO.2OK;$$

where $OK$ varies directly as $OP$ and therefore inversely as $CO$. Hence another proof that *the difference of the squares on two conjugate diameters is constant* (Art. 45).

### Corollary 3.

So long as the chord $RQr$ is drawn in a specified direction $Qr$ varies inversely as $QR$. If the chord be taken constantly *parallel to the asymptote* $CM$, so that the point $r$ recedes to $\infty$, it follows that $Q\infty$ varies inversely as $QR$, or directly as $CR$; and moreover, that if $QO$ be any finite portion of the chord, then $O\infty$ likewise varies as $CR$, and the rectangle $QO.O\infty$ varies as $CR.QO$. If $YOZ$ be a chord drawn in any other specified direction and meeting the chord parallel to the asymptote in $O$, then (Art. 16, Cor. 1) $OY.OZ$ varies as $QO.O\infty$, or as $CR.QO$; and in the special case in which $Q\infty$ is a fixed chord $OY.OZ$ varies as the length $QO$.

#### SCHOLIUM A.

If the hyperbola be defined as the locus of a point $P$ such that if $O$ be its projection upon one of two fixed straight lines $CL$, $CM$ (the asymptotes) by a straight line parallel to the other, $CO.PO = $ a constant $c^2$, we may proceed to investigate the properties of the curve as follows.

If $LM$ be drawn in a *specified* direction through any position of the tracing point $P$, it is evident that $PL.PM$ is constant, and also that in the case in which $LM$ becomes a tangent it is bisected at its point of contact $P$. In this case $CL.CM = 2CO.2PO = 4c^2$, and the triangle $CLM$ is of constant area. It may now be shewn that (with the notation of Art. 57) $QR = qr$; $CP$ bisects $Qq$ and all other chords parallel to the tangent at $P$; $QV^2$ varies as $CV^2 - CP^2$; and that the difference of the squares of any two conjugate diameters is constant (Art. 58, Cor. 2).

A straight line parallel to either asymptote $CM$ meets the curve in one point only, since (figure of Art. 60) if $CO$ be supposed constant, $CO.PO$ vanishes when $P$ is at $O$, and increases continuously up to $\infty$ as $P$ recedes from $O$, and is therefore equal to $c^2$ for one position of $P$ only. Hence at any point $P$ between the curve and its asymptotes $CO.PO$ is less than $c^2$. Moreover, for any assumed position of the intercept $LM$ it is evident that $PL.PM$ is a maximum, and therefore $PO.CO$ is a maximum, when $PL = PM$. Hence at the point of concourse $P'$ of any two tangents to the same branch of the curve $P'O.CO$ is less than $c^2$, and $P'$ therefore lies between the curve and its asymptotes, or the curve is *convex* to its asymptotes.

Lastly, if the hyperbola be regarded as the envelope of a straight line $LM$ which contains a triangle of constant area with two fixed straight lines $CL$ and $CM$, it may be shewn by the following method (which applies also to **Ex. 362**) that $LM$ is bisected at its point of contact. If $P$ be the point of concourse of the tangent line in any two positions $LM$, $L'M'$, the areas $LPL'$ and $MPM'$ are equal, and $PL.PL' = PM.PM'$; and therefore in the case in which $L'M'$ and $LM$ coalesce, $PL^2 = PM^2$, or $LM$ is bisected at the point $P$ of the envelope. The hyperbola may also be regarded as a special case of the envelope in Ex. 378, which makes intercepts $\dfrac{-d}{b}$ and $\dfrac{-d}{c}$ upon the fixed tangents, and therefore touches them at *infinity* when $b$ and $c$ vanish.

### PROPOSITION IV.

59. *Any tangent and its normal meet the asymptotes and the axes respectively in four points lying on a circle which passes through the centre of the hyperbola.*

The circle whose diameter is the intercept $Gg$ made by the axes on any normal passes through the centre, since the angle $gCG$ is a right angle.

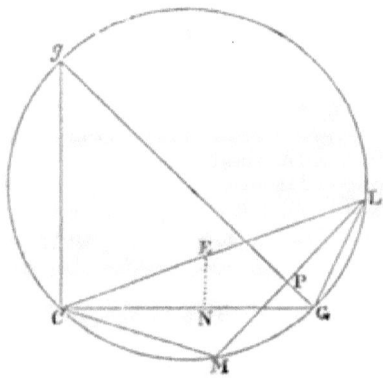

Let this circle meet the asymptotes in $L$, $M$, and let $LM$ meet $Gg$ in $P$. From any point $E$ in $CL$ draw $EN$ perpendicular to $CG$.

Then $\angle ECN = GCM = GLP$, in the same segment,

and $\angle CEN = ECg = LGP$, in the same segment;

and therefore the triangles $CNE$, $LPG$ are similar, so that the angle at $P$ is *a right angle*, and

$$PG : PL = EN : CN = CB : CA.$$

Similarly $\qquad PL : Pg = CB : CA.$

Hence $\qquad PG : Pg = CB^2 : CA^2,$

or $P$ is the **point at which** $Gg$ is **normal to the curve** (Art. 52, Cor. 2); and $LM$, which is at right angles to $Gg$, is the tangent at $P$.

### Corollary.

**From this** construction it appears again **the tangent** $LM$ is bisected at its point of contact; **and** that

$$PG : CD = CD : Pg = CB : CA,$$

where $CD$ is the semi-diameter **parallel to the tangent**.

### PROPOSITION V.

60. *The diameters of a hyperbola being regarded as terminated at the points in which they meet the curve or its conjugate, any two conjugate diameters are the diagonals of a parallelogram whose sides are parallel to and are bisected by the asymptotes, the tangents at their extremities meet on the asymptotes, and the difference of their squares is constant.*

(i) From a point $L$ on either **asymptote of** a pair of **conjugate hyperbolas** let a **tangent be drawn** to each, the one tangent meeting its curve in $P$ and the second asymptote in $M$, and the other meeting its curve in $D$ and the second asymptote in $M'$. Then will $CP$, $CD$ be conjugate semi-diameters, and $PD$ will be parallel to $MM'$, **and** will be bisected **at the point** $O$ in which it meets $CL$.

For since (Art. 57) the tangent $LM$ is bisected at $P$, and $LM'$ at $D$, therefore $PD$ *is parallel to the asymptote* $MM'$, and it also bisects $CL$.

Moreover (Art. 56, Cor.),

$$PO \cdot CO = \tfrac{1}{4}(CA^2 + CB^2) = DO \cdot CO,$$

or $PO$ is equal to $DO$; **that** is to say, $PD$ *is bisected by the asymptote* $CL$. But $PD$ likewise bisects $CL$, **and** therefore $CD$

152 THE ASYMPTOTES.

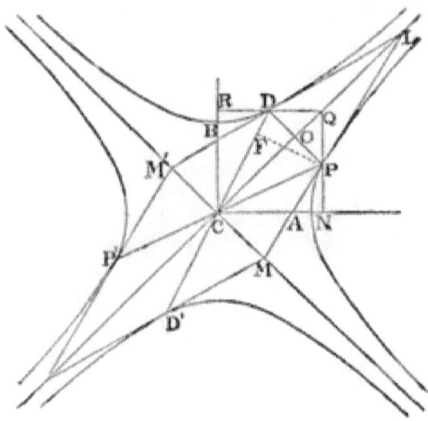

is parallel to the tangent at $P$ and is conjugate to $CP$. And if the parallel tangents touch the curves in $P'$ and $D'$, as in the diagram, the one will evidently pass through $M'$ and the other through $M$; and $P'D$ will likewise be parallel to one asymptote and bisected by the other.

(ii) Lastly, if $PN$ be the ordinate of $P$ to the transverse axis, and if it meet $CL$ in $Q$, it is easily seen that $OQ = OP$. And in like manner the ordinate $DR$ to the conjugate axis meets $CL$ in a point $Q'$ such that $OQ' = OD = OP = OQ$; that is to say, it meets it in the same point $Q$.

Hence $\qquad CQ^2 - CP^2 = QN^2 - PN^2 = CB^2$,

and $\qquad CQ^2 - CD^2 = QR^2 - DR^2 = CA^2$;

and therefore $\qquad CP^2 - CD^2 = CA^2 - CB^2$.

### Corollary.

To describe a pair of conjugate hyperbolas with given straight lines $CP$ and $CD$ as conjugate semi-diameters: draw $CO$ to the middle point of $DP$ and draw $CM$ parallel to $DP$; then will $CM$ and $CO$ be the asymptotes, and the axes will be the bisectors of the angles between them, and the foci will be the points in which the axes are cut by a circle whose radius $CS$ is a mean proportional to $PO$ and $2CO$.

## SCHOLIUM B.

CONJUGATE HYPERBOLAS are by no means to be regarded as organically related and together making up one continuous curve; but the one is a sort of auxiliary curve to the other, as the circle on its major axis is, for example, to the Ellipse.

(i) The two branches of a single hyperbola are to be regarded as constituting one continuous curve as was pointed out on p. 10, and as may be further illustrated in the following way. Let the hyperbola be considered to be traced by the extremity $P$ of a focal vector $SP$ (see fig. p. 145) moving round in the direction of the hands of a watch from the initial position $SA$. As $SP$ turns through an infinitesimal angle its extremity passes to a *consecutive* point on the curve, till at length by the continuous rotation of the focal vector the point $P$ recedes to infinity, $SP$ having become parallel to the asymptote $CE$: it then passes instantaneously to the *opposite* position at infinity, that is to say ($SP$ revolving gradually as before) the point $P$ passes at once from the extremity of the line $CE\infty$ to the extremity of the line $EC\infty$: at this infinitely distant point the curve *crosses its asymptote*, and $P$ proceeds to trace the opposite branch in the direction $p'A'$, and so forth. The two infinitely distant extremities of an asymptote or of any straight line may therefore be regarded as consecutive points, which likewise results from considering any straight line as (1) a circle of infinite radius in its own plane, or (2) as one of the great circles of a sphere whose radius has become infinite. Carrying on the latter illustration, we see that (since the length of a great circle on any sphere is constant) any finite straight line in a given plane together with its complement (p. 77) may be regarded as making up a *constant* infinite length; as was implicitly assumed in Chap. IV. Scholium C (p. 102), for if the bifocal property of the hyperbola,

$$HP - SP = AA',$$

be equivalent to $\quad H\infty P + SP = A\infty A',$

then $\quad HP + H\infty P = AA' + A\infty A'.$

(ii) It may be useful at this stage to give a conspectus of the several ways of viewing those diameters of the hyperbola which are not geometrically terminated by the curve.

*a.* By introducing the conception of *imaginary points* we may treat the hyperbola as a quasi-ellipse, and ignore the distinction between intersecting and non-intersecting diameters of the curve.

*b.* If we assign certain real magnitudes to the non-intersecting diameters (Art. 34)—arbitrarily, indeed, but in accordance with a partial analogy—we may then proceed to show (Art. 57) that any such diameter is equal to the intercept on the parallel tangent made by the asymptotes, and may prove as in Art. 58, Cor. 2 that the difference of the **squares of any two conjugate** diameters is of constant magnitude.

*c.* The non-intersecting diameters may also be treated as terminated by the conjugate hyperbola, as in **Arts. 45** and 60. The objection to this **mode of** treatment is that **it not** only proceeds upon an artificial analogy but **tends to obscure the** fact of the essential continuity and oneness **of the two branches of the** hyperbola.

*d.* Another method—very simple in practice, but presenting **of course** the same difficulties at the outset—is to start with the **Equilateral** Hyperbola (some of **the** properties of which can be **proved in** terms applicable likewise to the circle or Equilateral **Ellipse**) and **to** transfer the results thus obtained **to** the general **hyperbola** by the method of Orthogonal Projection.

(iii) APOLLONIUS, **in Lib. I.** prop. **14 of his Conics, defines the** two branches of a hyperbola as *Opposite* Sections ('Ἀντικείμεναι). At the end of the same **book** (prop. 56) he shews, quite independently of the asymptotes, how to construct two pairs of opposite sections with one and the same given pair of conjugate diameters, and he defines the curves so drawn as *Conjugate* (Συζυγεῖς). He afterwards proves in **Lib. II. prop. 15 that** opposite sections **have** the same asymptotes, **and in Lib. II.** prop. **17 that** conjugate opposite sections have **the same** asymptotes. **The** term *Conjugate* has also **been sometimes applied to** the **two branches** of a single hyperbola, **as for example in** prob. 36 of the *Arithmetica Universalis*, where **the words "conveniant ad** *conjugatam* Hyperbolam" refer to the further branch.

## *EXAMPLES.*

**391.** The eccentric circle of any point with respect **to a** hyperbola cuts the directrix at two points lying upon radii which are parallel to the asymptotes. Trace **the hyperbola by** the method of Art. 4, shewing **that the two** points in **which** the circle cuts the directrix correspond to the points **at infinity** upon **the asymptotes,** and the segment of the circle **beyond the** directrix to the further branch of the hyperbola.

**392.** The circle described about either **focus of a hyperbola** so as to bisect the semi-latus rectum cuts the **hyperbola at** points whose focal distances are parallel to the asymptotes; **and** the concentric circle which touches the asymptotes has its diameter equal to the conjugate axis.

393. Express the eccentricity of a hyperbola as a **function of the angle between its asymptotes**. If the eccentricity **and two points** on the curve be given, and if one asymptote **pass through a** third fixed point in the same straight line with **the** former two, the locus of the centre will be a circle.

394. If **the abscisses upon either asymptote** of any number of points **on a hyperbola be in** arithmetical progression, their ordinates **will be in harmonical progression, and** conversely.

395. **The** ordinates to either asymptote of the extremities **of any chord of** a hyperbola and the point of contact of a parallel **tangent** are proportionals.

396. The intercept made **by** the directrices **of a hyperbola on** either asymptote is **equal** to **the** transverse **axis**.

397. A hyperbola being regarded as the locus **of a point** whose distance from a given point is **equal to its distance from** a fixed straight line estimated in a given direction, **prove that** the given direction is that of an asymptote. Shew **also that** the straight line drawn from a focus to the nearer directrix parallel to an asymptote of a hyperbola is equal to the semi-latus rectum and is bisected by the curve.

398. **The** distance of any point on a hyperbola from either **focus is equal to the intercept** on either asymptote between the **ordinate of the point and the corresponding** directrix. Hence **prove in Art. 60 that if** $S$ **and** $H$ **be the foci,**

$$SP.HP = CQ^2 - CA^2 = CD^2.$$

Also prove that the difference **of the distances of the ends of two** conjugate radii of a pair of conjugate **hyperbolas** from their **nearer** foci is equal to **the** difference of **the semi-axes**.

399. Every chord drawn to a hyperbola from a fixed point on one asymptote **is** divided harmonically by that point and a fixed parallel to the said **asymptote,** and is bisected by a fixed parallel to the other.

400. The tangents at **the** vertices of a hyperbola meet its asymptotes **on the circumference** of the circle of which **the** straight line joining the foci is a diameter.

401. For what position of the tangent to a hyperbola is its intercept between the asymptotes a minimum?

402. The tangent to a hyperbola from the intersection of an asymptote with a directrix touches the curve upon a focal vector which is parallel to that asymptote.

403. The intercept on any tangent between the asymptotes subtends at the further focus an angle equal to half the angle between the asymptotes: it also subtends a constant angle at the intersection of the corresponding normal with either axis of the curve.

404. Every chord of a branch of a hyperbola which subtends at its focus an angle equal to the angle between the asymptotes touches a certain fixed parabola.

405. Find the relation between the intercepts made by any tangent to a hyperbola on two fixed straight lines parallel to the asymptotes.* If $OA$ and $OB$ be two straight lines given in position and $AB$ the intercept which they make on any tangent to a fixed conic which touches them, deduce from Ex. 378 that the locus of the point $P$ which completes the parallelogram $OAPB$ is a hyperbola whose asymptotes are parallel to $OA$ and $OB$; and examine the case in which the fixed conic is a hyperbola having its centre at $O$. Also find the locus of $Q$ if $A$ and $B$ be the points of contact of the given lines with any parabola which likewise touches a third given line.

406. The chords of intersection of any circle with the asymptotes of a hyperbola are equally inclined to either axis; the products of the segments of any two intersecting chords of the asymptotes are as the parallel focal chords; and if $O$ be any

---

* If $\xi$ and $\upsilon$ be the reciprocals of the intercepts made by a variable straight line on two fixed axes, the general condition that the variable line should envelope a conic is that $\xi$ and $\upsilon$ should be connected by an equation of the second degree. This system of "tangential coordinates" is fully developed in *A Treatise on some New Geometrical Methods* (vol. I., 1873) by the late Dr. James Booth, who had also given an account of his method in a separate tract published thirty years earlier. His discovery had however been anticipated by Prof. Plücker of Bonn, whose method given in Crelle's *Journal*, vol. VI. pp. 107—146 (1830), and dated Oct. 1829, is in reality identical with the above. See the obituary notice of Dr. Booth in the *Monthly Notices of the Royal Astronomical Society*, vol. XXXIX. pp. 219—225 (Feb. 1879).

point on a chord $PQ$ parallel to the radius $CD$ of the hyperbola, and $L$, $M$ the points in which it meets the asymptotes, then

$$OL.OM \sim OP.OQ = CD^2.$$

407. The tangent to a hyperbola at $P$ meets one asymptote in $T$ and $TQ$ is drawn parallel to the other to meet the curve in $Q$; prove that if $PQ$ meet the asymptotes in $L$ and $M$, the line $LM$ will be trisected at $P$ and $Q$.

408. The straight lines joining the points in which any two tangents to a hyperbola meet the asymptotes are parallel; and the intercepts which the tangents make upon the asymptotes are bisected by their chord of contact.

409. If one diagonal of a parallelogram whose sides are parallel to the asymptotes of a hyperbola be a chord of the curve, the other diagonal will lie upon the conjugate diameter; and conversely if the three sides of a triangle be taken as diagonals of three parallelograms whose sides are parallel to two given straight lines, their other three diagonals will pass through the centre of a hyperbola which circumscribes the triangle and whose asymptotes are parallel to the given lines.

410. In Art. 39, if $CM$ and $CN$ be the central abscissæ of the points in which the tangent meet the asymptotes, then

$$CV.CT = CM.CN = CP^2.$$

411. If the ordinate of a point on the hyperbola to a given diameter be equal to the conjugate semi-diameter, the product of the corresponding abscissæ will be equal to the square of half the given diameter.

412. Given the asymptotes of a variable hyperbola and a line parallel to one of them, if from the point in which it meets the curve a parallel to the other asymptote be drawn equal to either of the semi-axes, the locus of its extremity will be a parabola.

413. If an ellipse and a branch of a confocal hyperbola intersect in $P$ and $Q$, the asymptotes of the hyperbola pass through the points on the auxiliary circle of the ellipse which correspond to $P$ and $Q$.

414. A variable ellipse having its centre on a hyperbola and touching its asymptotes has in every position the maximum area: shew that its chord of contact with the asymptotes will envelope a similar hyperbola having the same asymptotes.

415. A parabola being drawn to touch the axes of a hyperbola at an extremity of each, prove that one asymptote is a diameter of the parabola and that the other is parallel to its ordinates.

416. If a parallelogram be formed by drawing two pairs of parallels to the asymptotes of a hyperbola, its sides will meet the curve at the extremities of two chords which intersect upon a diagonal of the parallelogram; and further, if any three hyperbolas have their asymptotes parallel, three and three, their three common chords will cointersect.

417. The tangents to an ellipse at $P$ and $Q$ being the asymptotes of a hyperbola, prove that a pair of their common chords are parallel to $PQ$, and that if the tangent to the hyperbola at an extremity of one of these chords pass through $P$ the tangent at its other extremity will pass through $Q$.

418. With two conjugate diameters of an ellipse as asymptotes a pair of conjugate hyperbolas are drawn; prove that if one of them touch the ellipse the other will touch it, and that the diameters through the points of contact will be conjugate.

419. If from any point $P$ on a hyperbola whose centre is $C$ straight lines $PM$ and $PN$ be drawn parallel to and terminated by the asymptotes, and if an ellipse be drawn having $CM$ and $CN$ for conjugate radii, the direction conjugate to $CP$ will be the same in both curves.

420. Given the base of a triangle and the difference of its base angles, or given the base of a triangle one of whose base angles is double of the other, it may be shewn that the locus of the vertex is a hyperbola. Determine the asymptotes and the eccentricity of each by supposing the vertex of the triangle to be at infinity.

421. If tangents be drawn to a hyperbola from any point on the conjugate hyperbola, their chord of contact will touch the opposite branch of the latter and be bisected at its point of contact.

422. The four normals to a hyperbola and its conjugate at points lying upon a perpendicular to either axis meet one another upon that axis.

423. Find the locus of the centre of gravity and the locus of the centre of the circumscribing circle of a triangle of constant area contained by one variable and two fixed straight lines.

424. A parabola and a hyperbola have a common focus and the asymptotes of the latter touch the former; prove that the tangent at the vertex of the parabola is a directrix of the hyperbola, and the tangents to the parabola where it meets the hyperbola pass through the further vertex of the latter.

425. Any two semi-diameters of a hyperbola contain the same area with the tangent at the extremity of either.

426. The asymptotes and any two conjugate diameters of a hyperbola divide any straight line harmonically.

427. The chords joining any point on a hyperbola to two given points on the same intercept a constant length on either asymptote; and the intercepts on a given parallel to an asymptote between the curve and two such chords are in a constant ratio.

428. If parallels to the asymptotes of a hyperbola be drawn from any point on the curve, any diameter will meet the parallels and either branch of the curve in three points whose central distances are in continued proportion.

429. If any two tangents to a hyperbola and their chord of contact intersect any parallel to either asymptote, the square of the intercept on the parallel between the curve and the chord of contact will be equal to the product of its intercepts between the curve and the tangents.

430. On a straight line drawn in a given direction to meet the three sides of a triangle a point is taken whose distances from the three sides are in continued proportion; prove that the locus of the point is a parabola or a hyperbola touching the two sides from which the extremes are measured at the extremities of the third side.

431. On a straight line drawn through a fixed point $C$ to intersect two given straight lines a length $CD$ is estimated a mean proportional to the intercepts between the fixed point and the two points of section; prove that the locus of $D$ is a hyperbola whose asymptotes are the parallels through $C$ to the fixed lines.

432. A diameter of a parabola and the tangent at its extremity being taken as the asymptotes of a hyperbola, what are the magnitudes to which the ordinate and abscissa of their point of concourse with respect to that diameter are a pair of mean proportionals? Conversely shew how to find a pair of mean proportionals to two given magnitudes.

433. The intercept on any parallel to an asymptote of a hyperbola (or to the axis of a parabola) between any point upon it and the polar of that point is bisected by the curve.

434. The intercept made upon any straight line through either vertex of a hyperbola by parallels drawn to the asymptotes through the other vertex is bisected at the point in which the straight line meets the curve again; the locus of the middle point of the intercept made upon any straight line through a fixed point by two given straight lines is a hyperbola to whose asymptotes they are parallel; and further, if the latter intercept be cut in any other constant ratio,\* the locus of the point of section will still be a hyperbola. In what case will the eccentricity of the locus be independent of the ratio in which this intercept is divided?

435. If $Q$ be a point on a hyperbola and $N$ a point on the nearer asymptote, and if $QE$ be drawn parallel to that

---

\* See the *Arithmetica Universalis*, prob. 25.

asymptote to meet the diameter conjugate to $QN$ in $E$, then will the area of the quadrilateral $CEQN$ be equal to half the triangle cut off by any tangent from the asymptotes; and if the diameter parallel to $QN$ meet $QE$ in $F$ and $QI$ be drawn in the conjugate direction to meet the same asymptote in $I$, the quadrilateral $CIQF$ will have the same constant magnitude.

436. If $O$ be any point in a chord $QQ'$ of a hyperbola parallel to the tangent at $P$ and $CE$ an asymptote meeting that tangent in $E$, and if $QR$ and $OT$ be drawn parallel to the asymptote to meet the diameter which bisects the chord, prove that

$$QO.OQ' : PE^2 = \text{quadrilateral } QRTO : \text{triangle } CEP.$$

437. If $P$ be any point on a hyperbola and $CD$ be conjugate to $CP$, shew that a pair of straight lines $PE$, $PF$ drawn parallel to the axes or to any other pair of conjugate diameters meet $CD$ in points $E$ and $F$ such that

$$CE.CF = CD^2.$$

438. A parabola which has an asymptote of a hyperbola for one of its diameters meets the hyperbola in general in three points such that the ordinates of two of them to that diameter are together equal to the ordinate of the third.

439. From any point $P$ on a hyperbola a parallel is drawn to one asymptote to meet the other in $M$, and an ellipse is drawn through $P$ and $M$ having its diameter which bisects $PM$ parallel to the latter asymptote and in a constant ratio to its conjugate diameter, viz. in the ratio of $PE$ to $PM$, where $PE$ is a perpendicular to the latter asymptote; prove that the ellipse meets the hyperbola again in three points such that the distances of two of them from the latter asymptote are together equal to the distance of the third point from the same.*

440. If two ellipses touch a hyperbola and have its asymptotes for conjugate diameters, any straight line whose pole with

---

* For Examples 425, 435—9 and others see De la Hire's *Sectiones Conicæ*, libb. IV., V. The references are given in detail in Walton's *Problems in illustration of the principles of Plane Coordinate Geometry*, pp. 276—292 (CAMBRIDGE, 1851).

respect to one of them lies **on the hyperbola** has it pole with respect to the other **on the hyperbola**.

441. If $ABCD$ be a convex quadrilateral, and $AD$ be produced to $K$ and $BC$ to $L$ so that $KL$ may be parallel to $AB$, then will $DL$ and $CK$ be parallel to the asymptotes of a certain hyperbola described about the quadrilateral; and if $\alpha B\beta$, $\beta C\gamma$, $\gamma D\delta$, $\delta A\alpha$ be the sides of a parallelogram and be parallel to the asymptotes, the straight lines drawn from $\alpha$, $\beta$, $\gamma$, $\delta$ to bisect $AB$, $BC$, $CD$, $DA$ respectively will cointersect at the centre of the hyperbola.

442. **If an ellipse pass through the centre** and have its foci on the asymptotes of a hyperbola, **and if the hyperbola passes through the centre of the ellipse, the axes of** each curve are a tangent and normal to the other, and the two axes which are normals are of equal length.

443. **If a diameter be** taken at right angles **to one asymptote of a hyperbola** and parallels **be drawn to the other** asymptote **from its extremities, any two supplemental chords from those points will make intercepts whose difference is constant** upon **the parallels**.

444. **The axes of the two parabolas which have a common focus and pass** through two given points are parallel to **the asymptotes** of the hyperbola which passes through the common **focus** and has the given points for foci.

445. Any circle which touches both branches of **a hyperbola** makes an intercept equal to the transverse **axis on either** asymptote; the tangents to it where it meets **the asymptotes pass** through one or other of **the foci, and** those which pass **through** the same focus contain a constant angle equal to the **angle between the asymptotes; and two** of the chords of intersection **of the circle with the** asymptotes are tangents **one to** each of two **fixed parabolas** whose foci are at the foci **of** the hyperbola.

446. If two conjugate diameters **of a** hyperbola be equal, every two conjugate diameters must be equal and the asymptotes must be at right angles.

447. If two parallel chords of a conic meet any tangent to the same in $T$ and $t$, and if any straight line meet the chords in $O$ and $o$ and the tangent in $L$, then

$$OT : ot = OL : oL.$$

Hence shew that the ratio of any two infinite parallel chords of a conic is finite, being a ratio of equality in the case of the parabola, and being equal to the ratio of the distances of the chords from the parallel asymptote in the case of the hyperbola.*

448. From two points $O$ and $o$ parallels are drawn to the asymptotes of a hyperbola, the parallels to one asymptote meeting the curve in $M$ and $m$ and the parallels to the other meeting the curve in $N$ and $n$; shew that if

$$OM : ON = om : on,$$

the points $O$, $o$ must lie either on one diameter or on a pair of conjugate diameters.

449. Prove by the Cartesian method or otherwise that if $CA$, $CB$ and $C\alpha$, $C\beta$ be semi-axes of a fixed and a variable confocal ellipse respectively, $P$ a point of contact of the latter with an ellipse drawn through the four extremities of the axes of the former, and $PN$ the principal ordinate of $P$, then

$$CN : PN = CA.C\alpha : CB.C\beta.$$

Deduce that the locus of $P$ is a hyperbola; and likewise determine its foci and asymptotes by considering special cases of the theorem.

450. Every ellipse drawn through the four extremities of the axes of a given ellipse is cut orthogonally by a hyperbola confocal with the given ellipse and having its equal conjugate diameters for asymptotes.†

---

\* See the notes on *Geometrical Evaluations* by R. W. Genese, M.A., in the *Messenger of Mathematics*, vol. IV. pp. 154—6 (1875).

† See Wolstenholme's *Mathematical Problems*, No. 1182 (ed. 2, 1878).

451. Find the locus of a point whose polar with respect to a conic cuts off a constant area from the space between two given conjugate diameters; and find the envelope of the polar of a point whose ordinates cut off a constant parallelogram from the same.

452. Having given the asymptotes of a hyperbola and a point on the curve, determine its foci and directrices.

453. Having given a focus and two points of a hyperbola and the direction of one of its asymptotes, or having given a focus and one point and the directions of both asymptotes, shew how to construct the curve.*

454. Given the centre of a hyperbola and three points on the curve, determine the directions of its asymptotes.

455. Having given the centre of a hyperbola and a self-conjugate triad, determine the directions of its asymptotes.

456. Having given four points and the eccentricity of a hyperbola, or four points and the direction of an asymptote, or three points and the direction of an asymptote and the eccentricity, shew how to construct the curve.

457. If three straight lines be drawn from three given poles and two of their intersections lie on fixed directrices, their third intersection will trace a curve of the second order. By the above system of radiants or otherwise describe a hyperbola having given one asymptote and three points or the directions of both asymptotes and three points on the curve.†

---

* Five data in general determine a conic. An asymptote is equivalent to two data, viz. to a tangent and its point of contact or two coincident points on the curve: having given the direction only of an asymptote we have one of the two points at infinity on the curve: a focus will be seen to be equivalent to two conditions. Compare the note at the end of Salmon's *Conic Sections*, "On the Problem to describe a Conic under Five conditions."

† See Leslie's *Geometry of Curve Lines*, Book II. props. 10, 21, 22.

458. Having given four points on any conic and one point on its director, or having given four tangents to an equilateral hyperbola, shew how to construct the curve.*

459. The area of the sector of a hyperbola made by joining any two points on it to the centre is equal to the segment cut off from the space between the curve and its asymptotes by the ordinates of the same two points to either asymptote; any other two ordinates in the same ratio as the former† cut off an equal segment; and the segment cut off by any two ordinates is bisected by the ordinate which is a mean proportional to them. Prove also that if two equal hyperbolas have two of their asymptotes coincident and the other two parallel, any parallel to their common asymptote will cut off from the space between two adjacent branches produced to infinity an area equal to the parallelogram contained by the said parallel and the three asymptotes.

460. If $O, P, Q, R...$ be any number of points on a branch of a hyperbola whose abscissæ $CK, CL, CM, CN...$ on either asymptote are in continued proportion, the hyperbolic sectors $OCP, OCQ, OCR...$ will be in arithmetical progression, and

---

\* The first case of Ex. 458 may be made to depend upon the second by reciprocating the conic with respect to the point on its director, as is done in Gaskin's *The Geometrical Construction of a Conic Section subject to Five Conditions of passing through given points and touching given straight lines, deduced from the properties of Involution and Anharmonic Ratio, with a variety of general Properties of Curves of the Second Order*, p. 53 (Cambridge, 1852). It is in this very able tract that the term Director seems to have been first used to denote the locus of intersection of tangents at right angles to a conic. The term is defined on p. 26, and in the Preface we read: "By a well known property of conic sections, the locus of the point of intersection of two tangents at right angles to one another is in general a circle concentric with the conic section, and when the curve is a parabola the locus is the directrix. There are several remarkable properties of this locus which, as far as the author is aware, have not been hitherto noticed, and he has found it convenient to denominate it the Director of the conic section, which in the case of the parabola coincides with the directrix."

† It is easily seen that the four ordinates to either asymptote of the extremities of any two parallel chords are proportionals, and that the ordinate of the point of contact of any tangent is a mean proportional to the ordinates of the extremities of any parallel chord.

the segments $OKLP$, $OKMQ$, $OKNR$... will be in arithmetical progression. Given any three terms of a geometrical series and the **logarithms** of two of them, shew how to determine geometrically the LOGARITHM of the third.*

---

\* For the first part of Ex. 460 see GREGORII A. S. VINCENTIO *Opus Geometricum* Quadraturæ *Circuli et* Sectionum *Coni*, lib. VI. prop. 125, p. 594 (Antverpiæ, 1647), and his *Opus Geometricum Posthumum ad Mesolabium*, prop. 24, p. 252 (Gandavi, 1668). The second part may be solved by taking hyperbolic segments in arithmetical progression to represent the logarithms of a corresponding series of abscissæ which are in geometrical progression, as was shewn by **Alf.** Ant. de Sarasa in a tract published (Antv. 1649) in vindication of Greg. de St. Vincent against some aspersions of Marinus Mersennus. Logarithms may also be represented by the "residual arcs" of a parabola (Booth's *New Geometrical Methods*, vol. I. p. 293).

# CHAPTER VI.

## THE EQUILATERAL HYPERBOLA.

**61.** The *Equilateral Hyperbola* is a hyperbola whose **latus rectum** is equal to its axis or latus transversum;* it is **also called** *Rectangular* since its asymptotes are at right angles. This curve and the circle, which is an equilateral ellipse, may be together designated the *Equilateral Conics*.

The properties of the equilateral hyperbola may for **the most part be** derived from **those** of the general hyperbola by **equating its axes to one another and to the latus** rectum, **or by supposing the angle between its asymptotes to become a right angle; but since** several of the special results thus obtainable may also be proved independently with peculiar ease, some of them in terms equally applicable to the circle also, we shall **here** treat the hyperbola in question to a great extent *ab initio*, **leaving it to be shewn** in the sequel how certain of the **properties of the** equilateral conics **may be** transferred to central **conics in general by the method** of Orthogonal Projection.

**62.** The latus **rectum being** supposed **equal to the axis,** it follows from Art. 33 that

$$PN^2 = AN \cdot A'N = CN^2 - CA^2,$$

which will however be proved independently in Art. 63.

**The axes** being equal, the radius of the director **circle** vanishes **(Art.** 40), or the equilateral hyperbola has no tangents at right angles except its asymptotes. Again, it follows from Art. 45 that every diameter is equal to its conjugate, which leads to many further simplifications; but **in this chapter we**

---

* In other words, this hyperbola is called equilateral because *the sides of the* FIGURE *upon its axis* (Schol. A, p. 82) *are equal*.

shall commence by *defining* any diameter which does not meet the curve as equal in length to its conjugate, in pursuance of the analogy between the equilateral hyperbola and the circle or equilateral ellipse.

It is to be noted at the outset (Arts. 35, 54) that *the eccentricity of the equilateral hyperbola is the ratio of the diagonal to the side of a* SQUARE, that the foot $X$ of the directrix bisects $CS$, and that

$$\tfrac{1}{2}CS^2 = CA^2 = 2CX^2 = 2SX^2.$$

### PROPOSITION I.

63. *The principal ordinate of any point on an equilateral hyperbola is a mean proportional to its abscisses.*

If $X$ be the foot of the $S$-directrix and therefore the middle point of $CS$, and if $PN$ be the principal ordinate of any point $P$ on the curve, then

$$PN^2 + SN^2 = SP^2 = 2NX^2,$$
and
$$CN^2 + SN^2 = 2CX^2 + 2NX^2;$$

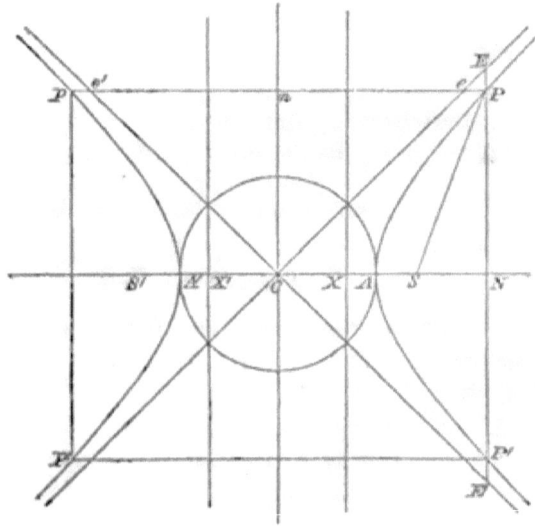

therefore $\qquad CN^2 - PN^2 = 2CX^2 = CA^2,$

or $PN^2$ is equal to $CN^2 - CA^2$ or $AN \cdot A'N$.

THE EQUILATERAL HYPERBOLA.

It is further evident from the figure that if $Pn$ be an ordinate to the conjugate axis,
$$Pn^2 = Cn^2 + CA^2,$$
as might also have been inferred from the consideration that the square of the conjugate semi-axis is $- CA^2$.

### PROPOSITION II.

**64.** *Any two conjugate diameters of an equilateral hyperbola make complementary angles with either axis and make equal angles with either asymptote.*

(i) If $V$ be any point on the directrix and $SZ$ be at right angles to $SV$, it is evident that $CV$ and $SV$ are equally inclined to the axis and $CV$ and $SZ$ make complementary angles with the axis. The proposition then follows at once as a special case of Art. 14; it may also be proved independently as below.

(ii) Let $Q$ and $q$ be any two points on the curve, $QM$ and $qm$ their principal ordinates, $O$ the middle point of $Qq$ and $OL$ its ordinate, $qK$ a parallel to the axis meeting $QM$ in $K$, and $n$ the point in which $Qq$ meets the axis.

Then since $QM^2 + CA^2$ is equal to $CM^2$ and $qm^2 + CA^2$ to $Cm^2$,

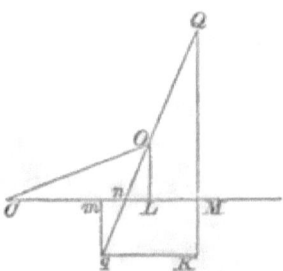

therefore $\qquad QM^2 \sim qm^2 = CM^2 \sim Cm^2,$

or $\qquad QM \sim qm : CM + Cm = CM \sim Cm : QM + qm;$

that is to say,
$$2OL : 2CL = qK : QK,$$
or the angle $OCL$ is equal to the angle $qQK$.

Hence if the chord $Qq$ be parallel to a fixed diameter, the locus of its middle point $O$ will be a second fixed diameter, and the inclinations of the two diameters to either axis will be complementary, and their inclinations to either asymptote will therefore be equal, and conversely.

*Corollary* 1.

**It is evident that** any two diameters which **are either conjugate** or at right angles must lie on opposite sides of an asymptote, and therefore that one of the two and one only meets the curve. It is likewise evident that if two equal diameters be **taken on** opposite **sides of** either axis, the one will be equal and at right angles **to the conjugate** to the other, and conversely that *any two diameters at right angles are equal.*

*Corollary* 2.

If the normal at $P$ meet the axes in $G$ and $g$ and the conjugate diameter in $F$, it is evident that $PCG$ is an isosceles

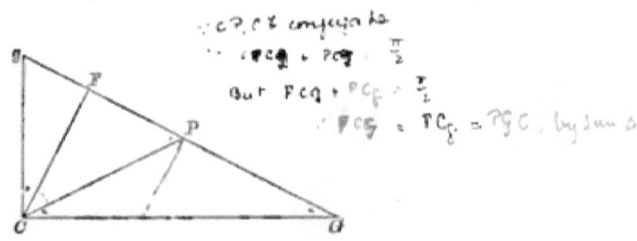

triangle having each of its angles **at $C$ and $G$** complementary to $FCG$, and hence that $PG = PC = Pg$, or $P$ is in this case the centre of the circle of Art. 59. Hence or by Art. 45, Cor. 1 **the** normal is also equal to $CD$.

*Corollary* 3.

*The angles between any **two diameters or chords** are equal to the angles between the diameters conjugate thereto.* For example, if $PQ$ and $PQ'$ be any two chords drawn from the **same** point $P$ on the curve and $P'Q$ and $P'Q'$ be the chords supplemental **to the** former from the further extremity $P'$ of the diameter **through** $P$, the angles between the former will be equal to angles between the latter (Art. 44), or *any chord QQ' of an equilateral hyperbola subtends at the extremities of any diameter PP' angles which are either equal or supplementary*. It will be seen that the angles subtended are *supplementary* when the diameter and the chord intersect WITHIN the curve (as in the figure of Prop. IV.) and *equal* when they intersect WITHOUT the curve.

## Corollary 4.

*The locus of the centre of an equilateral hyperbola circumscribing a given triangle is its nine-point circle*, since the diameters to the middle points of its sides contain two and two the same constant angles as the sides to which they are conjugate.* More generally it may be shewn that *the circumscribed circle of any self-conjugate triangle passes through the centre*, since the diameters to its angular points are conjugate to the directions of its opposite sides.

### PROPOSITION III.

65. *The projections of any two conjugate semi-diameters upon the axes are alternately equal to one another, and the triangle of which they are adjacent sides is of constant area.*

(i) If *CP* and *CD* be conjugate semi-diameters, and *PN* and *DR* be principal ordinates and *Dn* an ordinate to the

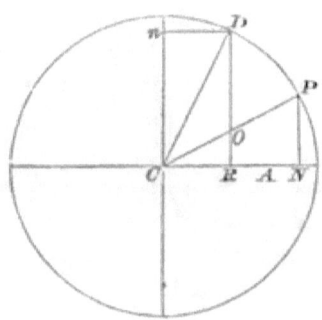

---

* Since each side and its perpendicular constitute a hyperbola (Art. 54), their intersections belong to the above locus: hence a fresh proof that the feet of the perpendiculars lie on the circle which bisects the sides.

conjugate axis, we have to shew that $Cn$ or $DR$ is equal to $CN$ and that $CR$ is equal to $PN$; which follows at once from the equality of the radii $CP$, $CD$ and of the angles $PCN$, $DCn$ (Prop. II.).

Hence also by Prop. I.,
$$CN^2 - CR^2 = DR^2 - PN^2 = CN^2 - PN^2 = CA^2.$$

(ii) The triangles $DCR$ and $PCN$ being equal, therefore
$$\triangle(DCR + DOP - COR) = PCN + DOP - COR,$$
or
$$\triangle PCD = DRNP = \tfrac{1}{2}(CN + CR)(CN - CR)$$
$$= \tfrac{1}{2}CA^2;$$

which is an equivalent of the theorem that the conjugate circumscribing parallelogram of an equilateral hyperbola is equal to $4CA^2$ (Art. 46).

### PROPOSITION IV.

66. *The base of a triangle and the sum or difference of its base angles being given, the locus of its vertex is an equilateral conic.*\*

(i) **If the base and the sum of** the base angles of a triangle be given, the third angle is constant **and** the locus of the vertex is a circle.

(ii) Let $P'CP$ be a fixed diameter of an equilateral hyperbola, $V$ any point in $CP$ produced, and $Q$ any point on the curve. Then since $QP$ and $QP'$ are supplemental chords, the sum of the acute angles which these make with the axis is equal to a right angle (Prop. II), and the sum of their inclinations to the fixed diameter $PP'$ is therefore constant.

The latter constant is at once seen, viz. by removing $Q$ to infinity, to be equal to *twice the angle which the nearer asymptote makes with $PP'$*.

---

\* This proposition forms **prob. 35 of** the *Arithmetica Universalis*, **and** was suggested by Eucl. III. 21, as **is shewn by** the preamble: " Ubi angulus ad verticem, sive (quod perinde est) ubi summa angulorum ad basem datur, docuit Euclides locum verticis **case** circumferentiam circuli; proposuimus igitur inventionem loci ubi differentia angulorum ad basem datur." NEWTON also stated the corollary given **above in the** text for the case in which the subtended angles are *equal.*

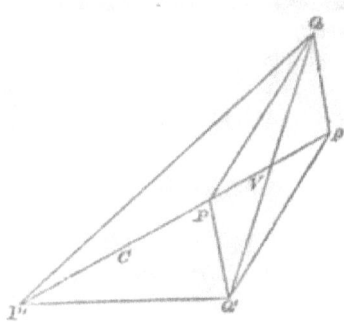

It follows that the angle at $P$ or **its supplement**, in the triangle $QPP'$, exceeds the angle at $P'$ or its supplement by a constant quantity; and conversely, that if the base $PP'$ of **a triangle** be given **and the** angle at $P$ or its supplement exceed the angle at $P'$ or its supplement by a constant quantity, the locus of the vertex $Q$ of the triangle will be an equilateral hyperbola whereof $PP'$ is a diameter, as was to be proved.

*Corollary.*

Hence it may be deduced that the angles which any chord of an equilateral hyperbola subtends at the extremities of any diameter are either equal or supplementary, as was shewn independently in Art. 64, Cor. 3.

PROPOSITION V.

67. *At any point of an equilateral* **hyperbola the** *ordinate to any diameter which meets the curve* **is a mean** *proportional to the abscisses on that diameter.*

Let $QV$ be the ordinate of any point $Q$ on the curve to the diameter $PP'$; then since the directions of $PV$ and $QV$ and likewise the directions of $PQ$ and $P'Q$ are conjugate, the angle $PQV$ is equal to $QP'V$ (Art. 64, Cor. 3), and the triangles $PQV$ and $QP'V$ are similar, so that

$$PV : QV = QV : P'V;$$

therefore $QV^2$ is equal to $PV.P'V$ or $CV^2 - CP^2$.

By changing the sign of $CP^2$ we obtain the corresponding property, viz.
$$QV^2 = CV^2 + CP^2,$$
of a diameter which does not meet the curve in real points.

### PROPOSITION VI.

**68.** *The product of the segments of any chord drawn through a fixed point to an equilateral hyperbola varies as the square of the parallel radius.*

Let $QQ'$ be any chord drawn through a fixed point $O$ and $V$ the middle point of $QQ'$, and let $CP$ be the semi-diameter parallel to the chord, $q$ the point in which $CO$ meets the curve or the conjugate rectangular hyperbola, and $qv$ the ordinate of $q$ to the diameter $CV$.

Then by Prop. v., taking for example the case in which $CP$ and $Cq$ are terminated by the curve,

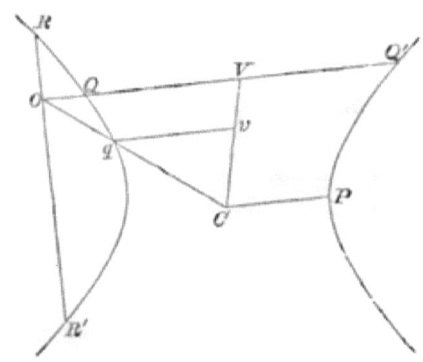

$$QV^2 - CP^2 : qv^2 - CP^2 = CV^2 : Cv^2 = OV^2 : qv^2.$$
Hence $OV^2 - QV^2 + CP^2 : CP^2 = OV^2 : qv^2 = CO^2 : Cq^2,$
which is a *constant* ratio since $O$ and $q$ are fixed points.

It follows that $OV^2 - QV^2$ or $OQ \cdot OQ'$ varies as $CP^2$, and if $ROR'$ be any second position of the chord and $CP'$ the radius parallel thereto,
$$OQ \cdot OQ' : OR \cdot OR' = CP^2 : CP'^2.$$

## PROPOSITION VII.

69. *If an equilateral hyperbola circumscribes a triangle it passes through its orthocentre, and conversely;* [*] *and every conic which passes through the four points of concourse of two equilateral hyperbolas is itself an equilateral hyperbola.*

If $ABC$ be any triangle and $AD$ one of its perpendiculars, any equilateral hyperbola which circumscribes the triangle will have its diameters parallel to $AD$ and $BC$ equal to one another.

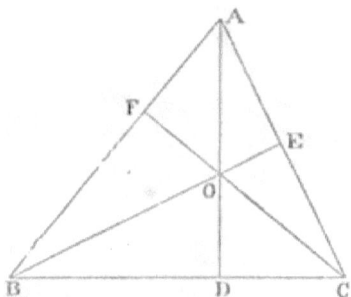

The hyperbola therefore meets $AD$ again in a point $O$ such that

$$AD.DO = BD.DC;$$

that is to say, it passes either through the orthocentre of the triangle or through the point in which $AD$ produced meets its circumscribing circle.

---

[*] This theorem was derived from **Pascal's hexagram** in a memoir by MM. Brianchon et Poncelet contributed to Gergonne's *Annales* (tome XI. pp. 205—220) at the commencement of the year 1821, under the title: *Recherches sur la détermination d'une Hyperbole Equilatère, au moyen de quatre conditions données*—viz. thus. Let a hyperbola be described through $A$, $B$, $C$, and the orthocentre $O$, and let $E$ and $F$ be the two points at infinity on the curve; let $H$, $I$, $K$ denote the three points of concourse $(AB, OE)$, $(EF, CB)$, $(AF, CO)$; then $HIK$ is a straight line *parallel to $BC$* (since $I$ is at infinity) or perpendicular to $AO$, whence it readily follows that $H$ is the orthocentre of the triangle $AOK$ and that $OE$ the direction of one asymptote is at right angles to $AK$ or $OF$ the direction of the other. The remainder of Prop. VII. follows independently from the fact that by adding together two equations of the form $a(x^2 - y^2) + bx + cy + d = 0$ we arrive at a third equation of the same form: the property of the orthocentre of any triangle is a *special case* of this latter theorem (Art. 54). See Prof. Cayley's *Note on the Rectangular Hyperbola* in the *Oxford, Cambridge, and Dublin Messenger of Mathematics*, vol. I. p. 77 (1862).

But it cannot pass through the latter point, for if so $AD$ and $BC$ would be equally inclined to its axes (Art. 16, Cor. 2) and parallel to its asymptotes, and either $B$ or $C$ would be at infinity: it therefore passes through the orthocentre.

Conversely, any conic which passes through the three angular points and the orthocentre of a triangle must be an equilateral hyperbola.

Moreover, if three of the points of intersection of any two equilateral hyperbolas be taken as the vertices of a triangle, both curves will pass through its orthocentre; and therefore every conic through their four points of concourse must likewise be an equilateral hyperbola.

From this proposition it is manifest that when three points of an equilateral hyperbola are given a fourth can be found; and hence that when four points are given the curve is in general determined.

### Corollary.

If $BAC$ be a right angle, the points $A$ and $O$ coalesce and $AD$ touches the curve at $A$. Hence the tangent at any point $A$ on an equilateral hyperbola may be determined by drawing any two chords $AB$ and $AC$ at right angles and drawing $AD$ perpendicular to $BC$. If $A$ be a fixed point, $BC$ is constantly parallel to the normal thereat.*

### PROPOSITION VIII.

70. *The product of the distances from the centre* **at which any** *tangent and the ordinate of its point* **of contact to any diameter meet** *the same is constant; and the product* **of the intercepts** *on any tangent between the curve and any* **two conjugate** *diameters is* **equal** *to the square of the parallel radius.*

(i) Let the tangent at $Q$ meet any diameter $CP$ in $T$, and let $QV$ be an ordinate to that diameter.

Then since $CP$ and $CQ$ are conjugate to $QV$ and $QT$ respectively, they contain equal angles (Art. 64, Cor. 3), so that

---

\* The fixed point on the normal (Ex. 279) through which $BC$ passes is otherwise seen to be at infinity since when $AB$ and $AC$ are parallel to the asymptotes $BC$ becomes the straight line at infinity.

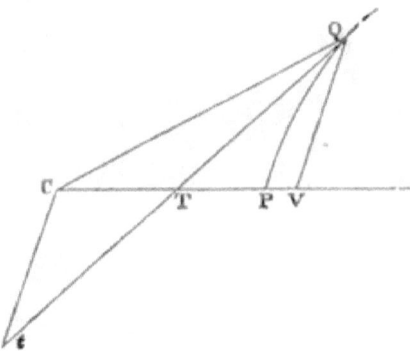

$$CV : QV = QV : VT.$$

Therefore $\quad CV.CT = CV^2 \sim CV.VT = CV^2 \sim QV^2$
$$= CP^2.$$

(ii) If the tangent at $Q$ makes intercepts $CT$ and $Ct$ on any two conjugate diameters, it may be shewn in like manner that

$$QT : CQ = CQ : Qt,$$
or $\quad QT.Qt = CQ^2 = CD^2,$

where $CD$ is the radius conjugate to $CQ$.

### SCHOLIUM.

An excellent machine for describing any number of RECTANGULAR HYPERBOLAS having the same asymptotes was constructed by Mr. H. H. S. Cunynghame, of St. John's College (1873), on the following principle. Let a fixed straight line meet the axis of a rectangular hyperbola at right angles in $H$; from any point $P$ on the curve draw $PM$ and $PN$ perpendicular to the fixed line and the axis; and on $CH$ produced take $HO$ equal to $CA$. Then

$$OM + PM = CN + PM = CH;$$

and conversely if $O$ be a fixed point and $MP$ a variable perpendicular to the fixed line $HM$, then provided that the length $OMP$ is constant the point $P$ will describe a rectangular hyperbola, and its centre $C$, which is determined by taking $HC$ equal to $OMP$, will be *independent of the distance OH*. The machine itself consists of a fixed bar $HM$ and a sliding cross bar placed in a horizontal plane: a string fixed at $O$ is kept stretched by a weight in the direction $OMP$: and a pencil attached at a point $P$ to the string traces an arc of a rectangular hyperbola by the motion of the cross bar. By varying the length $OH$ any number of rectangular hyperbolas having the **same asymptotes can be** traced with the same length of string.

In a note *On the Mechanical Description of the Cartesian*, by J. Hammond, *Bath, England* (*American Journal of Mathematics, pure and applied*, vol. I. no. 3, p. 283, 1878), the following, applicable to the HYPERBOLA, is given. Suppose two thin circular discs $A$ and $B$ rigidly attached to each other to rotate about **their** common centre, and suppose the opposite ends of a fine string (**which passes**

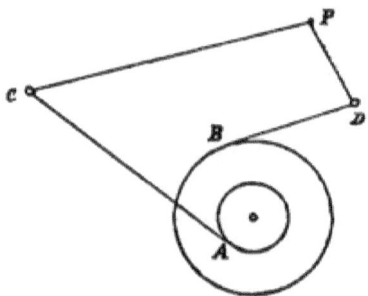

through small rings **at** $C$ **and** $D$ and is kept stretched by the point of a pencil at $P$) to be unwound from the two discs. Then will the increments of the lengths $CP$ and $DP$ be as the radii $a$ and $b$ of the discs, and $P$ will describe a curve having the property

$$a \cdot DP - b \cdot CP = \text{a constant},$$

which becomes a **hyperbola when the discs are equal.** If one end of the string be wound on to its disc whilst the other is unwound the curve traced will have the property

$$a \cdot DP + b \cdot CP = \text{a constant},$$

and will become an ellipse when the discs are equal.

The mechanical description of the ellipse by the property of **Ex. 219** was effected by Guido Ubaldi,[*] who was considered to have made an important discovery; but the property **is** mentioned by Proclus (on Eucl. I. def. 4) as was remarked in **the** first volume of the *Ærarium Philosophiæ Mathematicæ*, **auctore Mario Bettino, Lib. I.** pp. 38—45 (Bononiæ, 1648).

## EXAMPLES.

461. Trace the locus of the middle point of a straight line which cuts off **a constant area** from a corner of **a square.**

---

[*] Guidiubaldi *Planisphaeriorum Universalium Theorica*, Lib. II, end (Pisauri, 1579).

462. Place in a rectangular hyperbola a chord which shall be equal to and be bisected by a diameter of given length.

463. The chords connecting the ends of a fixed diameter of a circle and of any double ordinate of the same intersect upon an equilateral hyperbola.

464. In the rectangular hyperbola the diameter conjugate to the normal at any point is at right angles to the diameter through the point; any two diameters at right angles bisect chords at right angles, and conversely; and any chord subtends equal or supplementary angles at the extremities of a perpendicular chord.

465. The centre of an equilateral hyperbola circumscribing an equilateral triangle is upon the inscribed circle of the triangle, and the centre of the circle is on the hyperbola.

466. The tangents drawn from opposite foci of a hyperbola to any circle which touches both branches intersect upon one of two rectangular hyperbolas, each of them having one asymptote in common with the original hyperbola and having the line joining the foci of the latter for a diameter; and these two rectangular hyperbolas will coalesce if the original hyperbola be rectangular.

467. If two points $P$ and $Q$ move with equal velocities along the arms $AB$ and $BC$ of a right angle, the one starting from $A$ and the other simultaneously from $B$, and if $AA'$ be drawn equal to $AB$ and in the direction opposite to $BQ$, shew that $A'P$ and $AQ$ intersect upon a branch of a rectangular hyperbola, and determine its centre and asymptotes.

468. The circles described upon the six common chords of any two rectangular hyperbolas as diameters cut one another orthogonally in opposite pairs.

469. If a parallel to either asymptote of a rectangular hyperbola meet any principal double ordinate $PQ$ in $O$ and the curve in $R$, shew that

$$\triangle OCR = \tfrac{1}{4} OP . OQ.$$

470. Of two chords at right angles or conjugate in direction in an equilateral hyperbola one and one only is a chord of a single branch. **Explain** the apparent failure **of** the **proof of** Art. 16, Cor. 2 **which** arises from the equality **of** diameters which are conjugate **or at right** angles in the **equilateral** hyperbola;\*
**and shew** that no circle can intersect the **curve or its asymptotes** at the extremities of a pair of chords which are parallel **to two such** diameters.

471. The foci of an ellipse being situated at any two diametrically opposite points of a rectangular hyperbola, shew that **the tangents and** normals to the ellipse **at** the points in which it **meets the** hyperbola **are** parallel **to the asymptotes** of the latter; and shew that the tangents to the ellipse **from any** point **of the** hyperbola are parallel to conjugate diameters of the latter.

472. If $CA$ be a semi-axis of **a** rectangular hyperbola, **and** a perpendicular $CY$ be drawn to **the** tangent at $P$, the triangles $ACP$ and $ACY$ will be similar.

473. **Prove that the feet of** the perpendiculars of any triangle **are a conjugate triad with respect to any** equilateral hyperbola **which circumscribes the triangle; and shew** that **the** same result **may also be deduced from Example 76.**

474. Given a chord of an equilateral hyperbola and **the polar of a** given point on the chord, shew how to determine **two** other points on the curve.†

475. The circle described on the line joining **the foci of an** equilateral hyperbola as diameter meets **the asymptotes** at points lying upon the tangents **at the vertices;** and the circle described about any point **on** the **conjugate axis as** centre so as to pass

---

\* It is only in accordance with a convention which is not strictly accurate that such diameters are said to be equal. See Chap. IV, Scholium C, p. 101. If $ABC$ be a triangle simultaneously inscribed in a circle and an equilateral hyperbola, and if the perpendicular from $A$ to $BC$ meet the circle in $D$, the hyperbola in $E$, and $BC$ itself in $F$, then $FB.FC = FA.FD = -FA.FE$.

† On the given chord $AB$ as diameter describe a circle cutting the polar of the given point $O$ in $X$ and $Y$; then will the points $(AX, BY)$ and $(AY, BX)$ be the extremities of the chord through $O$ at right angles to $AB$.

through the vertices meets the curve again at the extremities of one of its own diameters.

476. If $PQ$ and $P'Q$ be any pair of supplemental chords of a rectangular hyperbola, and if the tangent at $Q$ and its ordinate to $PP'$ meet that diameter in $T$ and $V$, shew that the bisectors of the angle $PQP'$ are parallel to the asymptotes, the segments $CP$ and $TP'$ subtend equal angles at $Q$, and the circle around $CQT$ touches $QV$. Shew also that any chord subtends supplementary angles at its pole and the centre of the curve, and that the inclinations of any two tangents to their chord of contact are equal or supplementary to the angles which they subtend at the centre.

477. If a conic pass through the centres of the four circles which touch the sides of a triangle it must be a rectangular hyperbola, and its centre will lie on the circumscribed circle of the triangle.

478. The foci of all the ellipses which can be inscribed in a given parallelogram lie on a rectangular hyperbola passing through its four vertices.

479. The lines connecting the extremities of any two chords drawn through a focus parallel to conjugate diameters of an equilateral hyperbola pass through fixed points on the asymptotes. Examine the cases in which the focal chords coalesce or are parallel to the axis.

480. The axis of the rectangular hyperbola which touches an ellipse and has its axes for asymptotes is a mean proportional to the axes of the ellipse.

481. Construct a rectangular hyperbola having given the centre and a tangent and a point on the curve, or having given an asymptote and a tangent and its point of contact, or having given a diameter and one other point on the curve.

482. The common tangents to the circles described on any two parallel chords of opposite branches of a rectangular hyperbola as diameters subtend right angles at the extremities of the diameter which bisects the chords.

483. If two right angles revolve about opposite extremities of a diameter of a rectangular hyperbola so that the point of concourse of two of their arms is always a point on the curve, their other two arms will make equal intercepts on the normal at that point, and will themselves intersect upon the curve.

484. Tangents (or normals) are drawn in a given direction to a series of confocal conics: prove that the points of contact lie on a rectangular hyperbola passing through the foci and having an asymptote in the given direction.

485. The lines connecting the extremities of any chord and any diameter of a rectangular hyperbola intersect in two points which are concyclic with the extremities of the diameter: determine the condition that they may intersect on a *fixed* circle.

486. Find the points on an equilateral hyperbola at which the normal is parallel to a given chord.

487. The locus of the pole of any chord of a parabola which subtends a right angle at the focus is a rectangular hyperbola.

488. The subnormal at any point of an equilateral hyperbola is equal to the central abscissa; the tangent from the foot of the ordinate to the auxiliary circle is equal to the ordinate; the projection of the normal (terminated by either axis) upon either focal vector is equal to the semi-axis; and the intercept made on any tangent by the asymptotes subtends a right angle at the point in which the normal meets either axis.

489. Any two supplemental chords of a rectangular hyperbola form an isosceles triangle with either asymptote, and conversely.

490. Any two conjugate diameters of an equilateral hyperbola contain equal and similar triangles with the ordinates and abscissæ of their extremities to any other diameter.

491. The sum or difference of the inclinations of any two conjugate diameters of an equilateral conic to a fixed diameter is constant: distinguish between the several cases.

492. Any circle drawn through the extremities of a diameter of a rectangular hyperbola meets the curve again at the extremities of a diameter of the circle, and its tangents at those points are ordinates of the diameter of the hyperbola.

493. The circles described on parallel chords of a rectangular hyperbola as diameters have a common radical axis.

494. The ends of the equal conjugate diameters of a series of confocal ellipses lie on the confocal rectangular hyperbola.

495. The straight line joining the feet of the perpendiculars from any point of a rectangular hyperbola to two conjugate diameters is parallel to the normal at the point.

496. The opposite arcs cut off by any two diameters of a rectangular hyperbola subtend equal angles at any point on the curve.

497. Any two rectangular hyperbolas so placed that the axes of the one coincide with the asymptotes of the other intersect at right angles, and each of their common tangents subtends a right angle at the centre; and if two tangents to a pair of conjugate rectangular hyperbolas be at right angles, the straight line joining their points of contact subtends a right angle at the centre.

498. If on opposite sides of any chord of a rectangular hyperbola equal segments of circles be described, the four points in which the completed circles meet the hyperbola again will be the angular points of a parallelogram; and if parallels be drawn from any point on a rectangular hyperbola to the sides of an inscribed parallelogram, they will meet its opposite sides in two pairs of points lying on a circle.

499. The foot of the focal perpendicular upon any chord of a rectangular hyperbola which subtends a right angle at the focus lies on a fixed straight line.

500. The normal at any point $P$ of a rectangular hyperbola meets the curve again in $Q$, and $RR'$ is a chord parallel to the

normal: prove that $PR$, $QR'$ and $PR'$, $QR$ intersect on the diameter at right angles to $CP$.*

501. In any right angled **triangle inscribed** in an equilateral hyperbola the perpendicular upon **the hypotenuse** is the tangent **at the** right angle. Hence shew how to find **a third point on the curve** when two points and the tangent at one **of them are given**; and shew that the **curve is determined** when two points **and the** tangents thereat are given, or three points and the tangent at one of them, or two points and the tangent at one of them and a second tangent.

502. **Given the** middle **points and** the directions of two **chords of an** equilateral hyperbola, **the two** points and the intersection of the parallels through each **point to** the opposite chord determine **a circle** which passes **through** the centre of the hyperbola.

503. If through each of two points a parallel be drawn to the polar of the other **with respect to an equilateral** hyperbola, **the circle through the two points** and the intersection of the parallels will pass **through the centre of the hyperbola.**

504. **Given the centre of a** rectangular hyperbola and **a self-conjugate triad**, determine its asymptotes.†

505. **Two** equilateral hyperbolas can be inscribed in a given quadrilateral, and their centres are at the points in which the diameter‡ of the quadrilateral meets the circumscribed **circle** of the triangle formed by its three diagonals.

---

* Examples 471, 482, 484, 492—500 and others are from Wolstenholme's *Mathematical Problems*.

† If $C$ be the centre and $PQR$ the conjugate triad, let $CP$ meet $QR$ **in** $V$, and upon $QR$ take points $Q'$ and $R'$ such that $Q'V = R'V = CV$; then will $CQ'$ and $CR'$ be the asymptotes. The following method applies to the *general* hyperbola (Ex. 455). Draw $CP'$, $CQ'$, $CR'$ parallel to $QR$, $RP$, $PQ$, and find the two double lines of the involution determined by the pairs of conjugate rays $CP$, $CP'$; $CQ$, $CQ'$; $CR$, $CR'$.

‡ By a theorem of NEWTON (*Principia*, Lib. I. sect. V. lemma 25, cor. 3) the centres of all the conics inscribed in a quadrilateral lie upon the straight line (Ex. 372) which we have called the DIAMETER of the quadrilateral (p. 138). See also Ex. 513.

506. The three pairs of chords connecting any four points on an equilateral hyperbola intersect upon the circumference of a circle which passes through its centre.

507. The nine-point circles of the four triangles determined by four given points cointersect at the centre of the equilateral hyperbola which passes through the four points.*

508. Four points being taken at random in a plane, there exists in general one other point in the same and one only such that the lines radiating therefrom to the middle points of the six lines connecting the four points are inclined at the same angles as the lines which they severally bisect.

509. Given any two points in the plane of an equilateral hyperbola and the directions conjugate to the diameters passing through them, determine a circle on which the centre lies. If a chord and the direction of the polar of a point upon it be given, this circle passes through the point and bisects the chord and has its tangent at the middle point of the chord in the given direction.

510. Determine the centres of the four equilateral hyperbolas which pass through two given points and touch two given lines.†

511. Given two points of an equilateral hyperbola and two tangents to the same, determine the four positions of the chord of contact.‡

---

\* Three other circles may be determined by Ex. 502 and another by Ex. 506, making in all EIGHT, which pass through the same point.

† If $A$ and $A'$ be the given points, $C$ the intersection of the tangents, and $X$ and $Y$ the points in which they meet $AA'$, the points $A$, $A'$ and $X$, $Y$ determine an involution through one of whose foci $P$ or $Q$ the chord of contact of the two tangents must pass: let it pass through $P$, of which $CQ$ will be the Polar. Bisect $AA'$ in $I$ and $XY$ in $C$, and draw a circle through $P$ and $I$ having its tangent at $I$ parallel to $CQ$ (Ex. 509). Through the second intersection of $CI$ with the circle draw $Pz$ meeting $CX$ in $x$, and draw the tangent to the circle and let it meet $CQ$ in $y$: then the intersections of $xy$ with the circle determine two of the required centres, and the other two are determined by interchanging $P$ and $Q$. This construction is given by PONCELET in Gergonne's *Annales*, tome XII., where he corrects (p. 244) the misstatement of the joint article by Brianchon and Poncelet (XI. 218) that *the four centres lie on* ONE *circle*.

‡ Determine as before the point $P$ on the chord of contact and its polar $CQ$; find a third point $D$ on the curve (Ex. 474); and let $F$ and $F'$ be the foci of the involution determined by $A$, $D$ and the pair of points in which the tangents meet $AD$. Then will $PF$ and $PF'$ be two positions of the chord of contact.

512. Given that the centres of all the equilateral hyperbolas circumscribing a triangle lie on a circle, deduce the fundamental property of the nine-point circle of any triangle.

513. The three circles whose diameters are the diagonals of any quadrilateral belong to a coaxal system,* whose limiting points are the centres of the two inscribed equilateral hyperbolas.

514. The director circles of all the central conics touching the same four lines have a common radical axis,† which is also the directrix of the inscribed parabola; and if the conics touch but three lines, their director circles have a common radical centre.

515. The circumscribed circle of any triangle which is self conjugate with respect to a conic cuts its director circle orthogonally:‡ if the conic be an equilateral hyperbola the circle passes through its centre: if it be a parabola its directrix passes through the centre of the circle.

516. The base of an isosceles triangle being upon a fixed straight line and each of its equal sides passing through a fixed point, whereof one is on the fixed line, shew that the locus of the vertex of the triangle is an equilateral hyperbola passing through the fixed points and having an asymptote parallel to the fixed line.

---

\* See TOWNSEND's *Chapters on the Modern Geometry of the Point, Line, and Circle*, Art. 189 (vol I. p. 253).

† This follows from Prop. VII by reciprocation, as in the *Oxford, Cambridge and Dublin Messenger of Mathematics*, vol. I. p. 159. A direct proof by involution is given in vol. III. p. 31 of the same, by "W. K. C." [CLIFFORD.]

‡ It may be shewn that the circumscribed circle of the triangle formed by the three diagonals of a quadrilateral is orthogonal to the circles on its three diagonals as diameters. Ex. 515 then follows with the help of Ex. 514 by regarding the sides of any self conjugate triangle as the diagonals of a quadrilateral which envelopes the conic. This theorem is due to GASKIN, who proved it by the Cartesian method in his work (p. 33) already referred to in the note on Ex. 458. Eight years later the equivalent theorem: "*On donne un triangle conjugué à une ellipse...la tangente menée du centre de l'ellipse au cercle circonscrit au triangle est égale à la corde du quadrant d'ellipse*," was proposed by Cap. Faure as Quest. 524 in the *Nouvelles Annales*, tome XIX. p. 234 (1860). See also XIX. 290, 315; XX. 25, 77; V. 308 (2me série).

517. If through five concyclic points taken in fours five equilateral hyperbolas be drawn, their centres will lie on a second circle of diameter equal to the radius of the former.

518. The locus of the centres of all the conics which pass through four given points is a conic. Prove that the locus will reduce to a *circle* if any two of the conics through the four points be equilateral hyperbolas, and to an *equilateral hyperbola* if the four points lie on a circle.

519. The angular points and the centroid and orthocentre of any triangle determine ten triangles whose nine-point circles meet in a point; and this point lies on the circumference of the maximum ellipse that can be inscribed in the original triangle.*

520. Shew that the centre of any equilateral hyperbola inscribed in an obtuse angled triangle lies upon the circle with respect to which the triangle is self conjugate.

521. The angular points of a triangle and the extremities of any diameter of its circumscribing circle, taken four together, determine five equilateral hyperbolas whose centres lie on the nine-point circle of the triangle.

522. A variable triangle circumscribes an equilateral hyperbola and its nine-point circle passes through the centre of the curve: prove that the locus of the centre of its circumscribed circle is the hyperbola in question.

523. Prove that the opposite sides $AB$ and $CD$ of a parallelogram inscribed in a rectangular hyperbola subtend either equal or supplementary angles at any point $P$ on the curve; the circumscribed circles of the triangles $PAB$, $PBC$, $PCD$, $PDA$ are equal; and the product of the perpendiculars from $P$ to each pair of opposite sides of the parallelogram is the same.

524. With the extremities of any diameter of the circumscribed circle of a triangle as foci two parabolas are drawn

---

* See *Mathematical Questions, &c. from the* EDUCATIONAL TIMES, vol. IV. p. 89.

touching the sides of the triangle; prove that the tangents at their vertices are the asymptotes of one of the series of rectangular hyperbolas which pass through the vertices of the triangle.

525. Given the base of a triangle, prove that if the bisectors of its vertical angle be parallel to fixed lines, or if its two sides make equal angles with two fixed straight lines, the locus of its vertex will be a rectangular hyperbola whose asymptotes bisect the base of the triangle and are parallel to the bisectors of its vertical angle.

526. Given two fixed tangents to a variable parabola and a fixed point on its axis, prove that the locus of its focus is a rectangular hyperbola having its asymptotes parallel to the bisectors of the angle between the fixed tangents and its centre at the bisection of the line joining their point of concourse to the fixed point on the axis.

527. If a rectangular hyperbola has double contact with a parabola, the line joining the intersection of their common tangents with the centre of the hyperbola is bisected by the directrix of the parabola.

528. The circle described with any diameter of an equilateral hyperbola as radius meets the curve again in three points which determine an equilateral triangle; and conversely, the circumscribed circle of any equilateral triangle inscribed in an equilateral hyperbola has one of its radii coincident with a diameter of the hyperbola. If $OA$ and $OB$ be the bounding radii of a circular arc $AB$, shew that a point of trisection of the arc lies upon the rectangular hyperbola which has $OA$ for a diameter and passes through the point of concourse of $OB$ with the tangent at $A$ to the circle. Deduce from the above that the problem, to trisect a given angle, admits of three solutions. Prove also that that if points $P$ and $Q$ be taken on $AB$ such that

$$\text{arc } AP = 2 \text{ arc } BQ,$$

the intersection of $AP$ and $OQ$ will lie on the hyperbola.

529. A pair of mean proportionals to two given magnitudes $m$ and $n$ may be found as follows. Describe a parabola of **latus rectum** equal to $m$, and with its axis and the tangent **at its vertex as** asymptotes describe a hyperbola whose semi-latus rectum is a mean proportional to $m$ and $2n$; then will the distances of **their point of concourse from** the asymptotes be the two **mean proportionals which were to be found.**\*

530. The circle described on any radius of a rectangular hyperbola as diameter meets the curve in two points whose distances from the asymptotes are in continued proportion; and conversely, the hyperbola drawn through the point of concourse of two sides of a rectangle so as to have the other two sides for asymptotes meets the circle circumscribing the rectangle in a second point whose distances from the asymptotes are a pair of mean proportionals to the sides of the rectangle. Hence shew how to find a pair of mean proportionals to two given magnitudes.†

531. The difference of the ordinates of the points in which any tangent to an equilateral hyperbola meets the directrices is to the difference of their distances from the centre as the diagonal to the side of a square; and their distances from the centre are to one another as the focal perpendiculars upon the tangent.‡

---

\* This construction also (cf. Art. 20, Cor.) is ascribed to Menaechmus.

† The *Delian* problem of the DUPLICATION OF THE CUBE (*i.e.* the construction of a cube of twice the volume of a given cube), which so exercised the ancient geometers, was reduced by Hippocrates of Chios to the problem of finding a pair of mean proportionals to two given magnitudes (Art. 20, Cor. and Exx. 432, 529, 530). See Reimer's *Historia Problematis de* CUBI DUPLICATIONE (Gottingæ, 1798); Walton's *Problems in illustration of the principles of Plane Coordinate Geometry*, p. 157; Bretschneider's *Die Geometrie und die Geometer vor* EUKLIDES, §78. The method of Ex. 530 is employed in Grégoire de St Vincent's *Opus Geometricum Quadraturæ Circuli* (Lib. VI. prop. 138, p. 602), and elsewhere. The TRISECTION OF THE ANGLE (Exx. 308, 390, 528) like the former problem is equivalent to the solution of a cubic equation, and either may be effected by the intersection of a circle with a parabola as was proved, in the third book of his *Geometria*, by DES CARTES; who further shews that solid problems in general can be reduced to the same two constructions, and gives his reasons for concluding *a priori* that " *Problemata Solida construi non possint absque Sectionibus Conicis, nec quæ magis composita sunt sine aliis lineis, magis compositis.*"

‡ Exx. 531—7 are from Booth's *New Geometrical Methods*, i. 291—2 and i. 343; Exx. 538—40 from Gregory St. Vincent's *Opus Geom. Quadrat. Circuli*, Lib. VI. props. 146, 156, 166 (pp. 606—16).

532. The auxiliary circle of an equilateral hyperbola is the envelope of the lines joining the points in which any two diameters at right angles meet the curve and its directrices respectively.

533. If tangents be drawn to an equilateral hyperbola from a point on one of its directrices and their chord of contact be produced to meet the directrix, the intercept upon it between the chord and the point will subtend a right angle at the centre; and if the tangents be drawn from any point not on the directrix, the focal distance and the polar of the point will intercept on the directrix a length which subtends a right angle at the centre.

534. The intercepts on either directrix of an equilateral hyperbola between any chord and the tangents at its extremities subtend equal angles at the centre.

535. The chords drawn from any two fixed points on an equilateral hyperbola to a variable point on the same intercept on either directrix a length which subtends a constant angle at the centre, the constant angle being a right angle in the case in which fixed points are the vertices; and the angles subtended at the centre by the intercepts on the two directrices are together equal to the angle subtended by the chord joining the fixed points.

536. If a right angle revolve about the centre of an equilateral hyperbola, the abscissa of any point on either arm varies inversely as the abscissa of the point in which its polar meets the other arm.

537. If a diameter of a parabola meet the curve in $P$ and the directrix in $M$ and a length $MPQ$ be taken on it equal to the normal at $P$, the locus of $Q$ will be a rectangular hyperbola having its centre at the vertex of the parabola. If $M'P'Q'$ be any second position of $MPQ$, shew that the hyperbolic area $QMM'Q'$ is equal to the product of the arc $PP'$ of the parabola and its semi-latus rectum.*

---

\* When the diameters are consecutive the distance between them is to the arc $PP'$ as the subnormal at $P$ to the normal, whence the required result readily follows. Thus the QUADRATURE of the Hyperbola is reduced to the RECTIFICATION of the Parabola.

**538.** A hyperbola having for asymptotes the axis of a parabola and the tangent at its vertex cuts the parabola in $O$, and $APQ$ is drawn from the vertex of the parabola to meet it in $Q$ and to cut the hyperbola in $P$; prove that if the ordinate of $Q$ cut the hyperbola in $R$, the segment $AOP$ is equal to one-third of the segment $APR$; and if from the latter segment $AO'P$ be cut off equal to one-third of its area, then will $AO'$ and the ordinate of $O'$ meet $QR$ and $AQ$ respectively on a parallel through $O$ to the axis of the parabola.

**539.** If from any two points $Q$ and $Q'$ on the above hyperbola parallels be drawn to its asymptotes meeting the curve in $M, M'$ and $N, N'$, the areas $OQM$, $OQN$, $OQ'M'$, $OQ'N'$ will be proportionals.

**540.** If through the point $Q$ a second parabola be drawn having the asymptotes for its axis and the tangent at its vertex, the arcs of the two parabolas will trisect the area $QMN$.

### NOTE ON THE NINE-POINT CIRCLE.

The property of the Nine-point Circle was stated and proved by Brianchon and Poncelet in Gergonne's *Annales*, XI. 215 (1821). See above, p. 175, note. The property may be verified as suggested in Ex. 512, viz. thus. Each of the six chords connecting a triad $ABC$ and its orthocentre $O$ (Art. 69) is a diameter of one of the series of equilateral hyperbolas which can be drawn through $A, B, C$: these six chords are therefore bisected by the locus of centres (a circle), which also contains the three intersections $D, E, F$ of the chords taken in opposite pairs (Art. 54 and p. 171, note). A short proof by inversion of the theorem (Salmon's *Conic Sections*, Art. 131, Ex.), that *the nine-point circle of a triangle touches its inscribed and exscribed circles*, was given by Mr. J. P. Taylor, Fellow of Clare College, in the *Quarterly Journal of Mathematics*, vol. XIII. p. 197. The same nine-point circle touches the SIXTEEN inscribed and exscribed circles of the four triangles determined by a triad and its orthocentre.

# CHAPTER VII.

## THE CONE.

**71.** An unlimited straight line which passes through **a fixed point in space and moves round the circumference of a fixed circle generates a surface which is called a** *Cone*.\* The line in **any** of its positions is called a *Side* **or a** *Generating Line* of the cone; **the fixed point is called the Vertex,** and the straight line **joining it to the centre of the circle is called the** *Axis* of the cone.†

When the axis is at right angles to the plane of the circle the surface generated is a *Right Circular* cone: in other **cases** the cone is called *Oblique* or *Scalene*. **In** this chapter we shall shew that the curve of intersection of **a cone with a plane is a parabola, an ellipse, or a hyperbola;** and **we shall derive their** elementary properties **from the** cone itself, confining our attention **in** general, **for the sake** of simplicity, **to the** right circular cone.

In **the particular case** in which the section **of a right circular cone is taken** at right angles to its axis, it is evident that the section **is a** circle. Any circular section may be regarded as the *Base* of the cone.

The *Focal Spheres* of any plane section **of a right circular** cone are the spheres which can be inscribed **in the cone so as** to touch the plane of section. Their points **of contact** may be defined as the *Foci*, and the intersections **of** their planes of **contact** with the **plane of** the section **as** the *Directrices* of the

---

\* The **complete cone consists of two** infinite portions **on** opposite sides of **the** vertex. The (right) cone **as** defined by EUCLID (Book XI. def. 18) **is the finite** figure (p. 193) **described by the** revolution of a right-angled triangle **about one of** the sides containing **the right** angle.

† The cone and its axis are thus defined by APOLLONIUS at **the beginning of his** Περὶ Κωνικῶν (p. 13, ed. Halley). In the oblique cone, which has two sets of **circular** sections, this definition gives two lines, either of which may be called the "**axis.**" In analytical **treatises on** Solid Geometry the **term axis** is *not* used as above.

section. We shall show that these points and lines are identical with the foci and directrices as hereinbefore defined.

In what follows suppose a plane through the axis and at right angles to the base of the cone to be taken as the *Plane of Reference* and the *Section* to be made by a plane at right angles thereto.

## THE ORDINATE.

### PROPOSITION I.

**72.** *The square* of the principal ordinate in any section varies as *the product of the corresponding abscissæ.*

(i) Let $A$ and $A'$ be the vertices of the section, $PP'$ a principal double ordinate meeting $AA'$ in $N$, **and let the plane of circular section through $PP'$ meet $OA'$ in $L$, and $OA$ in $M$,** the point $O$ being the vertex **of the cone.**

Then in the circle $PN^2$ **is equal to** $LN.MN$. And as $LM$ moves parallel to itself, $MN$ varies as $AN$ and $LN$ varies as

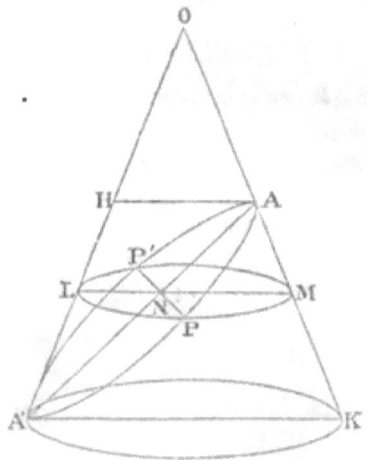

$A'N$. Therefore $PN^2$ varies as $AN.A'N$, or the square of the ordinate varies as the product **of** the abscisses.

When the section cuts all the generating lines on the same side of the vertex it **is an** *Ellipse,* and when it cuts both branches of the cone (fig. **p.** 199) **it is a** *Hyperbola.*

o

(ii) If the axis $AN$ of the section be *parallel* to the side $OL$ of the cone, then, in the figure of Art. 74, since the length $LN$ is constant whilst $NR$ varies as the abscissa $AN$, therefore $PN^2$ (or $LN.NR$) varies as $AN$, and the section is a *Parabola*.

Hence it appears that **whatever be the vertical angle of the cone the section is a parabola, a hyperbola or an ellipse** according as the angles $LOA$ and $OAN$ are together **equal to or greater or less than two right** angles.

### Corollary 1.

Since in the former part of the proposition
$$PN^2 : AN.A'N = LN.MN : AN.A'N = AH.A'K : AA'^2,$$
where $AH$ and $A'K$ are the **diameters of** circular sections, it follows that *the conjugate axis of the section is a mean proportional to the diameters of the circular sections through its vertices,* and the semi-axis conjugate is a mean proportional to their radii or to the perpendiculars from the vertices of the section upon the axis of the cone.

### Corollary 2.

Hence it readily follows that **the orthogonal projection of the section upon a plane of circular section is a conic having a focus upon the axis of the cone.**\*

#### SCHOLIUM A.

MENÆCHMUS (or Menechmus) is said to have been the **discoverer** of the conic sections, which have been accordingly called after him the *Menæchmian Triads*. Thus Proclus in the second book of his commentaries on the First Book of Euclid, **writing on Def. 4,** states upon the authority of Geminus: "But **with** respect to these sections, the conic were invented by Mœnechmus (*sic.*), which also Erastosthenes relating says, *Nor in a cone Mænechmian ternaries divide*" (Thos. Taylor's *Proclus*, I. 134); and see the end of the letter of Erastosthenes to Ptolemy, given **by** Eutokius in his commentary on Archimedes, De *Sphær. et Cyl.* (Archim. *Op.*, p. 146, ed. Torelli), where the same **verse,**

μηδὲ Μενεχμείους κωνοτομεῖν τριάδας,

---

\* This property was given by W. H. Talbot, of Cambridge, in Gergonne's *Annales*, XIV. 126.

appears in its original context. The authorities are given as above in Bretschneider's, *Die Geometer und die Geometrie vor Euklides*, § 116, p. 155.

The parabola, the ellipse, and the hyperbola were anciently regarded as the sections of right circular cones of different angles by planes *at right angles to their sides*, and were accordingly known as the sections of the right-angled, the acute-angled, and the obtuse-angled **cones respectively.** APOLLONIUS shewed that they could all be **cut from one and the same right or scaleno** cone, and he gave **them their** names *Parabola, Ellipse, Hyperbola*, for the reason **assigned above in** Chap. IV. Scholium A, p. 82. See Pappi Alex. *Collectio*, lib. VII. § 30 (p. 672, ed. Hultsch); and J. H. T. Müller's *Beiträge* zur *Terminologie der* **Griechischen** *Mathematiker*, p. 25 (Leipzig, 1860). ARCHIMEDES is sometimes wrongly supposed to have employed the term *Parabola*, for the reason that one of his treatises came to be known by the title, ʼΑρχιμήδους τετραγωνισμος Παραβολῆς, whereas throughout the treatise the author uses only the periphrasis, ἡ τοῦ ὀρθογωνίου κώνου τομά. In like manner he calls the ellipse ἡ τοῦ ὀξυγωνίου κώνου τομά, and the exceptional **occurrence** of the term *Ellipse* itself in **his work** *De Conoid. et Sphæroid.* (lib. I. cap. 9, &c.) is **rightly attributed to an** error of transcription.

Eutokius, at the commencement of **his** commentary on the *Conics* of APOLLONIUS (p. 9, ed. Halley) explains the names of the three conics as follows.\* Let $LOR$ be the angle of the cone, $AN$ the axis of the section supposed *at right angles to the side* $OR$, and $A$ **the** vertex of the section, which will be a Parabola, a Hyperbola, or an Ellipse, according as the angle of the cone is equal to, or greater or less than a right **angle.** The Parabola is accordingly said to be so called because $AN$ **is** *parallel* **to** $OL$: the Hyperbola because the **angles** $LOA$ **and** $OAN$ **together** *exceed* **two** right angles, **or because** $NA$ **falls beyond the vertex and meets the** side $LO$ produced: **and** the **Ellipse because the angles** $LOA$ and $OAN$ **are** together *less than* two right angles, **or because it is a defective** circle (κύκλον ἐλλειπῆ). If however **the** names **in** question **were** first introduced by APOLLONIUS, it is clear that they are **to** be **explained as** on p. 82. The property of the ordinate there given is **used by him to** discriminate between the three conics and forms the **actual basis of** his investigations, so that having once obtained **it he makes** in reality very little further use of the cone.

## THE ASYMPTOTES.

### PROPOSITION II.

73. *The sections of* **a cone** *by parallel planes are similar curves; and the asymptotes of the hyperbolic sections made by*

---

\* The passage is given in the Greek at the end of Walton's geometrical *Problems* (see above, Ex. 530, note).

parallel planes are parallel *to the sides of the cone which lie on the parallel plane through its vertex.*

If any fixed straight line through the vertex $O$ of the cone meet a pair of **parallel planes** in $M$ and $N$, and if a variable **plane through** $OMN$ meet the sections made by those planes in $P$ and $Q$, then

$$MP : NQ = OM : ON;$$

or the parallel vectors $MP$ and $NQ$ are in a constant ratio, and the sections are therefore similar.

If $M$ and $N$ be the centres of a pair of hyperbolic sections the vectors $MP$ and $NQ$ become infinite together: hence the asymptotes of any two **parallel** hyperbolic sections are parallel to one another, and therefore also **to the sides of the cone** determined by the parallel plane **through** the vertex, since **this** is a limiting position of one of the planes of section.

### Corollary.

The angle between the asymptotes of a hyperbolic section cannot exceed the vertical angle of the cone; and conversely in order to cut a hyperbola of given eccentricity from a cone we must take a cone whose vertical angle is not less than that between the asymptotes.

## THE FOCAL SPHERES.

### PROPOSITION III.

74. *The distance of any point of a section* **from the** *point of contact of its plane with either focal sphere* **is in a** *constant ratio to the distance of the point from the plane of contact of the sphere with the cone,* **or to its distance** *from the line in which that plane meets the plane of section.*\*

Let $S$ be the point in which the plane of the section touches

---

\* The reader who prefers to define a conic as the section of a cone by a plane may define its foci and directrices by means of the focal spheres (p. 192), as Pierce Morton (Schol. B) proposed to do. The proposition will then take the form that "The distance of any point on a conic from either focus is in a constant ratio to its distance from the corresponding directrix."

# THE CONE.

either of its focal spheres, and $MX$ the line in which it meets the plane of contact of the sphere with the cone.

Take any point $P$ on the section, and let $Q$ be the point in which the side of the cone through $P$ touches the sphere, and let $PM$ be supposed parallel to the axis of the section.

(i) Then the tangent $PS$ to the sphere is equal to the tangent $PQ$, and the perpendicular from $P$ to the plane of contact varies as $PQ$, and likewise as $PM$; and therefore $SP$ varies as that perpendicular, and likewise as $PM$.*

Hence the point of contact $S$ and the line $MX$ are a *Focus* and *Directrix* in accordance with their definition on p. 1.

(ii) This result is usually obtained, rather less directly, as follows.

Having made the same construction, let the side of the cone through the vertex $A$ of the section touch the sphere in $E$

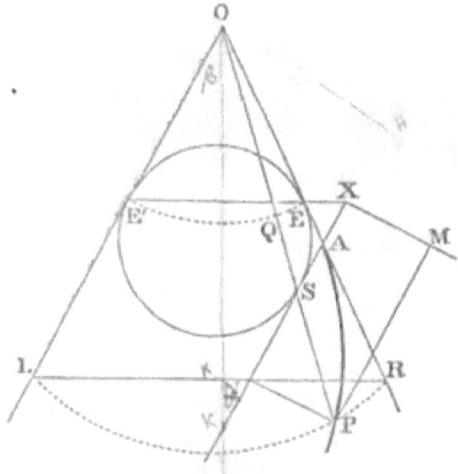

and meet the plane of circular section through $P$ in $R$; let $PN$ be the ordinate of $P$ to the axis $AN$ of the section, and let $X$ be supposed to lie in the plane of reference.

---

* If $a$ and $\beta$ be the inclinations of the axis of the cone to the axis of the section and to a side of the cone respectively, then $SP = \cos a . \sec \beta . PM$; or *the eccentricity* is equal $\cos a . \sec \beta$, and is therefore limited by the vertical angle of the cone and *cannot exceed* $\sec \beta$.

Then since *SP* is equal to *PQ*, and *PQ* is equal to *RE*, therefore *SP* is equal to *RE*.

Hence and by parallels,

$$SP : NX = RE : NX = AE : AX$$
$$= AS : AX,$$

or *SP* is to *PM* or *NX* in the constant ratio of *SA* to *AX*.

In the case of a bifocal conic the second focus and directrix are determined in like manner, as is indicated in the diagrams of Art. 75.

### PROPOSITION IV.

**75. *The sum* or *difference of the distances* of any point on a section from the points of contact *of its plane* with the focal spheres is constant, being equal to the intercept made by the planes of contact of the spheres upon any *side of the* cone.**

Let *S* and *H* be the foci, or points of contact of the focal spheres, and *Q* and *R* the points in which the spheres meet the generating line through any point *P* of the section.

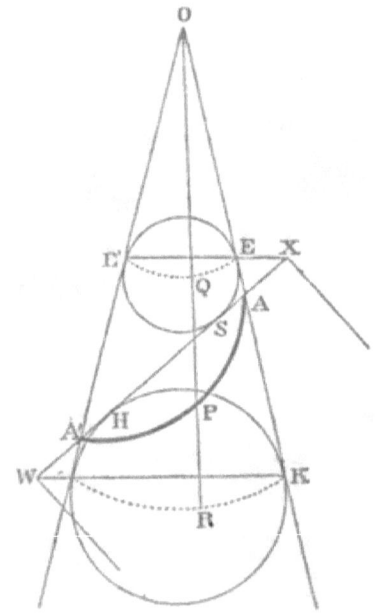

THE CONE.    199

(i) Then since the tangents from $P$ to either sphere are equal, therefore by addition in the case of the ellipse,

$$SP + HP = PQ + PR = QR,$$

which is the same for all positions of $P$ on the section.

(ii) And by subtraction in the case of the hyperbola,

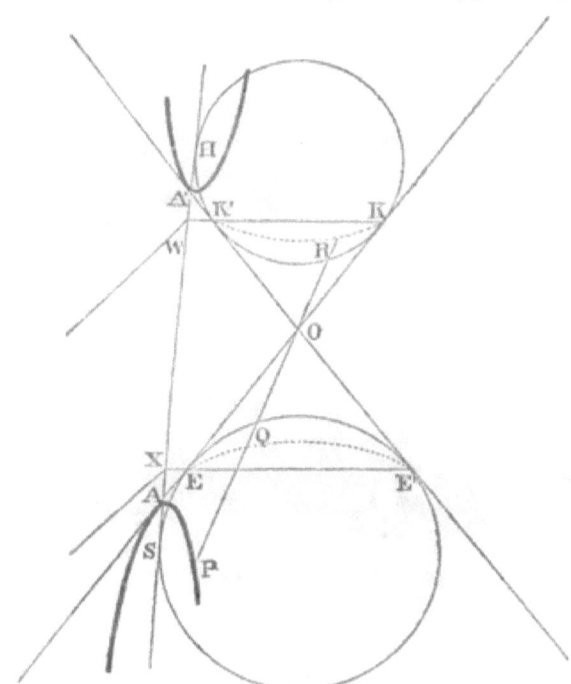

$$SP \sim HP = PQ \sim PR = QR,$$

which is constant, as in the former case.

*Corollary.*

In the first figure if $OA$ and $OA'$, drawn from the vertex of the cone to the vertices of the section, touch the $S$-sphere in $E$ and $E'$, then

$$OA' - A'S = OE' = OE = OA - AS,$$

or $OA' \sim OA$ is equal to $SH$. In the second figure it may be shown in like manner that $OA' + OA$ is equal to $SH$. Hence

200 THE CONE.

the eccentricity is the ratio of $OA' \pm OA$ to $AA'$, and the distance of the foci from the centre is $\tfrac{1}{2}(OA' \pm OA)$.

### PROPOSITION V.

**76.** *The tangent at any point of a section makes equal angles with the focal distances and with the side of the cone.*\*

Let $TPt$ be the tangent at any point $P$ to the section, and let the side $OP$ of the cone meet the focal spheres in $p$ and $p'$.

Then since the tangents $PS$ and $Pp$ to the $S$-sphere are equal, and likewise the tangents $TS$ and $Tp$, therefore the

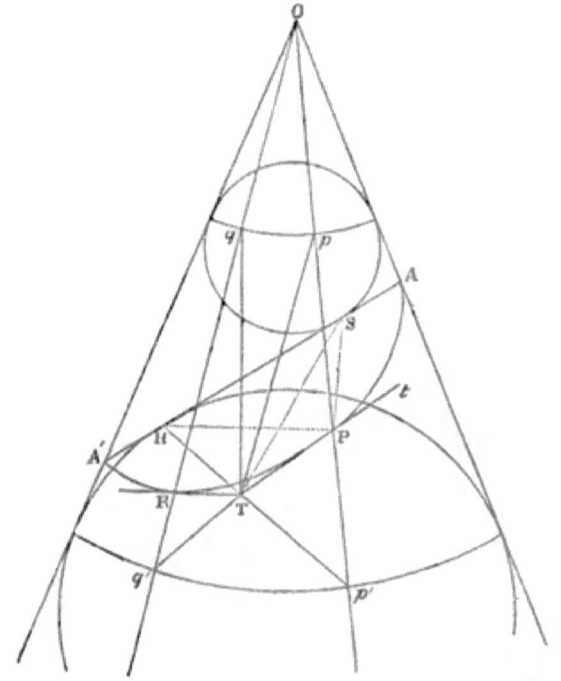

---

\* This property and its applications were pointed out by me in an article on *An Angle-property of the Right Circular Cone* contributed in June 1871 to the *Messenger of Mathematics* (vol. I. p. 67), and in subsequent articles. The same methods were employed in Booth's *Treatise on Conics* published some years later in the second volume of his *New Geometrical Methods*; but from the introduction to that volume we learn (p. x) that the substance of the treatise had been read before the Royal Irish Academy in 1837, although not published till forty years after.

triangles $TSP$ and $TpP$ are equal in all respects, having their angles at $T$ equal and their angles at $P$ and their supplements equal.

In like manner it may be shewn that the angles at $T$ and those at $P$ in the triangles $THP$ and $Tp'P$ are equal.

Hence $\angle SPt = pPt = p'PT = HPT$,

or the tangent $TPt$ makes equal angles with the focal distances $SP$ and $HP$ and with the side $OP$ of the cone.*

### PROPOSITION VI.

**77.** *If tangents be drawn* to a section *from any point in its plane,* **and** *a side of* the cone *be drawn* **through** *either point of contact, the intercept upon it* **between** *the focal spheres subtends at the point of concourse of the tangents an angle equal to the angle between* **them.**

It may be shewn as in Art. 76 that the angles $STP$ and $pTP$ are equal, where $T$ is any point on the tangent at $P$; and in like manner that the angles $HTP$ and $p'TP$ are equal.

Hence $\angle pTp' = STP + HTP = STH + 2STP$.

If $TR$ be the second tangent from $T$ to the section, and if the side of the cone through $R$ meet the spheres in $q$ and $q'$, it may be shewn in like manner that

$$\angle qTq' = STR + HTR = STH + 2STR.$$

And since the triangles $pTp'$ and $qTq'$ have their sides which touch the spheres equal and their bases $pp'$ and $qq'$ equal, their angles at $T$ are equal. Hence a fresh proof that the angle $STP$ is equal to $HTR$ (Art. 50); and it follows that

$$\angle PTR = pTp' = qTq',$$

as was to be proved.†

### *Corollary.*

If $PTR$ be a right angle, $pTp'$ is a right angle and $T$ lies on a certain sphere. The locus of $T$ is therefore the section

---

\* This may also be proved by the method of Art. 48 (1), since $OP \sim SP$ is constant.

† Observe that the triangle $Tpp'$ is identically equal to the triangle $STH'$ of Art. 50 (Cor. 2 and Scholium D).

of a sphere by a plane; that is **to say, it is the** circle which is called the Director Circle.

### PROPOSITION VII.

**78.** *The conjugate axis of any section **is a mean proportional** to the diameters of its focal spheres, and its **latus rectum** varies **as the** perpendicular to the plane of section from the vertex of the cone.*

**Let** $AA'$ be the axis of the section and $f$ and $F$ the centres **of its** focal spheres.

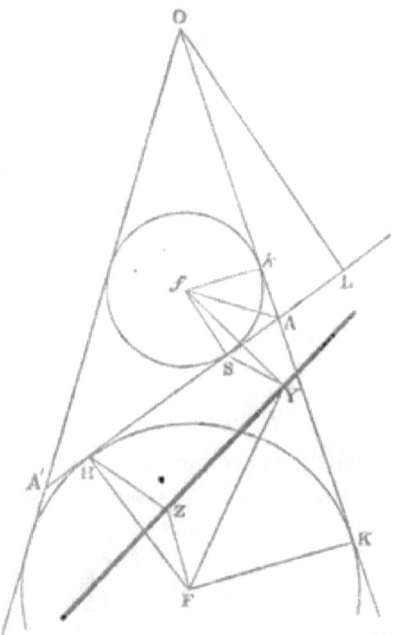

(i) Then since $fA$ and $FA$ bisect the supplementary angles between $AA'$ and **the side of the** cone through $A$, therefore by similar triangles $fSA$, $FHA$,

$$fS : AS = AH : FH,$$

and therefore $\quad fS \cdot FH = AS \cdot AH = CB^2,$

**or** $CB$ is a mean proportional to the radii of the spheres, and $2CB$ to their diameters.

# THE CONE.

(ii) **Draw** $OL$ perpendicular to $AA'$, and draw $fk$ and $FK$ to the points of contact of the spheres with the side $OA$ of the cone.

Then since $OK$ is **equal to** the semi-perimeter of the triangle $OAA'$,

$$fk \cdot OK = \triangle OAA' = \tfrac{1}{2} OL \cdot AA',$$

where $OK$ varies as the radius $FK$.

Therefore $OL \cdot AA'$ varies as $fk \cdot FK$ or $CB^2$; that is to say, $OL$ is in a constant ratio to the latus rectum.

### Corollary.

**If a** sphere be described about **the vertex** of the cone **as** centre, the latus rectum of **the section made** by any **plane** touching it will be constant, **and will be equal to the diameter** of the circular sections **whose planes touch the sphere.**

### PROPOSITION VIII.

79. *The sphere of which the line joining the **centres of the** focal spheres of any section is a diameter contains the auxiliary circle of the section.*

(i) **Since** $fF$ (in Art. 78) subtends right angles at $A$ and $A'$, the **sphere on** $fF$ **as** diameter cuts **the** plane of section in the **circle on** $AA'$ **as diameter, which is the** auxiliary circle of the **section.**

The annexed **duplicate proof further establishes the relation** between the auxiliary circle and **the tangent.**

(ii) **Through any tangent** $YZ$ **to the section draw a** plane **through** $f$ and likewise **a** plane through $F$. These bisect the supplementary angles between the plane of section and the tangent **plane** through $YZ$ to the cone, and are therefore *at right angles.*

If $SY$ and $HZ$ be the focal perpendiculars upon the **tangent,** $fY$ is at right angles to $YZ$ and to the plane $FYZ$.

Hence $fY$ is at right angles to $FY$, and the sphere **on** $fF$ as diameter passes through $Y$, **and** its trace on the plane of section is a circle, whereof $AA'$ **is evidently a diameter.**

*Corollary.*

The right angled triangles $fSY$ and $FHZ$ being similar,

$$SY \cdot HZ = fS \cdot FH = CB^2,$$

or the product of the focal perpendiculars upon the tangent is constant.

SCHOLIUM B.

The constructions for the Foci and Directrices of the sections of the cone are due to Hamilton, Dandelin, and others.

HUGH HAMILTON of Dublin, in Lib. II. prop. 37 of his work entitled *De Sectionibus Conicis Tractatus Geometricus in quo ex natura ipsius Coni Sectionum Affectiones facillime deducuntur methodo nova* (Londini, 1758), establishes the following properties. In the figures of Art. 75 (supposing the spheres to be omitted) if $S$ be a focus and $AE$ be taken equal to $AS$, then (1) the $S$-directrix is determined by the intersection of the plane of the conic with the plane of circular section $EQE'$; and (2) the vector $SP$ to any point $P$ of the conic is equal to the segment $PQ$ cut off by the same circular section from the side $OP$ of the cone; and (3) when the conic is bifocal two circular sections are thus determined which intercept on any side of the cone a length $QR$ equal to the transverse axis. Having thus established the equality of $AS$, $AE$ and of $A'S$, $A'E'$, as well as the equality of $OE$ and $OE'$, he had virtually proved that the focus $S$ might be determined as the point of contact of $AA'$ with the inscribed circle or one of the escribed circles of the triangle $OAA'$, or in other words as the point of contact of a Focal Sphere with the plane of section. He did not however state his conclusion in this form, but presupposed the determination of $S$ by the relation $AS \cdot A'S = CB^2$, and then proved $MX$ to be the directrix by shewing that $CS : CA = SA : AX$.

QUETELET contributed a *Mémoire sur une nouvelle Théorie des Sections Coniques considérées dans le Solide* (presented Dec. 23, 1820) to the *Nouveaux Mémoires de l'Académie Royale des Sciences et Belles-lettres de Bruxelles* (tome II. pp. 123—153, 1822), in which he shewed *inter alia* (1) that the foci of a section are determined by the relation $OA \pm OA' = SH$; and (2) that in an elliptic section $OP - SP$ is constant and equal to $OB - CA$. These results, so far as they go, are identical with Hamilton's; but Quetelet (unlike Hamilton) gives no construction for the directrices. In the course of the above *Mémoire* he refers to his tract on the *Curva Focalis*, or "Focale" (Gandavi, 1819).

DANDELIN, in a *Mémoire sur quelques propriétés remarquables de la Focale Parabolique* published in the same volume of the *Nouveaux Mémoires* (II. 171—202), begins by inscribing the FOCAL SPHERES and thus determining the foci of the sections. In *tome* III. (1826) he extends the same construction to the Hyperboloid of revolution, but in neither case does he make any mention of **the directrices.** See also *tome* IV. 77; Quetelet's *Correspondance mathématique et physique,* I. 82; Gergonne's *Annales,* XV. 392.

A complete determination of the Foci and Directrices of the sections of the cone by means of the Focal Spheres was at length proposed by PIERCE MORTON (B.A., 1825) before the *Cambridge Philosophical Society* in 1829 (*Transactions,* vol. III. pp. 185—190, 1830; and see pp. 228—9 of the anonymous *Geometry, Plane Solid and Spherical,* **in the Library of** Useful Knowledge, London, 1830). From his **introductory remarks it would seem** that he was not acquainted **with the investigations of Hamilton and** Dandelin.

## THE SEGMENTS OF CHORDS.

### PROPOSITION IX.

**80.** *A chord of a cone being divided at any point, to determine the rectangle contained by its segments.*

Upon the surface of a right **or scalene cone take** *any* **two points** $A$ **and** $A'$ (figures **of Art. 75), and in the line** $AA'$ or its complement take any point $X$: **it is required to determine** the magnitude of the rectangle $XA \cdot XA'$.

Take any fixed circular section $KRK'$, and from the vertex $O$ of the cone draw a parallel to $AA'$ to meet its plane in $Z$; and let the plane $AOZA'$ meet the fixed circular section in $K$ **and** $K'$ and the parallel section made by a plane through $X$ in $E$ **and** $E'$.

**Then by similar triangles** $AEX$, $ZKO$ **and** $A'E'X$, $ZK'O$,
$$XA \cdot XA' : XE \cdot XE' = ZO^2 : ZK \cdot ZK',$$
where (1) the **latter ratio depends only upon** *the direction of* $AA'$, since when this **is given the point** $Z$ is known and the product $ZK \cdot ZK'$ **is determined, and** (2) the magnitude of $XE \cdot XE'$ depends only upon *the position of* $X$ in space.

**By** taking $A$ and $A'$ on a given plane **of section we deduce the** results of Art. 16.

### SCHOLIUM C.

The method of Prop. IX. is frequently attributed to Hamilton, in whose treatise it holds a prominent place (*Sectiones Conicæ,* 1758, Lib. I. props. X. XI.). He supposed that he had thereby settled the old controversy about the cone and the plane in favour of the ancients, who derived the "sections" **from the** cone; not being aware that the property in **question could be** proved *in plano* with at least equal ease (Art. 16). Hamilton's proposition had appeared in the preceding century in the *Tractatus* XXIV. *De Sectionibus Conicis*, props. 48, 49 (pp. 419—20) appended to EUCLIDES ADAUCTUS &c. *auctore D. Guarino Guarino* (Augustæ Taurinorum, 1671); where in prop. 48 the case **of** two **parallel** chords cut by a single chord is considered, and the property of **two pairs of** parallel chords is deduced in prop. 49. See also *Synopsis* **Palmariorum** *Matheseos: or A New Introduction to the Mathematics*, by W. **Jones** (London, 1706), Part II. § 69 (3), p. 255 : "In any Conic Section, if **two** Parallels are cut by two others, and all terminate at the **Curve**, the Rectangles of the Segments shall be Proportional." **These** references are given by Abr. Robertson, *Sectionum Conicarum libri septem*, **p.** 348 (Oxon. 1792).

## EXAMPLES.

541. The latus rectum **of a** parabola **cut from a** given right cone varies as **the** distance between **the vertices** of the section and of the cone.

542. The foci of all similar sections of **a** given right cone lie upon two other right cones.

543. **Prove by means of** Art. 75, Cor. that the parallel sections **of a cone have the same eccentricity; and** give a construction **for cutting an ellipse of given eccentricity** from a given right cone.

544. The eccentricity of a section is a ratio of majority or minority according **as** the acute angle between the axes **of the cone and of** the section is less or greater than half the vertical angle **of** the cone.

545. If two or more plane sections **have the same directrix,** the corresponding foci lie **on a straight line** through the vertex of the cone.

546. Shew how **to cut from a given** right cone an ellipse whose axes shall **be of given lengths, or** whose latus rectum and area shall be of given magnitude.

547. **The area of** *the triangle through the axis*\*, in the right

---

\* APOLLONIUS supposed his sections to be made by planes at right angles to the plane drawn through the axis of the cone (defined as on p. 192) at right angles to its base. The *triangle through the axis* (viz. of the cone) was the triangle determined by the vertex of the cone and **the trace** of the conic upon the plane through the axis. See Chasles, *Aperçu Historique*, p. 18 (ed. 2, Paris, 1875).

cone, varies as the square of the minor **axis of the section,** and the volume cut off from the cone by the plane of the section varies as the cube of its minor axis, and is of constant magnitude so long as the minor axis is constant. Is the same true of **the** scalene cone?

548. If the minor axis of a **section be given** in length and direction, the **locus of the centre of the section is a** hyperbola.

549. **Given a** right cone and **a point** within it, construct **the two sections** which have the point for a focus; and shew **that their planes** make equal angles with the straight line **joining the** point to the vertex of the cone.

550. The vertical angle of a right cone being a right angle, the perpendicular from the vertex to any plane is equal to the semi-latus **rectum of the section made by that plane.**

551. Shew how **to cut a section of maximum eccentricity** from a given cone. Under what conditions **is it possible to** cut a rectangular hyperbola from a cone?

552. Shew how to place a given section (when possible) in a given cone.

553. If from the centre $C$ of a hyperbolic or elliptic section a line $CVV'$ **be drawn at** right angles to the axis to meet the **sides of the triangle** through the axis, **the square** of the semi-axis conjugate **is equal to** $CV.CV'$; **and the** semi-axis conjugate is equal to the distance of $C$ from the vertex of the cone in the case in which the transverse axis **of the section is parallel** to the axis of the cone.

554. Two right **cones** of supplementary vertical angles **being placed** with their axes at right angles **and their vertices coincident,** shew how to cut from them a pair of conjugate* hyperbolas. Show also that if the two cones be cut by a plane perpendicular to their common generating line, the directrices of one of the sections will pass the foci of the other.

---

\* That is to say, conjugate in form and dimensions, but *not* lying in the same **plane.**

555. The locus of the centres of the elliptic sections whose major axes are equal is a prolate or oblate spheroid.

556. If $V$ be the vertex of a right angled cone, and $PN$ the ordinate of $P$ in a parabolic section whose vertex is $A$, shew that the semi-latus rectum is equal to $VP - AN$.

557. An ellipse and a hyperbola lying in planes at right angles are so situated that the foci of each are at the vertices of the other. Shew that if $S$ be the vertex and $A$ the focus and $P$ any point upon a branch of the hyperbola, and if $Q$ be any point on the ellipse, then
$$AS + PQ = AP + SQ.$$

558. If two cones be described touching the same two spheres, the eccentricities of their sections by identical planes are in a constant ratio.

559. The vertex of a right circular cone which contains a given ellipse lies on a certain hyperbola, and its axis touches the hyperbola, and conversely.

560. Two parallel tangents to a section of a right cone meet in $M$ and $M'$ a plane which touches the focal spheres in points $Q$ and $Q'$ on a generating line: shew that a circle goes round $MQM'Q'$.

561. Determine the asymptotes of a given section of a scalene cone.

562. Assuming that one focus of the shadow of a sphere standing on a horizontal plane and exposed to the light of the sun is its point of contact with the plane; find the envelope of the corresponding directrix, and the locus of the remaining focus, for a given day and place.

563. By properties of the cone, or otherwise, find the locus of the extremity of the shadow of a vertical gnomon erected on a horizontal plane, on a given day and in a given latitude.

564. The centre of a sphere moves in a room a vertical plane which is equidistant from two candles of the same height from the floor: determine its locus if the shadows upon the ceiling be always in contact.

565. If a point move in a plane in such a way that the sum or difference of its distances from two fixed points, one of which lies in the plane and the other without it, is constant, its **locus will be a plane section of a right circular cone whose vertex is at the external given point.**

566. Prove in Art. 75 that $SE$ **and** $SE'$ **bisect** the diameters of the circular sec**tions through** $A'$ **and** $A$ respectively; and that $SO$, $AE'$, $A'E$ **cointersect, and hence that** $S$ **and** $E'E$ produced divide $AA'$ harmonically.

567. **Prove** from the cone that the **intercept** on any tangent **to a conic** between the curve and a directrix subtends a right **angle at the** corresponding focus.

568. Prove **also** that the **tangents** from **any point to a** section subtend **equal or supplementary angles at either focus.**

569. A tangent to a **right cone being drawn, there may** always be drawn through it **pairs of planes cutting the cone in** sections which have equal **parameters.**

570. The section of maximum parameter which can **be** drawn through a given point on a right cone has its plane at right angles to the generating line through the point, and **has its tangent at that** point parallel **to** the base of the cone.

571. **Through a given point on a right** cone there may be drawn any number of planes making sections which have equal parameters; and the envelope of these planes is another right cone, having its **vertex at the given point and its axis coincident** with the side of the original cone through **that point.**

572. **In a** given right cone, **the locus of the foci of all** equal **parabolas** is a circle whose plane is parallel to **the** base; **the locus of the foci** of all the parabolas whose planes are parallel is a straight line through the vertex of the cone; and the locus of the foci **of all** the parabolas that can be drawn in the cone is another right cone having the same axis.

573. The sphere inscribed **in** a right cone so as to pass through a vertex of a **section** intercepts upon the axis of the section a length equal to its latus rectum.

P

574. If sections of a right cone be made having one of their vertices at a fixed point on the cone, their circles of curvature at that vertex lie upon a sphere inscribed in the cone.

575. The sum or difference of the tangents from any point on a conic to the circles of curvature at its vertices is constant.

576. Prove from the right cone that a conic section may be regarded as the locus of a point such that the sum or difference of the tangents therefrom to two fixed circles is constant.

577. Prove from the right cone that a conic section is the locus of a point such that the tangent therefrom to a fixed circle is in a constant ratio to its perpendicular distance from a fixed straight line; and prove that in the case in which the straight line cuts the circle it is the chord of real *double contact* of the circle with the conic;* and prove that the above-mentioned constant ratio or *modulus* is equal to the eccentricity of the conic.

578. Two circles have double internal contact with an ellipse,† and a third circle passes through the four points of contact. If $t$, $t'$, $T$ be the tangents from any point on the ellipse to these three circles, prove that $T^2 = tt'$.

579. Notice the forms assumed by the several properties of the acute-angled cone when its vertical angle is diminished indefinitely, so that the surface becomes a cylinder.

580. An oblique cone or cylinder being described upon a circular base, show that its subcontrary sections are likewise circular.‡

---

\* Hence it appears that *a focus of a conic may be regarded as an evanescent circle having double contact with the conic* at the two imaginary points in which it is intersected by the corresponding directrix; and it may be inferred that the lines joining the focus to the two imaginary points at infinity through which all circles pass are tangents to the conic, and hence that *all conics which have the same two foci may be regarded as inscribed in the quadrilateral which has the two foci and the two circular points at infinity for its opposite vertices.* See SALMON'S *Conic Sections*, chap. XIV, on *Methods of Abridged Notation*.

† If $CM$ be the central abscissa of a point of contact, and $CN$ the abscissa of any point on the conic, the tangent from that point to the circle is equal to $\dfrac{CM}{e} - e \cdot CN$.

‡ Two sections are said to be *subcontrary* when the traces of their planes upon the plane of reference are inclined to the sides of the cone or cylinder at angles which are alternately equal.

581. Extend the theorems of Art. 72 to the oblique cone; and shew that if through any point on a side of the triangle through the axis there be drawn two planes of circular section and any other plane *between* them, the third plane will cut the cone in an ellipse, having its *minor* *axis* in the plane of reference.

582. If $AA'$ be the axis of a section of a scalene cone, and $AD$ and $Ad$ be diameters of its circular sections through $A$, shew that the square of the distance between the foci is equal to $A'D \cdot A'd$. Shew also in the right cone, with the construction of Art. 72, that the circle drawn with the middle point of $AA'$ as centre to bisect $AH$ and $A'K$ passes through the foci.

583. If a scalene cone be cut by any plane at a given distance from its vertex, the latus rectum of the section will be constant.*

584. The *transverse* *axis* of a section of an oblique cone being supposed to lie in the plane of reference, prove that the circle which touches the axis and passes through the centres of the two circular sections which can be drawn through either vertex determines the nearer focus.†

585. If the *conjugate axis* lie in the plane of reference, and if two circular sections be drawn through either of its

---

* This extension of Art. 78, Cor. may be proved as follows. Supposing the transverse axis to lie in the plane of reference, let $DD'$ be the diameter (in that plane) of the section which is parallel to the base and equidistant with the plane of the conic from the vertex $O$ (fig. Art. 72) of the cone. Draw $OYZ$ parallel to $AA'$ to meet $AH$ in $Y$ and $DD'$ in $Z$. Then it may be shewn that $AY = OZ$, and hence that $DD' : AH = A'K : AA'$. The latus rectum is therefore equal to $DD'$. This theorem is due to James Bernouilli. See the Leipzig *Acta Eruditorum*, ann. 1689, pp. 586—8; and Chasles' *Aperçu Historique*, p. 19.

† Exx. 584 and 585 are from Chasles' *Aperçu Historique*, Note IV, p. 285, where it is added that the *Eccentricity* is a mean proportional (Ex. 582) to the distances of the centre of the conic from the centres of the two circular sections through either of its vertices. It is to be noted that (before the directrix came into general use) the eccentricity was sometimes defined (1) as the *distance* of the foci from the centre, or (2) as the ratio of that distance to the semi-axis, or (3) as the ratio $\dfrac{b-a}{a}$, where $b$ denotes the semi-latus rectum and $a$ the perihelion distance of the orbit (Euler's *Theoria Motuum Planetarum et Cometarum*, prob. VIII. cor. 1, p. 36).

extremities, the circle upon the line joining their centres as diameter passes through the foci.

586. Every tangent plane to a cone cuts the cyclic planes in a pair of lines making equal angles with the line of contact; and every plane through two sides of the cone cuts the cyclic planes in two lines which make equal angles with those sides.

587. The sum or difference of the angles which any tangent plane makes with the cyclic planes is constant.

588. The sum or difference of the lines drawn from the vertex of a right cone to the extremities of any diameter of a section is double of the line from the vertex to an extremity of the conjugate axis.

589. If a sphere drawn through the vertex of an oblique cone cuts the cone in a circle, the plane of a section subcontrary thereto cuts the planes of any two great circles of the sphere in a pair of lines inclined at the same angles as their planes.

590. The lines of intersection of any tangent plane to a scalene cone with its two cyclic planes are such that the product of the tangents of the angles which they make with the intersection of the cyclic planes is constant.

591. The product of the sines of the angles which any side of a cone makes with the two cyclic planes is constant.

592. If a section of a cone be made by a plane which cuts the planes of two subcontrary circular sections and the sphere containing them in two right lines and a circle respectively, and if a chord be drawn to the circle from any point of the section; the product of the segments of the chord is to the product of the perpendiculars from the assumed point to the subcontrary planes in a constant ratio.*

---

* For this very general theorem and its corollaries see the article by Mr. JOHN WALKER on *Geometrical Propositions relating to Focal Properties of Surfaces and Curves of the Second Order* in the *Cambridge and Dublin Mathematical Journal*, vol. VII. pp. 16—28 (1852). The special case of Ex. 594 is also proved in the *Messenger of Mathematics*, vol. IX. pp. 33, 34. See also Mr. S. A. Renshaw's treatise on *The Cone and its Sections* (London, 1875).

**593.** The above-mentioned constant ratio **is the ratio of the product** of the sines of the angles which the cyclic planes **make** with the plane of the section to the product of the sines **of the** angles which they make **with any** side of the cone.

**594.** A sphere being drawn through two subcontrary circular sections of a cone, and the planes of those sections being produced to meet; **prove that** their **line of** intersection is a directrix of the **section made by either plane drawn** through it to touch **the sphere, and that the point of** contact **is the** corresponding **focus.**

**595.** From the above construction deduce that any tangent **to** a conic (from the curve to either directrix) subtends a right angle at the corresponding focus.

**596.** All the right cones which **have the** same **conic section for their base have their vertices upon another conic section,** lying in a plane **at right angles to that of the former, the** foci of each curve being at the vertices of **the other.**

**597.** If a hyperboloid of revolution and its asymptotic cone be cut by a plane, their two sections will be similar.*

**598. If** two spheres be inscribed in a conoid† so as to **touch a given plane** of section, the two points of contact will **be the foci of the section.**

**599. All sections of a conoid made by** planes through one its foci have **that point for one of** their foci, and they have the intersections **of those planes with** the directrix plane of the conoid for their corresponding **directrices.**

**600.** The cone whose **vertex is** at a focus and whose base **is any** plane section of a conoid is a right cone.

---

\* For proofs of the theorems of Exx. 597—600 see Hutton's *A Course of Mathematics*, composed for the use of the Royal Military Academy, vol. II. pp. 196—203 (12th edition, ed. Thomas Stephens Davies, 1843); and see Besant's *Conic Sections treated geometrically*, chap. XII.

† A *conoid* is the surface generated by the **revolution of a** conic about one **of** its axes.

# CHAPTER VIII.

## CURVATURE.

**81.** If $PQ$ be a small arc of a curve and $PT$ its tangent at $P$, the angle $QPT$ is called the *Angle of Contact* of the arc $PQ$.

If $PQ'$ be an arc of a second curve touching the former at $P$, and if $QQ'$ meet the common tangent in $T$; then will the curvature of $PQ$ at $P$ be equal to or greater or less than the curvature of $PQ'$ at $P$ according as the limiting ratio of the angle of contact $QPT$ to the angle of contact $Q'PT$ (when $PQ$ and $PQ'$ are diminished indefinitely) is a ratio of equality, majority, or minority.

The *Circle of Curvature* of a conic at any point $P$ is the circle which has the same curvature as the conic at $P$: it is therefore the limiting position of the circle which touches the curve at $P$ and meets it again at an adjacent point which ultimately coalesces with $P$: it is also the limiting position of the circle which meets the curve at $P$ and at two other points which ultimately coalesce with $P$. The centre, radius and diameter of this circle are called the *Centre of Curvature*, the *Radius of Curvature*, and the *Diameter of Curvature* of the conic at $P$, and its chord in any direction through $P$ is called the *Chord of Curvature* of the conic in that direction.

It is easily seen that a circle which cuts a conic must cut it in two or four points, and hence that the circle of curvature at any point $P$ of a conic will in general *cut* as well as touch it at $P$, and will also cut it in one other point. Any other circle touching the conic at $P$ must lie wholly within or without the former, and since it cannot cut the conic at $P$ it is easily seen that it cannot pass *between* the curve and its circle of curvature at that point. The circle of curvature is therefore the circle of closest contact with the conic at $P$, and is called its *Osculating Circle* at that point.

A circle which touches a conic at an extremity of either axis will in general meet it again at the two extremities of a chord parallel to the other axis; whence it readily follows that the circle of curvature at an extremity of an axis is to be regarded as meeting the conic in *four* coincident points. It may also be regarded as having double contact with the conic at two coincident points,* and it does *not* cut the curve at its point of contact. It is easy to determine the points in which any other circle touching the conic at the same point meets it again, and hence to shew that no such circle can pass between the conic and its circle of curvature at that point.

### PROPOSITION I.

82. *The focal chord of curvature at any point of a conic is equal to the focal chord of the conic parallel to the tangent at that point.*

Let $PSP'$ be any focal chord of a conic, $PT$ the tangent at $P$, and $RSR'$ the focal chord parallel to $PT$.

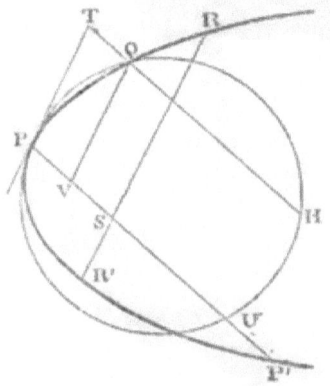

---

* By regarding the centre of curvature at the vertex as the ultimate position of the foot of the normal at $P$ when $P$ coalesces with $A$, we deduce from the property $SG = e \cdot SP$ (Art. 10) that the radius of curvature at $A$ is equal to $AS(1 + e)$, or to the semi-latus rectum. By like considerations it may be shewn that the radius of curvature at $B$ in the ellipse is equal to $\dfrac{CA^2}{CB}$.

Describe a circle touching the conic at $P$ and cutting it at an adjacent point $Q$, and let $TQ$ be taken parallel to $PP'$, and let it be produced to meet the circle again in $H$ and the conic in $Q'$. Then by Art. 16 and by a property of the circle,

$$TQ.TH : TQ.TQ' = TP^2 : TQ.TQ' = RR' : PP',$$

or $\qquad TH : TQ' = RR' : PP'.$

Let $PS$ meet the circle again in $U$; and let the point $Q$ coalesce with $P$, so that the circle becomes the circle of curvature at $P$.

Then $T$ likewise coalesces with $P$, and $TQ'$ with $PP'$, and $TH$ with $PU$; and therefore (from the above proportion) the chord of curvature $PU$ (being the limit of $TH$) is equal to the focal chord $RR'$ parallel to $PT$, as was to be proved.*

### Corollary 1.

It follows that in a central conic the chord of curvature at $P$ through *either* focus is equal to $\dfrac{2CD^2}{CA}$ (Art. 36, Cor.): in the parabola the chord of curvature through the focus, or parallel to the axis,† is equal to $4SP$: and in the general conic the focal chord of curvature is equal to $\dfrac{2PG^2}{L}$ (Art. 15, Cor. and Ex. 45).

### Corollary 2.

Given the chord of curvature at $P$ in any one direction, the chord in any other direction can be determined. For let $PU$ be the given chord and $PV$ any other, and let a parallel to the tangent at $P$ meet $PU$ in $K$ and $PV$ in $F$; then it is evident from the circle that

$$PV : PU = PK : PF,$$

---

\* This proof is due to Professor TOWNSEND (Salmon's *Conic Sections*, Art. 397): a variation of it will be given in Prop. II., where the circle of curvature is regarded as the limit of a circle which cuts the conic in three points, two of which ultimately coalesce with the third.

† Draw a circle touching the parabola in $P$ and cutting it in $R$ (fig. Art. 32), and let $MR$ meet the circle again in $K$. Then $MR.MK = MP^2 = 4SP.MR$ (Art. 28, Cor. 2), or $MK = 4SP$. Hence another proof of the result given above in the text. It is to be noted that any two chords of the circle of curvature equally inclined to the normal at $P$ are equal.

# CURVATURE.

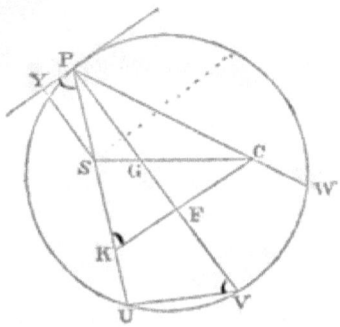

or the ratio of $PV$ to the given chord $PU$ is known. If $K$ be supposed to coincide with $S$, then

$$PV.PF = SP.PU = SP.RR',$$

which gives an expression for the chord of curvature $PV$ in any direction in terms of the focal chord $RR'$ (of the **conic**) parallel to the tangent at $P$. For example, in the parabola it follows at once that the *Diameter of Curvature* is equal to $\dfrac{4SP^2}{SY}$, where $SY$ is the **focal perpendicular** upon the tangent. In the central conics, supposing $KF$ to pass through $C$, we deduce that the *Diameter of Curvature* is equal to $\dfrac{PU.CA}{PF}$ ($PU$ being drawn through $S$), or $\dfrac{2CD^2}{PF}$, or $\dfrac{2CD^2}{CA.CB}$, or $\dfrac{2PG^3}{L^2}$ × (used by Newt

## PROPOSITION II.

**83.** *At any point of a conic* **the chord of curvature in** *any direction is to* **the chord of the conic in the same direction as the focal chords** *(of the conic) parallel to the tangent* **and to** *the chord of curvature.*

Let a circle meet a conic in three adjacent points $Q$, $P$, $Q'$,* and let $PU$ be a chord **of the** circle, and let it meet $QQ'$ in $V$ and the conic again in $P'$.

---

\* Complete the chord $QVQ'$ in the figure of Prop. I.

218    CURVATURE.

Then by a property of the circle and by Art. 16,

$$PV.VU : PV.VP' = QV.VQ' : PV.VP' = q : p,$$

where $p$ and $q$ are the focal chords of the conic parallel to $PP'$ and $QQ'$.

Therefore $\quad\quad VU : VP' = q : p.$

Hence, if $Q$ and $Q'$ (and therefore also $V$) be supposed to coalesce with $P$, so that the circle becomes the circle of curvature and $QQ'$ becomes the tangent at $P$, it follows that **the chord of curvature $PU$ is to the chord of the conic $PP'$ in the same direction as its focal chord parallel to the tangent at $P$ to its focal chord parallel to $PP'$, as was** to be proved.*

### PROPOSITION III.

**84.** *To determine the length of the central* **chord of curvature** *at any point of an ellipse or hyperbola, and likewise the length of the chord of curvature drawn in any other direction.*

(i) Let $PCP'$ be **any diameter of a central conic**, $QV$ a double **ordinate of that diameter adjacent to its** extremity $P$: we have **to evaluate the central chord of** curvature of the conic at $P$.

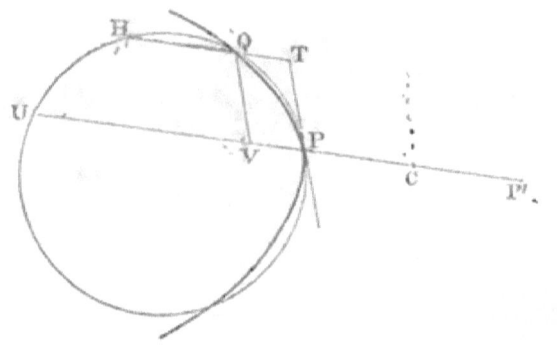

---

* If any focal chord meet the tangent at $P$ in $T$, it follows at once by Ex. 79 (cf. Ex. 447, note) that the chord of curvature at $P$ parallel to it is equal to $\dfrac{SP.RR'}{ST}$, as was virtually shewn in Art. 82, Cor. 2. It will be observed that Prop. II. alone completely **determines** the curvature of a conic at any point, but it seemed desirable to regard the subject from different points of view.

## CURVATURE. 219

Draw a circle touching the conic at $P$ and cutting it at $Q$, and let it meet $PP'$ again in $U$, and let its chord $HQ$ parallel to $PP'$ meet the tangent at $P$ in $T$.

Then since $\qquad QV^2 = TP^2 = TQ \cdot TH,$

therefore $\qquad TQ \cdot TH : PV \cdot VP' = CD^2 : CP^2,$

or $\qquad TH : VP' = CD^2 : CP^2,$

where $CD$ is the semi-diameter conjugate to $CP$.

By making $Q$ coalesce with $P$, so that the circle becomes the circle of curvature at $P$, we deduce that

$$PU : PP' = CD^2 : CP^2,$$

or $\qquad PU \cdot CP = 2CD^2,$

where $PU$ is the central chord of curvature at $P$.

(ii) More generally, if $QH$ and $PU$ be parallel chords of the circle drawn in any direction, and if $QV$ meet $PU$ in $R$ and $CD$ in $E$, then since $QR : QV$ is a ratio of equality when $Q$ coalesces with $P$,[*] it is easily seen, as in the former case, that ultimately

$$PR \cdot TH : PV \cdot VP' = CD^2 : CP^2;$$

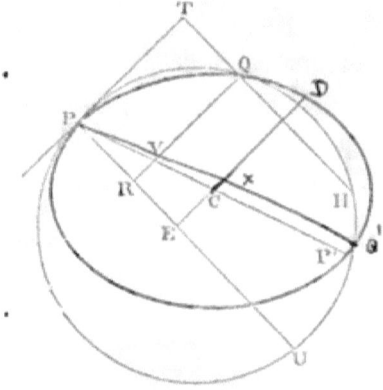

and it follows by parallels that

$$PE \cdot TH : CP \cdot VP' = CD^2 : CP^2,$$

---

[*] See NEWTON's *Principia*, **Lib.** I. Sect. I. lemma 7, cor. 2. Also, upon curvature in general, see lemma 11 with the notes in the edition by Mr. P. T. Main (after J. H. Evans) of Sections I–III, **IX, XI**; and see the *Appendix*, pp. 131–141, in that edition (Cambridge, 1871).

or ($TH$ being now coincident with the chord of curvature $PU$)

$$PU \cdot PE = 2CD^2.$$

The *Diameter of Curvature* at $P$ is equal to $\dfrac{2CD^2}{PF}$, where $PF$ (defined as in Art. 46) is equal to the central perpendicular upon the tangent.*

### PROPOSITION IV.

85. *To evaluate the common chord of a conic and its circle of curvature at any point.*

If three of the four points in which a circle meets a conic coalesce at $P$ and their fourth common point be $Q$, it follows from Art. 16, Cor. 2 that the chord $PQ$ and the tangent at $P$ are equally inclined to the axis; that is to say, *the common chord of a conic and its circle of curvature at any point and their common tangent at that point are equally inclined to the axis.*

(i) In the Parabola, let the common chord $PQ$ meet the axis in $R$, and let the common tangent at $P$ meet the axis in $T$. Then it is easily seen that the second tangent $TP'$ from $T$ is parallel to $PQ$, and that $PP'$ is a double ordinate to the axis, and that $PQ$ is equal to four times $PR$ or $TP$.†

(ii) In the Ellipse, let the diameter $CD$ parallel to the tangent at $P$ meet the common chord $PQ$ in $X$; then will $PX \cdot PQ$ be equal to $2CD^2$.

Take a circle of radius $CA$, and take any diameter of the same, and draw the chord $PQ$ making the same angle therewith as the tangent at $P$, and let the diameter parallel to that tangent meet the chord in $X$.

---

* The radius of curvature of the general conic may be evaluated by regarding the centre of curvature as the ultimate intersection of two consecutive normals, and assuming that $SG = e \cdot SP$ (Art. 10), and $PK = \tfrac{1}{2}$ latus rectum (Art. 11). The expression $\dfrac{PG^3}{L^2}$ (Art. 82, Cor. 2) for the radius of curvature has been obtained in this direct manner by Professor ADAMS (*Oxf. Camb. Dubl. Messenger of Mathematics*, vol. III. pp. 97—99).

† The chord of curvature in any other direction may be deduced. For example, it may be shewn by angle-properties that the circle meets $PS$ produced in a point $V$ lying on the diameter through $Q$ in the parabola, and hence that $PV$ is equal to $4SP$.

Then if $PC$ meet the circle again in $P'$,

$$PX \cdot PQ = PC \cdot PP' = 2CA^2.$$

It follows by Orthogonal Projection that **in the** Ellipse (the **same letters** being used) $PX \cdot PQ$ is equal to twice the square of the semi-diameter parallel to $PQ$: therefore, **the diameters** parallel to $PQ$ and the tangent at $P$ being equal,

$$PX \cdot PQ = 2CD^2.$$

It is left to the reader to obtain the same result by a method applicable to the hyperbola also.*

### SCHOLIUM.

APOLLONIUS in treating of maxima **and minima takes a point on** the axis of a conic at a distance equal **to the semi-latus rectum from** its vertex (*Conics*, Lib. v. props. 4—6, &c.), which is in fact the centre of curvature at the vertex, although he does not in any direct manner touch upon the subject of curvature. Cf. Vincentio Viviani's treatise, *De Maximis et Minimis geometrica* ***divinatio in*** *quintum Conicorum Apollonii Pergæi adhuc desideratum* (Florent. 1659); and see Abr. Robertson's *Sectionum Conicarum libri septem*, p. 372 (Oxon. 1792).

HUYGHENS came very near to the subject of curvature when he propounded his theory "De linearum curvarum evolutione et dimensione" (***Horologium*** *Oscillatorium*, Pars III. Paris. 1673), since the evolute of a curve is the envelope of its normals and the **locus of its centres of cu**rvature. The first case of rectification of a curve was that of the cubical **parabola by** William NEIL:† the **cycloid was rectified soon** after **by Christopher** WREN. Huyghens had previously (**1657**) **reduced the rectification of** the parabola [Ex. 535, note] to the **quadrature of the hyperbola** (*Horolog. Oscill.* pp. 72, 77).

---

* The proof in the text is given to indicate **the applicability of** Orthogonal Projection (chap. IX.) to the treatment of curvature. If any **two lines (as** $PQ$ and the tangent at $P$) equally inclined to the axis of the ellipse **be projected** on to its auxiliary **circle, the projected** lines will be equally inclined **to the** axis; or if the term *equally* inclined **be** restricted **to** parallels, we must say that **if** two lines make equal or supplementary **angles** with the axis, their projections will make **equal** or supplementary angles with the **axis**.

† This was suggested by the *Arithmetica Infinitorum* of WALLIS (Oxon. **1656**): Van-Heuraet seems to have rectified the same curve independently (1659) **very soon** after Neil. See Montucla's ***Histoire des*** *Mathématiques*, Part IV. Liv. VI. § II, tome II. p. 303. Huyghens awards the palm **to the** later discoverer, considering that Neil's investigation was incomplete (*Horolog.* p. **72**).

The *Osculating Circle* received its name from LEIBNITZ: see *Meditatio nova de natura Anguli Contactus*\* *& Osculi* by G. G. L. in the *Acta Eruditorum* for 1686 (pp. 289—292, by misprint 489—292). See also pp. 292—300, and the *Acta* for 1692, pp. 30—35, 110—116, 440—6. The subject of curvature was considered by NEWTON in the seventh and eighth chapters of his *Geometria Analytica*.

## *EXAMPLES.*

601. Prove the following construction for the centre of curvature at any point $P$ of a conic. From the point in which the normal meets the axis draw a perpendicular to the normal meeting $SP$ in $Q$, and from $Q$ draw a perpendicular to $SP$ meeting the normal in $O$. Then $O$ is the centre of curvature at $P$.

602. The circle which touches a conic at $P$ and intercepts upon the diameter through $P$ a length equal to its parameter† is the circle of curvature at $P$.

603. The circle of curvature at any point of a conic being the circle through that point and two others which ultimately coalesce with it, shew that the centre of curvature may be regarded as the point of ultimate intersection of two consecutive normals to the conic.

604. The circle of curvature at any point of a conic may be regarded as touching three consecutive tangents to the conic ‡

605. Determine the position of the common chord of a parabola and its circle of curvature at an extremity of the latus rectum.

606. The diameter at either extremity of the latus rectum of a parabola passes through the centre of curvature at its other extremity.

---

\* For earlier controversies on the nature of the angle of contact see WALLIS, *De Angulo Contactus et Semicirculi disquisitio geometrica* (Oxon. 1656).

† The parameter of any diameter of a central conic is defined as a third proportional to that diameter and its conjugate.

‡ For proofs of Exx. 603–4, see the section on curvature in Main's NEWTON, *Appendix* (see above p. 219, note).

607. The radius of curvature at any point of a parabola is double of the portion of the normal intercepted between the curve and the directrix.

608. At any point of a parabola the intercept made by the circle of curvature upon the axis is a third proportional to the latus rectum and the parameter of the diameter to the point.

609. At any point $P$ of a parabola, if $PY$ be the projection of $SP$ upon the tangent, the chord of curvature through the vertex is a third proportional to $AP$ and $2PY$.

610. If $R$ be the middle point of the radius of curvature at $P$ in a parabola, $PR$ subtends a right angle at $S$.

611. If the normal to a parabola at $P$ be produced to any point $O$, express the radius of curvature at $P$ in terms of the line $OP$ and its inclinations to the tangents from $O$.

612. The radius of curvature at an extremity of the latus rectum of a parabola is equal to the projection upon the directrix of the focal chord parallel to the tangent at that point.

613. Given a circle and a straight line, it is required to find a parabola (see Ex. 607) having the line for its directrix and the circle for a circle of curvature.

614. The envelope of the common chords of a parabola and its several circles of curvature is a parabola, and the locus of their middle points is a parabola.

615. The tangent from any point of a parabola to the circle of curvature at its vertex is equal to the abscissa of the point.

616. If tangents $TQ$ and $TQ'$ be drawn to a parabola from any point $T$ on the fixed diameter which meets the curve in $P$ and the directrix in $M$, the centre of the circle round $TQQ'$ lies at a constant distance $2PM$ from the directrix. If $O$ be the centre of the circle, $D$ the projection of $O$ upon the fixed diameter $V$ the middle point of $QQ'$, and $X$ the foot of the directrix, shew that
$$OD : TM = OD : DV = MX : SX.$$
Shew also that the radius of the circle varies as $ST$.

**617.** If from the vertex of a parabola chords $AR$ and $AR'$ be drawn equally inclined to the axis, the normals at the extremities of any chord parallel to $AR$ intersect upon the normal at $R'$;* and the centre of the curvature at the extremity of the diameter which bisects $AR$ lies upon the normal at $R'$.

**618.** Triads of points can be found on a parabola such that the normals thereat cointersect. The circle through any three such points passes through the vertex of the parabola, and the centroid of the triangle which they determine lies on the axis.

**619.** In Ex. 616 shew that the ordinate of the point of concourse of the normals at $Q$ and $Q'$ is to the ordinate of $V$ as the product of the ordinates of $Q$ and $Q'$ to the square of the semi-latus rectum. Hence determine the ordinate of the centre of curvature at $P$, and likewise the length of the radius of curvature.

**620.** Give a geometrical method of drawing normals to a parabola from a point on the curve.

**621.** Determine a point on a given conic at which the circle of curvature is of given magnitude; and in the case of the ellipse determine the limits of its magnitude.

---

* Let $QQ'$ in Ex. 616 be supposed parallel to $AR$. Produce $TO$ to meet the circle again in $N$ (the point of concourse of the normals at $Q$, $Q'$), draw the perpendicular $NH$ to the fixed diameter, and take $HK$ equal to $TM$ upon $MH$ produced. Then it is easily seen that $K$ is a fixed point (the position assumed by $N$ when $QQ'$ passes through $S$), and $NH : HK = 2MX : SX$. Therefore $N$ lies on a fixed straight line through $K$, which may be proved (from the property of the subnormal) to be the normal at $R'$. For the above I am indebted to the Rev. A. F. Torry, Fellow of St. John's College. The following method may also be suggested: since the angle $OST$ is a right angle (Ex. 146), we have an intercept $TO$ between two fixed straight lines $DO$ and $DT$ at right angles subtending a right angle at a fixed point $S$, and we have to shew that the extremity $N$ of $TO$ produced to double its length lies upon a fixed straight line. Or again, taking three positions of $P$ in Art. 26—$P$ and $P'$ on one side and $P''$ on the other side of the axis, and supposing the normals at the three points to cointersect at a point whose projection upon the axis is $Z$, we easily prove that $PN.ZG = P'N'.ZG' = P''N''.ZG''$; whence it follows, after some reductions, that $PN + P'N' = P''N''$. On the tetrads of concurrent normals to a central conic see Ex. 286.

# EXAMPLES.

622. If the osculating circle at a vertex of an ellipse passes through the further focus, determine the eccentricity.

623. The circle of curvature at a point on an ellipse passes through a vertex: find the point.

624. The circle of curvature at an extremity of one the equal conjugate diameters of an ellipse passes through its other extremity. Explain the corresponding result in the hyperbola.

625. The circles of curvature at the extremities of two conjugate radii $CP$ and $CD$ of an ellipse meet the curve again in $Q$ and $R$: shew that $PR$ is parallel to $DQ$.

626. Find the points on a central conic at which the diameter of curvature is a mean proportional to the axes.

627. Express the chord of curvature perpendicular to the axis at any point of an ellipse in terms of the ordinate of the point.

628. From the point in which the tangent to an ellipse at $P$ meets the axis a straight line is drawn bisecting one of the focal distances and meeting the other in $Q$. Prove that $PQ$ is one fourth of the focal chord of curvature at $P$.

629. If the circle of curvature at $P$ in an ellipse passes through a focus, then $P$ lies midway between the minor axis and the further directrix, and the parallel to the tangent through the focus divides the diameter at $P$ in the ratio of three to one. The circle of curvature cannot pass through either of the foci if the semi-axis exceed the distance between them.

630. Shew that a conic can be described with a given focus so as to have a given circle of curvature at a given point.*

631. The foci of all the ellipses which have a common maximum circle of curvature at a given point lie on a circle.

632. The tangent at $P$ in an ellipse meets the axes in $H$ and $K$, and $CP$ is produced to meet the circle round $CHK$ in $L$: prove that $2PL$ is equal to the central chord of curvature at $P$, and that $CL \cdot CP$ is constant.

---

* This is a limiting case of the problem to describe a conic of which a focus and three points are given.

633. A hyperbola which touches an ellipse and has a pair of its conjugate diameters for asymptotes has the same curvature* as the ellipse at their points of contact.

634. At any point $P$ of a rectangular hyperbola, if $PN$ be a perpendicular to an asymptote, the chord of curvature in the direction $PN$ is equal to $\dfrac{CP^2}{2PN}$.

635. At any point of a rectangular hyperbola the radius of curvature varies as the cube of the radius of the curve.

636. At any point of a rectangular hyperbola the normal chord is equal to the diameter of curvature.

637. At any point of a rectangular hyperbola the diameter of the curve is equal to the central chord of curvature.

638. If from a given point on an ellipse there be drawn a double ordinate to either axis, and if to the diameter through its further extremity a double ordinate be applied from the given point, it will be a chord of the circle of curvature at the given point.

639. The normal chord which divides an ellipse most unequally is a diameter of curvature,† and is inclined at half a right angle to the axis.

640. A chord of constant inclination to the arc of a closed curve divides its area most unequally when it is a chord of curvature.

641. Every chord of a conic which touches the circle of curvature at its vertex is divided harmonically by that circle and the tangent at the vertex.‡

642. The radius of curvature at any point of a parabola is bisected by the circle which touches the parabola at that point and passes through the focus.

---

* Curvature is measured by the reciprocal of the radius of curvature.

† The normal in two consecutive positions must cut off equal areas, and must be bisected at the centre of curvature.

‡ Exx. 641–3 are from the *Mathematician*, vol. I. 290 (London, 1845).

643. The radius of curvature at any point of a central conic is cut harmonically by the two circles which touch the conic at that point and pass one through each focus.

644. If $T$ be the point of concourse of the common tangents of an ellipse and its circle of curvature at $P$, and if $O$ be the centre of the circle, $C$ the centre and $S$ either focus of the ellipse; then $T$ lies on the confocal hyperbola through $P$, and $OC$ bisects $PT$, and $SP$ and $ST$ are equally inclined to $OS$.

645. A circle through the vertex of a parabola cuts the curve in general in three other points, the normals at which cointersect. If the point of cointersection of the three normals describe a coaxal conic having its centre at the vertex of the parabola, the locus of the centre of the circle will be a conic having its centre at the focus of the parabola, and the ratio of its axes will be twice as great as the ratio of the axes of the former conic.

646. Find a point $P$ on an ellipse at which the common tangents of the ellipse and its circle of curvature are parallel. Shew that their common tangents and common chord are parallel to the asymptotes of the confocal hyperbola through $P$; and that their finite common tangent is bisected by the common chord.

647. The common chord of an ellipse and its circle of curvature at any point and their common tangent thereat divide their further common tangent harmonically.

648. On the normal at $P$ to an ellipse $PO$ is measured outwards equal to the radius of curvature: shew that $PO$ is divided harmonically at the points $Q$ and $Q'$ in which it meets the director circle, and that

$$\frac{1}{PQ} \sim \frac{1}{PQ'} = \frac{2}{OP}.$$

649. The angle between the normals at adjacent points $P$ and $Q$ of an ellipse whose foci are $S$ and $S'$ is equal to $\frac{1}{2}(PSQ + PS'Q)$. Deduce that, if $PR$ be the focal chord of curvature at $P$,

$$\frac{4}{PR} = \frac{1}{SP} + \frac{1}{S'P}.$$

650. Three points can be found on an ellipse whose osculating circles meet at a given point on the curve:* the three points determine a maximum inscribed triangle, and the four points lie on a circle.

---

* Let $BCD$ be a maximum inscribed triangle, and $A$ the point in which the osculating circle at $B$ meets the ellipse again. If $B'$ be a consecutive position of $B$, the triangles $BCD$ and $B'CD$ must lie between the same parallels. Hence the tangent at $B$ is parallel to $CD$, which is therefore equally inclined to the axis with $AB$ (Art. 85), so that a circle goes through $ABCD$ (Art. 16, Cor. 2). Ex. 650 (except as regards the area of the triangle) is due to STEINER, who stated the theorem without proof in Crelle's *Journal*, vol. XXXII. 300 (1846): it was proved in vol. XXXVI. 95, by JOACHIMSTAL, who further shewed that the centroid of the three points is at the centre of the ellipse, and that the normals at the three points cointersect.

# CHAPTER IX.

## ORTHOGONAL PROJECTION.

**86.** THE foot of the perpendicular from any point in space upon a plane is called the *Orthogonal Projection*, or briefly the *Projection*, of the point upon the plane. The projection of any line or figure upon a plane is the aggregate of the projections of its several points thereupon. The term *Plane Projection* is defined in Art. 90.

The *Parallel Projection* of any figure upon a plane is the aggregate of the points in which a system of parallels from all points of the figure meet the plane: when the parallels are taken at right angles to the plane of projection we come back to the special case of orthogonal projection, to which our attention will be in the first instance confined.

### PROPOSITION I.

**87.** *The projections of parallel straight lines upon any plane are parallel straight lines, and each line or segment is in the same ratio to its projection.*

(i) The projection of a straight line upon a plane is the common section of that plane with the plane drawn at right angles to it through the line, since their common section evidently contains the projections of all points on the line. The projections of parallel straight lines, being the common sections of the plane of projection with a system of planes at right angles thereto, are themselves parallel straight lines.

(ii) If any two parallel straight lines $AP$ and $BQ$ meet a plane in $A$ and $B$ (fig. Art. 89), and if $AC$ and $BD$ be their projections upon it, then by similar triangles

$$AP : AC = BQ : BD,$$

or each of the parallels is in the same ratio to its projection, and in like manner it may be shewn that any segment of either is in the same ratio to its projection. This ratio is the trigonometrical secant of the angle between any of the parallels and its projection.

Hence it is evident that parallel straight lines (and the segments of one and the same straight line) are to one another as their projections upon any plane.

If two or more systems of parallels be projected orthogonally, each line in any system will be in a constant ratio to its projection, but this ratio will change as we pass from one system to another.

### PROPOSITION II.

88. *The points of concourse of straight lines or curves correspond to the points of concourse of their projections: tangents project into tangents: and a curve of any degree or class projects into a curve of the same degree or class.*

(i) Since figures are projected by projecting their several points, if two or more figures have a point in common their projections must have the projection of the point in common. Thus if a variable line pass through a *fixed* point, the projection of the line will pass through a fixed point.

(ii) If $P$ and $Q$ be adjacent points on a curve, and $p$ and $q$ be the feet of the perpendiculars from them to the plane of projection, then if $Q$ be made to coincide with $P$ the perpendicular $Qq$ must coincide with $Pp$, and the point $q$ with $p$; in other words, of $PQ$ become the tangent at $P$, then in the projection $pq$ will become the tangent at $p$.

(iii) It is hence evident that a curve and any straight line in its plane intersect in the same number of points as their projections, and hence that a curve and its projection must always be of the same degree or order; and further, that the number of tangents which can be drawn from any point to a curve is the same as the number which can be drawn from the projection of the point to the projection of the curve, and hence that a curve and its projection are always of the same class.

## PROPOSITION III.

89. *The areas of all figures in one plane are in the same ratio to the areas of their projections on a given plane.*

Let $AB$ be the common section of the primitive plane and the plane of projection, and let $PA$ and $QB$ be drawn from

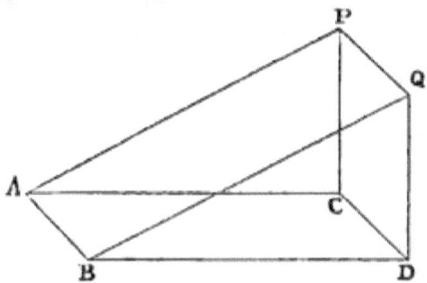

any two points $P$ and $Q$ in the former plane at right angles to $AB$, and let $PC$ and $QD$ be perpendiculars to the plane of projection. Then since $AP$ and $BQ$ are in a constant ratio to their projections $AC$ and $BQ$, the rectilinear figure $APQB$ is in the same constant ratio to its projection $ACDB$.

In like manner, if $PP'$ be any segment of $AP$ and $QQ'$ any segment of $BQ$, the rectilinear area $PP'Q'Q$ is in a constant ratio to its projection; and hence the aggregate of any number of figures as $PP'Q'Q$ is in the same ratio to the aggregate of their projections.

But any rectilinear area in the primitive plane may be divided into elements as $PP'Q'Q$ by drawing planes at right angles to $AB$, and any curvilinear area may be regarded as the limit of a rectilinear area. Hence every area is to its projection in a constant ratio, which is easily seen to be equal to the secant of the angle between its plane and the plane of its projection.

90. *Plane Projection.*

The above propositions all follow from the constancy of the ratio of $AP$ to its projection, where $PA$ is the perpendicular from any point $P$ in the primitive plane to its common section with the plane of projection. The relations between the locus of $P$ and its projection will evidently still be of the same kind if the former plane be turned about $AB$ into coincidence with the latter, the situation of $P$ in its own plane being supposed to remain undisturbed. The new position of $P$ will lie in $AC$ produced, and will be such that its projection $C$ now divides $AP$ in a constant ratio. Since all points and their projections are thus brought into one plane, we shall speak of the point $C$ which divides $AP$ in a constant ratio as the orthogonal *Plane Projection* of $P$.

The relations between a figure and its plane projection may of course be obtained directly by plane geometry. Thus if $PN$ be an ordinate to a fixed axis, and if it be divided (externally

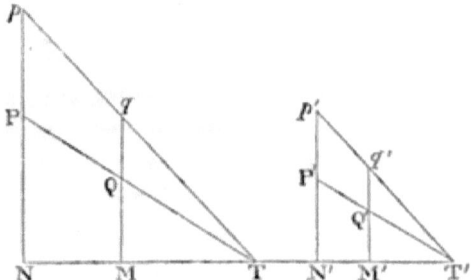

or internally) in a constant ratio at $p$, it is easily proved by similar triangles that if the locus of $P$ be a straight line meeting the axis in $T$, the locus of $p$ will be a straight line meeting the axis in the same point: parallels as $PT$, $P'T'$ correspond to parallels $pT$, $p'T'$: and the results of Props. I.—III. in general may be similarly established *in plano*.

It may be proved also that if **two straight lines make** *supplementary* **angles with the axis their projections make** supplementary angles with the axis.

The following examples will serve to illustrate the method of orthogonal projection.

### 91. *The Ellipse.*

A cylinder being regarded as a cone whose generating lines are parallel, it follows from Art. 72 that the plane sections of a right circular cylinder are ellipses, which may be of any eccentricity. It follows that *an ellipse of any eccentricity may be projected orthogonally into a circle* having its plane at right angles to the axis of the cylinder. This circle will evidently be equal to the minor auxiliary circle of the ellipse.

An ellipse may also be projected orthogonally *in plano* into a circle by producing its principal **ordinates in the ratio of** $CA$ to $CB$, or by diminishing **the** ordinates to its minor axis in the ratio of $CB$ to $CA$; and conversely a circle may be projected *in plano* into an ellipse by increasing or diminishing every ordinate $pN$ in a constant ratio. [Art. 33.

*a.* Let an ellipse and a circle be projectively related. Let $PQ$ be any one of a system of parallel chords in the ellipse: $pq$ the corresponding chord of the circle, which will itself belong to a system of parallels.

The middle point $O$ of $PQ$ corresponds to the middle point $o$ of $pq$; and if the locus of the one be a straight line, the locus of the other will be a straight line. But the locus of $o$ in the circle is a straight line, and therefore in the ellipse the locus of the middle points of any system of parallel chords is a straight line. From this it is further evident that conjugate diameters in the ellipse correspond to diameters *at right angles* in the circle.

*b.* Next let $O$ be a fixed point and $PQ$ any chord of the ellipse passing through it: then $pq$ in the circle will pass through the corresponding fixed point $o$.

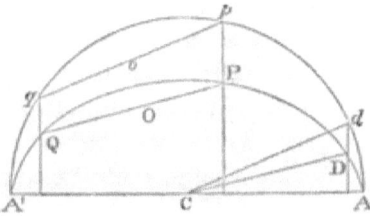

Let $CD$ be the radius of the ellipse parallel to $PQ$, and $Cd$ the corresponding radius of the circle, which will be parallel to $pq$. Then by Prop. I.

$$OP : OQ : CD = op : oq : Cd.$$

Therefore $OP.OQ : CD^2 = op.oq : Cd^2,$

or ($op.oq$ and $Cd$ in the circle being constant) $OP.OQ$ varies as the square of the parallel semi-diameter $CD$.

c. Since the ordinates $PN$ and $pN$ in the second figure of Art. 33 are as $CB$ to $CA$, the area of any segment or sector of the ellipse is to the area of the corresponding segment or sector of the circle as $CB$ to $CA$, and the area of the whole ellipse is to that of the circle in the same ratio. The area of the ellipse is therefore equal to $\pi.CA.CB$, and it is a mean proportional to the areas of its major and minor auxiliary circles.

d. Any *maximum* or *minimum* or *constant* area related to the ellipse projects into an area similarly related to the circle. Thus, the greatest triangle which can be inscribed in a circle being equilateral, the greatest triangle which can be inscribed in an ellipse may be shewn to be that which has its sides parallel to the tangents at its angular points and its centroid at the centre of the ellipse.

92. *The Equilateral Conics.*

Any hyperbola can be projected into an equilateral hyperbola in the same way that an ellipse can be projected into a circle (Art. 33); and thus from certain properties of the equilateral conics (Art. 61) we can deduce corresponding properties of central conics in general. For example, having proved for the equilateral hyperbola (Art. 65) as well as for the circle, that the conjugate parallelogram is equal to $4CA^2$, we infer that in any central conic the conjugate parallelogram is equal to $AA'.BB'$. In like manner other projective properties which have been established for the two equilateral conics are at once seen to be capable of extension to central conics in general.

93. *Properties of Polars.*

Using the same letters as in Art. 41 but supposing the curve to be a *circle*, we have the angles at $o$ and $T$ right angles, and therefore

$$CO.CT = Co.Ct = CA^2,$$

or $T$ is a fixed point and $tT$ a fixed straight line.

In other words, if a chord of a circle pass through a fixed point $O$ the tangents at its extremities intersect on a fixed straight line $tT$, and conversely. The same follows at once for the ellipse by orthogonal projection.

### 94. *Property of a Quadrilateral.*

If $ABCD$ be a fixed quadrilateral inscribed in a circle, $PQ$ a variable diameter of the circle, $Pa$, $Pb$, $Pc$, $Pd$ perpendiculars to $AB$, $CD$, $AC$, $BD$ respectively; then by Euc. VI. C the rectangle $Pa.PQ$ is equal to $PA.PB$, and thus

$$Pa.Pb = \frac{PA.PB.PC.PD}{PQ^2} = Pc.Pd.$$

Project the circle into an ellipse. Then (still using the same letters) $Pa.Pb$ in the ellipse *varies as* $Pc.Pd$, and the lines $Pa$, $Pb$, $Pc$, $Pd$ meet the sides of $ABCD$ at constant angles, and therefore vary severally as the perpendiculars from $P$ to those sides.

The theorem is inferred to be true for the hyperbola regarded as a quasi-ellipse (Scholium C, p. 101), as also for the parabola. Hence *the products of the perpendiculars from any point on a conic to the sides of a fixed inscribed quadrilateral, taken in opposite pairs, are in a constant ratio;* and conversely, if a point be thus related to a fixed quadrilateral, its locus will be a conic circumscribing the quadrilateral.

It follows by supposing two opposite sides of the quadrilateral to vanish that *the product of the perpendiculars to any two fixed tangents to a conic from a variable point on the curve varies as the square of the perpendicular from that point to their chord of contact.*

### 95. *Curvature.*

The common chord of an ellipse and its circle of curvature at any point and their common tangent at that point make supplementary angles with the axis (Art. 85), and may therefore be projected into a chord and tangent at a point of a circle making supplementary angles with a given diameter.

[Art. 90.

Given a point $O$ on a circle, three positions may be found on the curve of the point $P$ such that $OP$ and the tangent at $P$ make supplementary angles with a given diameter, and the three positions of $P$ determine an equilateral triangle.

Hence it may be deduced that there are three points on an ellipse, lying at the vertices of a maximum inscribed triangle, whose osculating circles cointersect at a given point on the circumference of the ellipse. [Ex. 650.

## PARALLEL PROJECTION.

96. The parallel projections of straight lines, being the common sections of parallel planes drawn through them with the plane of projection, are themselves parallel straight lines; and it is easily deduced that Props. I.–III. in general are applicable to parallel projection. Moreover it is to be noted that in the figure of Art. 90 the ordinate $PN$ may be supposed to meet the axis at *any constant angle* and not necessarily at right angles. We may therefore "project" a point obliquely also *in plano* by increasing or diminishing its ordinate (drawn in any given direction) in a constant ratio.

Although parallel projection is not essentially more general than orthogonal projection, it enables us to effect some constructions with greater ease and directness, as will appear from the next article.

97. *Any two angles of a triangle may be projected into angles of given magnitude.*

(i) Given the angles $SHP$ and $HSP$, it is required to project them into angles equal to $SHg$ and $HSg$ respectively. This may be effected most simply by parallel projection (*in plano* or in space) as follows.

Take $SH$ as the axis of projection, and let it meet $gP$ in $G$. Then (1) if $gG : PG$ be taken as the projecting ratio, the triangle $SPH$ may be projected *in plano* by ordinates parallel to $PG$ into the triangle $SgH$; or (2) if the triangles be situated in different planes (having $SH$ for their common section), the one may be projected into the other by lines parallel to $Pg$.

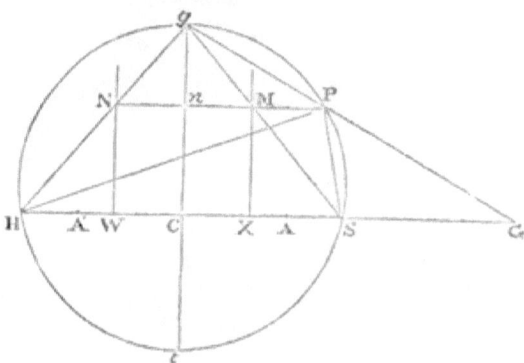

It is hence evident that *any triangle may be projected by parallel lines into an equilateral triangle;* and in like manner that any parallelogram may be projected into a square.

(ii) To project any triangle *SPH orthogonally* into an equilateral triangle, let an ellipse be described about it so as to have its centre at the centroid of the triangle. Then if the ellipse be projected into a circle, *SPH* will at the same time project into an equilateral triangle inscribed in the circle, since each side of the projected triangle will be parallel to the tangent at the opposite vertex.

## LAMBERT'S THEOREM.*

98. *The intercept on any focal vector of an ellipse between the curve and any chord parallel to the tangent at its extremity is equal to the diametral sagitta of the arc cut off by the corresponding chord in the auxiliary circle.*

In an ellipse take any focal vector $SP$, let the diameter conjugate to $CP$ meet $SP$ in $K$, and let any chord $QQ'$ parallel to $CK$ meet $SP$ in $O$.

Produce the ordinates of $P$, $Q$, $Q'$ outwards to meet the auxiliary circle in $p$, $q$, $q'$, and let $m$ and $o$ be the middle points of $QQ'$ and $qq'$, which will lie on a common ordinate $omM$.

---

* This name is given to the theorem in elliptic motion deduced below in the Scholium from Art. 99.

Then $PO : PK = Pm : PC = po : pC$

by parallels; and $PK$ is equal to $pC$ (Art. 38, Cor. 3); and therefore $PO$ and $po$ are equal, as was to be proved.

If $O'$ and $o'$ be any second positions of $O$ and $o$, it follows that $OO'$ is equal to $oo'$: thus by taking $O'$ at $S$ we infer that *SO is equal to the altitude of the triangle $Sqq'$.* If the parallel from $S$ to $qq'$ meet $Cp$ in $s$, it is thus evident that every segment of the line $KSOP$ is equal to the corresponding segment of the radius $Csop$. Hence

$$KS . KO = Cs . Co = CS . CM,$$

since the angles at $s$ and $M$ are right angles.

99. *If in two ellipses described on equal major axes there be taken equal chords $QQ'$ such that $SQ + SQ'$ is of the same magnitude in both ellipses, the areas of the two focal sectors $SQQ'$ will be in the subduplicate ratio of the latera recta of their ellipses.*

In each of two ellipses having equal major axes make the same construction as in Art. 98, with the proviso that $SP$* and $PO$ shall be of the same lengths in both ellipses; then will $QQ'$ and $SQ + SQ'$ be of the same lengths in both ellipses, and the areas $SQQ'$ will be in the subduplicate ratio of the latera recta of their ellipses, and conversely.

(i) For the semi-diameter $CD$ parallel to $QQ'$, being a mean proportional to $SP$ and $AA' - SP$, is of the same length in

---

* Take $SP$ at random in the less eccentric ellipse, and find an equal $SP$ in the other.

both ellipses; and, the sagittæ $po$ in the two circles being equal (Art. 98), the chords $qq'$ are equal; whence it follows in virtue of the projective relation

$$QQ' : CD = qq' : CA,$$

that the chords $QQ'$ are equal in the two ellipses.

(ii) And since $CM$ is the abscissa of the middle point of $QQ'$, therefore

$$SQ + SQ' = AA' - 2\frac{CS}{CA}.CM,*$$

where $CS.CM$, or $KS.KO$ (Art. 98), is the same in both ellipses; and therefore also $SQ + SQ'$ is of the same magnitude in both.

(iii) It is further evident that the segments $qpq'$ are equal and the triangles $Sqq'$ are equal in both circles, and hence that the two sectors $Sqq'$ are equal. It follows by projection that (the major axes of the two ellipses being equal) the focal sectors $SQQ'$ are in the ratio of the minor axes of their ellipses, or in the subduplicate ratio of their latera recta. Conversely, if two focal sectors $SQQ'$ of any pair of ellipses described on equal major axes be thus related, then will $QQ'$ and $SQ + SQ'$ be of the same lengths in both ellipses.

100. *The area of any focal sector $SQQ'$ of a central conic divided by the square root of its latus rectum may be expressed in terms of $QQ'$, $SQ + SQ'$, and the tranverse axis, and is independent of the magnitude of the conjugate axis.*

This may be inferred from Art. 99 for the ellipse: or it may be proved by the following method, which will be seen to be applicable to the hyperbola as well as to the ellipse.

In an ellipse whose axes are $AA'$ and $BB'$ take any diameter $PCP'$, and let $PS$ meet the conjugate diameter $CD$ in $K$ and the tangent at $P'$ in $H$, so that $PK$ is equal to $CA$ and $PH$ to $AA'$.

---

\* Assume that at any point $P$ on the curve $SP = CA \pm e.CN$. Notice also in Art. 100 (ii) that $SK = PK - SP = CA - SP = e.CN$.

240  ORTHOGONAL PROJECTION.

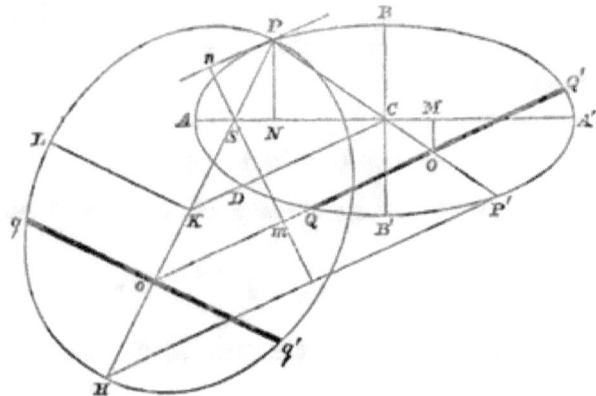

With $S$ as focus and $P$ and $H$ as vertices describe a second ellipse, and let $KL$ be its semi-axis conjugate. Then if $S'$ be the further focus of the first ellipse,
$$CD^2 = SP.PS' = SP.SH = KL^2.$$

(i) Let $QOQ'$ be any double ordinate to $PP'$, let it meet $PH$ in $o$, and let $qoq'$ be a double ordinate to $PH$ in the second ellipse. Then by the nature of ordinates,
$$QO^2 : CD^2 = PO.OP' : CP^2 = Po.oH : KP^2$$
$$= qo^2 \quad : KL^2,$$
or $QO$ is equal to $qo$ and $QQ'$ to $qq'$.

(ii) And since $SK : CN = CS : CA = CS : PK$,
and $\qquad CN : CM = CP : CO = PK : Ko$;
therefore $\qquad SK : CM = CS : Ko$,
or in terms of the eccentricities $e$ and $e'$,
$$e.CM = e'.Ko;$$
and therefore
$$SQ + SQ' = AA' + 2e.CM = PH + 2e'.Ko$$
$$= Sq + Sq'.$$

(iii) **Let a line through $S$ at right angles to $QQ'$ meet $QQ'$ in $m$ and the tangent at $P$ in $n$; then will the equal chords $QQ'$ and $qq'$ in any two consecutive positions cut off elements of area which, being as their own breadths, are** in the *constant* ratio of $Sm$ to $So$; and this is also the ratio of the triangles $SQQ'$ and $Sqq'$ on equal bases.

Therefore the whole segments $QP'Q'$ and $qHq'$, as also the sectors $SQQ'$ and $Sqq'$, are as $Sm$ to $So$, or as $Sn$ to $SP$, or as $CB$ to $CD$ or $KL$; that is to say (the transverse axes being equal) the sectors $SQQ'$ and $Sqq'$ are in the subduplicate ratio of the latera recta of their ellipses,

Lastly (calling the second ellipse described as above the *auxiliary ellipse*) if any two ellipses having equal major axes be placed so as to have a common focal vector $SP$, they will have the same auxiliary ellipse; and it is inferred by comparing each of the two ellipses with their auxiliary ellipse that their corresponding focal sectors $SQQ'$ will be in the subduplicate ratio of their latera recta provided that $QQ'$ and $SQ + SQ'$ be equal in both. Hence, when the three lengths $QQ'$, $SQ+SQ'$, $AA'$ are given the area of the focal sector $SQQ'$ divided by the square root of the latus rectum is determined.

### SCHOLIUM.

The substance of the preceding section may be found in LAMBERT's *Insigniores Orbitæ Cometarum Proprietates*\* (pp. 102–125), where it is deduced that *the time in any arc $QQ'$ of an elliptic orbit described about $S$ is a function of $SQ + SQ'$, the chord $QQ'$, and the major axis*, and does not depend upon the minor axis. For an investigation of the expression for the time the reader may consult the *Messenger of Mathematics*, vol. VII. p. 97: the result may also be obtained in the same form by the following process, which has the advantage of giving a more direct geometrical interpretation to some of the symbols employed.

Assuming the results of Arts. 98, 99, and noticing further that $sf$ is equal to $CD$ (Ex. 651), let $\alpha$ denote the angle $oCq$, and $\beta$ the angle $sCf$. Then

$$\cos\alpha \cdot \cos\beta = \frac{Co \cdot Cs}{CA^2} = \frac{CM \cdot CS}{CA^2} = 1 - \frac{SQ + SQ'}{AA'},$$

$$\sin\alpha \cdot \sin\beta = \frac{\tfrac{1}{2}qq' \cdot sf}{CA^2} = \frac{\tfrac{1}{2}qq' \cdot CD}{CA^2} = \frac{QQ'}{AA'};$$

and therefore $\quad 1 - \cos(\beta \pm \alpha) = \dfrac{SQ + SQ' \pm QQ'}{AA'}$.

---

\* See above, p. 57; and for a further account of the work see the *Messenger of Mathematics*, vol. IX. p. 63.

Let $n$ denote the mean **motion in the ellipse**: $t$ the time in the arc $QQ'$, which **is proportional to the** sectorial area $SQQ'$ (or $\frac{CB}{CA}$ sect.$Sqq'$) **divided by the square root of the** latus rectum (Newton's *Principia*, Lib. I. Sect. III. prop. 14, theor. 6). **Then** since in the circle

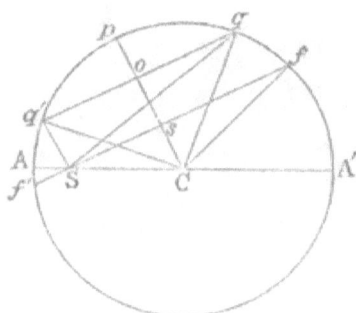

sect.$Sqq'$ = sect.$Cqq'$ − $\triangle$ ($Cqq'$ − $Sqq'$) = $CA^2$ ($\alpha$ − $\sin\alpha \cdot \cos\beta$),

it may be deduced that

$$nt = \{\beta + \alpha - \sin(\beta + \alpha)\} - \{\beta - \alpha - \sin(\beta - \alpha)\},$$

where $\beta + \alpha$ and $\beta - \alpha$ **are** known functions of $AA'$, $QQ'$, and $SQ + SQ'$.

The corresponding **theorem in** parabolic motion had been given **by Euler** in the *Miscellanea Berolinensia*, tom. VII. pp. 19, 20 (1743). **Lambert** (1761) extended it to the central conics generally; although **Lexell** in his elaborate article (*Nova Acta Academiæ Scientiarum Imperialis Petropolitanæ*, tom. I. pp. (140)—(183), 1787) overlooks Lambert's references in Preface and text (§§ 57, 95, 213, &c.) **to** the hyperbola. On the parabola see Examples 691 and 692.

## EXAMPLES.

651. The diameter parallel **to any focal** chord of an ellipse **is equal to** the projectively **corresponding** chord of its major **auxiliary circle.**

652. **The** orthogonal **or parallel** projection of a parabola is a parabola; and an ellipse or hyperbola may **be projected** into an ellipse or hyperbola of any eccentricity.

653. Shew how **to** project a parabola and a given point within it so that the projection of the point may be **the** focus

of the projection of the curve:\* and thus deduce (from a property of the directrix) that the tangents to a parabola at the extremities of any chord drawn through a fixed point within it intersect upon a fixed straight line.

654. Any two similar and coaxal ellipses may be projected into concentric circles. Hence shew that **a chord of** an ellipse which **always touches a similar and coaxal ellipse** is bisected at its **point of contact, and** that **it cuts off** a constant area **from the outer** ellipse; **and** shew that the portions of any **chord intercepted** between the two curves are equal.

655. **Shew** how to project any two coaxal **ellipses** or **hyperbolas** into confocal ellipses or hyperbolas; **and** extend the theorem of Ex. 382 **to any pair of coaxal** conics **of the same** species.

656. **If** a chord of **an ellipse and the tangents at its extremities contain a constant area, the chord cuts off a constant** area from the ellipse and touches a similar ellipse, **and the** tangents at its extremities intersect on another similar ellipse.

657. If $CP$ and $CD$ be conjugate radii of an ellipse, and if the lines connecting $P$ and $D$ with opposite ends of the major **axis intersect in** $O$; then will $DOP$ and an extremity of the **minor axis determine a** parallelogram. When is its area least **or greatest?**

658. The least triangle **circumscribing a given** ellipse has its sides bisected at the points **of contact.**

659. The greatest ellipse which **can be inscribed in a given parallelogram** is that which bisects its sides.

660. An $n$-**gon** described about an ellipse so **as to have** its sides bisected **at their points of** contact is of constant area, and the $n$-gon formed by joining every two successive points of contact is of constant area.

---

\* Let $TP$ and $TQ$ be the tangents whose chord of contact $PQ$ passes through **and** is bisected by the given point; and let $TPQ$ be projected into an isosceles triangle **right-angled** at the projection of $T$ (**Art. 97**).

661. If a triangle be inscribed in an ellipse, the parallels through its vertices to the diameters bisecting the opposite sides meet in a point.

662. Parallel chords drawn to an ellipse through the extremities of conjugate diameters meet the curve again at the extremities of conjugate diameters.

663. Two ellipses of equal eccentricities and parallel major axes can have only two points in common; and the common chords of three such ellipses meet in a point.

664. $P$, $Q$, $R$ being three points on an ellipse, a diameter $ACA'$ bisects $PQ$ and meets $RP$ in $N$ and $RQ$ in $T$; prove that $CN.CT = CA^2$.

665. Through the centre of an ellipse and the points of concourse and contact of any two tangents a similar and similarly situated ellipse can be drawn.

666. $A$, $B$, $C$ are similar and similarly situated ellipses, and $B$ is concentric with $A$ and passes through the centre of $C$. Shew that the common chord of $A$ and $C$ is parallel to the tangent to $B$ at the centre of $C$.

667. Through a given internal point draw a straight line cutting off a minimum area from a given ellipse.

668. If the tangent at the vertex $A$ of an ellipse meets a similar coaxal ellipse in $T$ and $T'$, any chord of the former drawn from $A$ is equal to half the sum or difference of the parallel chords of the latter through $T$ and $T'$.

669. Determine the greatest triangle that can be inscribed in an ellipse with one angular point at a given point on the circumference.

670. Given an ellipse, if a parallelogram be inscribed in it so that one side divides the diameter conjugate thereto in a constant ratio; or if a parallelogram be described about it so that one of its diagonals is divided in a constant ratio by the curve; the area of the parallelogram will be constant.

671. The tangents to an ellipse at $P$ and $P'$ are parallel, any two conjugate diameters meet them in $D$ and $D'$, and any third tangent meets them in $T$ and $T'$; shew that

$$PD : PT = P'T' : P'D'.$$

672. A triangle $ABC$ inscribed in an ellipse has its centroid at the centre of the ellipse; shew that the tangents at the opposite extremities of the diameters through $A$, $B$, $C$ form a triangle similar to and four times as great as triangle $ABC$.

673. If two conjugate hyperbolas having a pair of conjugate diameters of an ellipse for asymptotes cut the ellipse at points lying on four diameters 1, 2, 3, 4 taken in order: then will 1, 3 and 2, 4 be conjugate in the ellipse, and 1, 4 and 2, 3 in the hyperbolas.

674. The locus of the middle point of a chord of an ellipse drawn through a fixed point is a similar ellipse, having its centre midway between the fixed point and the centre of the given ellipse.

675. Determine the greatest isosceles triangle that can be inscribed in an ellipse with its base parallel to either axis.

676. Any two radii of a circle and any pair of the radii at right angles thereto determine equal triangles: what is the corresponding property of the ellipse?

677. Any double ordinate to a given diameter of an ellipse being divided into segments whose product is constant, the point of section traces a similar coaxal ellipse.

678. If the tangents at $T$ and $T'$ to an ellipse meet on a similar coaxal ellipse having $AA'$ for major axis, shew that the loci of the points $(AT, A'T')$ and $(AT', A'T)$ are ellipses similar to the former.

679. If $PP'$ be a diameter of an ellipse and $Q$, $R$ be any two points on the curve, shew that the line joining the intersections of $PQ$, $P'R$ and $PR$, $P'Q$ passes through the intersection of the tangents at $Q$ and $R$ and is conjugate in direction to $PP'$.

680. In a given ellipse shew how to place a chord passing through a given point in the axis which shall contain with the radii to its extremities a maximum triangle.

681. From the extremities of any diameter of an ellipse straight lines are drawn to either focus; shew that the radii parallel thereto in the auxiliary circle meet the circle at points lying on a chord through the focus.

682. A triangle of constant area being inscribed in an ellipse, find the locus of its centroid.

683. If $P$ and $Q$ be projectively corresponding points on an ellipse and its major auxiliary circle, the radius of the ellipse in the direction $CQ$ is equal to the perpendicular from $C$ to the tangent at $P$.

684. $AA'$ being a diameter of a conic and $Q$ any point on the tangent at $A$, shew that the line $A'Q$ cuts the conic in a point $B$ such that the tangent thereat bisects $AQ$.

685. If $CP$, $CP'$ be semi-diameters of an ellipse, $Q$ and $Q'$ the points in which the conjugate diameters meet the tangents at $P'$ and $P$ respectively; the triangle determined by a pair of its semi-axes is a mean proportional to the triangles $CPP'$ and $CQQ'$.

686. If a triangle inscribed in an ellipse have its sides parallel to the diameters $b'$, $b''$, $b'''$ and the focal chords $c'$, $c''$, $c'''$; and if $a$, $b$ be the axes and $p$ the parameter of the ellipse, and $D$ the diameter of the circle circumscribing the triangle; shew that

$$D = \frac{b'b''b'''*}{ab} = \left(\frac{c'c''c'''}{p}\right)^{\frac{1}{2}}.$$

687. The square root of the continued product of the six focal chords of an ellipse parallel to its six chords of intersection with a circle is equal to the parameter of the ellipse multiplied by the square of the diameter of the circle.

---

* The theorem is due to MAC CULLAGH. (Salmon's *Conic Sections*, Art. 369).

# EXAMPLES.

**688.** Shew that $S'B$ (fig. p. 79) is perpendicular to the tangent from $X$ to the auxiliary circle. If $Z$ be the point of contact of this tangent and $G$ its intersection with $S'B$, determine the eccentricity of the ellipse so that the lines $CZ$, $SG$, $XB$ may meet in a point.

**689.** Two particles describe the same ellipse subject to the same force at the centre: shew that their directions of motion at any time intersect on a similar ellipse.

**690.** A given triangle may be orthogonally projected from an equilateral triangle; or it may be orthogonally projected into an equilateral triangle. Determine by a geometrical construction the magnitudes of these equilateral triangles.

**691.** Prove by the method of Art. 100 that the area of any focal sector $SMN$ of a parabola divided by the square root of its latus rectum may be expressed in terms of $SM + SN$ and $MN$.

**692.** Prove also that if $PV$ be the diametral sagitta of the arc $MN$, then

$$\frac{SM + SN \pm MN}{2} = \{\sqrt{(SP)} \pm \sqrt{(PV)}\}^2;$$

and from this last result deduce that

$$\frac{3}{\sqrt{(AS)}} \text{ sector } SMN = \left\{\frac{SM + SN + MN}{2}\right\}^{\frac{3}{2}} - \left\{\frac{SM + SN - MN}{2}\right\}^{\frac{3}{2}}.*$$

**693.** If in any two ellipses or hyperbolas on equal transverse axes there be taken equal focal vectors $SP$, and if chords $QQ'$ be drawn in them parallel to the tangents at $P$ and so as to make equal intercepts on the equal focal vectors; shew that the chords thus drawn are equal in the two conics, and that $SQ + SQ'$ is of the same length in both, and that the

---

\* This is here suggested as perhaps the most direct way of establishing geometrically the theorem in parabolic motion commonly ascribed to LAMBERT but previously discovered by EULER (Scholium, p. 242).

focal sectors $SQQ'$ are in the subduplicate ratio of the latera recta of their ellipses or hyperbolas.*

694. Given, in a conic orbit described about the focus $S$ under the action of a given force, a chord $QQ'$, the parallel focal chord, and the sum of the terminal focal vectors $SQ + SQ'$; the time of describing the arc $QQ'$ is constant.†

695. Given the directions of two sides of a triangle inscribed in a given ellipse, determine the envelope of its third side.

---

\* The proof is very much as in Art. 100; but the two conics are thus compared directly and without the interposition of an auxiliary conic in which the chord $qq'$ of the sector is at right angles to the principal axis.

† This extended enunciation of LAMBERT's theorem (as I am informed by Prof. TOWNSEND) was given in a lecture in the year 1845 by the late Prof. MAC CULLAGH of Dublin, who proved that the sectorial area $SQPQ'$ and the square root of the latus rectum of the orbit vary severally as the sine of the angle between $SP$ and $QQ'$, having previously shewn that $OP$ is constant, and that every chord parallel to $QQ'$ which cuts $SP$ in a constant ratio is of constant length.

# CHAPTER X.

## CROSS RATIO AND INVOLUTION.

**101.** A SET of points on a straight line constitute a *Range*\* or row, of which the line is called the **base** or axis; and a set of straight **lines** radiating from **a point** constitute a *Pencil*, of which the point is called the *Vertex* or the centre and the lines are called *Rays* or radiants.

The ratio of the ratios in which the straight line joining two points is divided by any other two points in its length is called a *Double Section Ratio*, or an *Anharmonic Ratio*, or a *Cross Ratio* of the four points. If two ranges of four points have one cross ratio of the one equal to one cross ratio of the other, it may be shewn that every cross ratio of the one is equal to the corresponding cross ratio of the other: we may therefore in practice speak of a tetrad as having one cross ratio only, although its four points determine six segments to each of which belong a reciprocal pair of cross ratios.†

**102.** If a line $BD$ be divided at $A$ and $C$, the ratio $AB.CD : AD.CB$ (which is a *cross-multiple* of the ratios $\dfrac{AB}{AD}$ and $\dfrac{CB}{CD}$) may be taken as the cross ratio of the points $ABCD$, and it is expressed by the notation $\{ABCD\}$. It may

---

\* Collinear points are (or were) sometimes said to "range on a straight line"—for an example see Booth's *New* **Geometrical** *Methods*, vol. I. §48.

† For a further exposition of the principles of Anharmonic Section see TOWNSEND's *Modern Geometry of the Point Line and Circle*, vol. II. The purpose of this chapter is not so much to establish these principles as to apply them to conics.

be shewn that *this ratio is unaltered by the interchange of any two points of the range provided that the remaining two points be at the same time interchanged*. In other words, it may be shewn that

$$\{ABCD\} = \{BADC\} = \{CDAB\} = \{DCBA\}.$$

The cross ratio of a pencil of four rays is that of the range which it determines on any transversal: this ratio will be shewn to be *independent of the position of the transversal* (Art. 103). The cross ratio of four radiants $ABCD$ is expressed by $\{ABCD\}$, and the cross ratio of four radiants $Oa$, $Ob$, $Oc$, $Od$ by $O\{abcd\}$ or $\{O.abcd\}$.

Tetrads of points and rays are said to be *equal* or *equianharmonic* when their cross ratios are equal; and ranges and pencils of any number of constituents are said to be *homographic*\* when every tetrad of the one system and the corresponding tetrad of the other have equal cross ratios. The notation

$$\{ABCDE \&c.\} = \{A'B'C'D'E' \&c.\}$$

is used to express that two systems whose constituents correspond in pairs are homographic. More briefly, if $A$ and $A'$ be a variable pair of homologous (*i.e.* corresponding) constituents of two homographic systems, the equation $\{A\} = \{A'\}$ may be used to signify that every four positions of $A$ and the corresponding positions of $A'$ have equal cross ratios. The corresponding notation for a pencil having its vertex at $O$ is $O\{A\}$.

Thus far we have spoken only of collinear points and of concurrent lines, but the following definitions may be added in anticipation of Art. 113. The cross ratio of *Four Points on a Conic* is that of the pencil which they subtend at any fifth point on the curve; and the cross ratio of *Four tangents to a Conic* is that of the range which they determine on any fifth tangent thereto. In the special case of the *Circle* it is at once evident

---

\* This term is sometimes confined to systems of more than four constituents, but it may be used with equal fitness for the special case of tetrads, in place of the longer word *equianharmonic*. The still shorter term *equal* (although not in general so used) may often be employed with advantage as an abbreviation for "equal as regards cross ratio"—for which the term *equicross* is suggested by Prof. TOWNSEND.

that these ratios are *independent of the position of the fifth point or tangent*, since four given points on a circle subtend a pencil of *constant angles* at any fifth point on the curve, and since the segments of any fifth tangent by four fixed tangents subtend *constant angles* at the centre. It will be shewn in Art. 113 that the cross ratios of four given points on (or tangents to) any conic are likewise constant.

We proceed to establish certain Lemmas relating mainly (1) to Cross Ratio in general, and (2) to the special case of homography called Involution, with a view to their eventual application to some of the fundamental properties of conics.

## CROSS RATIO.

103. *The cross ratio of the points in which four fixed radiants are met by a variable transversal is constant.*\*

(i) Let four lines radiating from $O$ be met by any transversal in the points $A$, $B$, $C$, $D$, and let the parallel through $B$ to $OD$ meet $OA$ in $a$ and $OC$ in $b$.

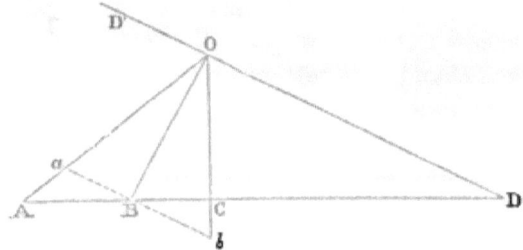

Then by similar triangles,

$$AB : AD = aB : DO$$

and

$$CD : CB = DO : Bb.$$

Hence

$$\{ABCD\} = \frac{AB.CD}{AD.CB} = \frac{aB}{bB},$$

which is constant for all positions of $ab$ parallel to $OD$; and therefore $\{ABCD\}$ is constant for all positions of $ABCD$ without restriction.

---

\* It is to be noted that this very important theorem of the ancients implicitly contains the method of Conical Projection.

We have thus proved that
$$\{ABCD\} = \{aBb\infty\},$$
where $\infty$ denotes the point at infinity in which $aBb$ meets $OD$.

It is to be noted that the transversal may be drawn so as to cut the radiants (produced through $O$) upon either side of $O$. Thus in the figure of Art. 107 (i), considering the pencil whose vertex is $O$ and supposing $QO$ to meet $AB$ in $E$ and $CD$ in $F$, we have
$$\{DMAQ\} = \{BPAE\} = \{BRCQ\} = \{DPCF\},$$
the rays of the pencil being taken in the same order in every case.

(ii) It is further evident that different pencils have equal cross ratios when their corresponding angles are equal or supplementary.

For example, if $SR$ and $ST$ turn about $S$ (Art. 9, Cor. 1) so as to be always at right angles, the pencils $S\{R\}$ and $S\{T\}$ are equiangular and therefore homographic. Or again, if $TS$ and $TH$ in Art. 50 remain fixed whilst the conic varies (so that the tangents from $T$ always make equal or supplementary angles with $TS$ and $TH$ respectively), the pencils $T\{P\}$ and $T\{Q\}$ are homographic.

**104.** *Condition that a variable straight line should pass through a fixed point.*

Four rays determine equal ranges upon any two transversals (Art. 103); and conversely, if two equal tetrads of points on different axes be such that three of the lines joining them in homologous pairs meet in a point, the join\* of the fourth pair must pass through the same point.

In particular, if $A$ be a point on one of the four rays, they determine equal ranges $\{ABCD\}$ and $\{AB'C'D'\}$ on any two transversals through $A$.

---

\* This term is used by some writers as an abbreviation for the *line joining* two points. See for example HENRICI's *Elementary Geometry, Congruent Figures*, § 47 (London, 1879). The "join" of two lines is their common point.

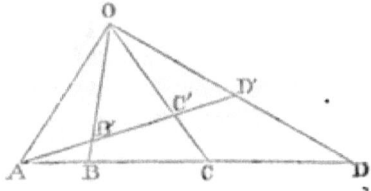

Conversely, if $BB'$ and $CC'$ be any two assumed positions of a variable straight line, $O$ their point of concourse and $A$ that of $BC$ and $B'C'$, and if $OA$ be a third position and $DD'$ any other position of the variable line, the condition that $DD'$ should pass through the fixed point $O$ is that $\{ABCD\}$ should be always equal to $\{AB'C'D'\}$.

**105.** *Condition that a variable point should lie on a fixed straight line.*

A range of four **points subtends equal pencils at any two vertices**; and conversely, if **two equal pencils having different vertices** be such that three pairs of their corresponding **rays intersect on one straight line, the fourth pair must intersect on the same straight line.**

In particular, if $O$ and $O'$ be the vertices of two pencils which have $OO'A$ for a common ray, and if $B, C, D$ be the intersections of their remaining three pairs of rays, then will $BCD$ be a straight line provided that

$$O\{ABCD\} = O'\{ABCD\}.$$

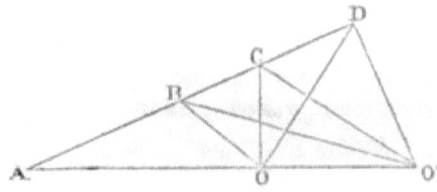

If now (supposing the rays of the two pencils to be variable) the intersections of two pairs of homologous rays be at **given** points $B$ and $C$, the remaining intersection $D$ will lie on the fixed **straight** line $BC$.

This result and that of Art. 104 may be together expressed as follows:

*If two tetrads of points or rays $ABCD$ and $A'B'C'D'$ be equianharmonic, then if one pair $(A, A')$ coalesce the remaining three pairs $(B, B'), (C, C'), (D, D')$ are in perspective.*

**106.** *Condition that a tetrad of points or rays may be harmonic.*

(i) If $BD$ be divided harmonically at $A$ and $C$, and if the signs $\pm$ be used to denote opposite directions, then it is easily seen that
$$\frac{AB}{AD} \cdot \frac{CD}{CB} = \frac{AD}{AB} \cdot \frac{CB}{CD} = \frac{CB}{CD} \cdot \frac{AD}{AB} = -1,$$
or $\quad\{ABCD\} = \{ADCB\} = \{CBAD\} = -1.$

The condition that the range $ABCD$ should be harmonic is therefore that the points $A$ and $C$ or the points $B$ and $D$ should be *separately* interchangeable. Each pair of interchangeable points are said to be *harmonically conjugate* to each other with respect to the remaining pair.

(ii) A pencil $O\{ABCD\}$ will be harmonic if a transversal drawn parallel to one of its rays $OD$ be cut in *equal* and opposite segments $Ba$ and $Bb$ (fig. Art. 103) by the remaining three rays. For example, since $DP$ in Art. 60 is parallel to one asymptote and is bisected by the other, it follows that *the asymptotes and any two conjugate diameters of a hyperbola constitute a harmonic pencil.*

(iii) A pencil will also be harmonic if two of its rays be the bisectors of the angles between the other two. For example, *the tangent and normal and the focal distances at any point of a conic constitute a harmonic pencil.*

**107.** *Properties of a complete quadrilateral.*

Let $ABCD$ be a quadrilateral, and let $O, P, Q$ be the intersections of $(AC, BD), (AB, CD), (AD, BC)$ respectively. Then the figure as thus completed is called a Complete Quadrilateral, and $PQ$ is called its third Diagonal.

(i) Let the ray $PO$ meet $AD$ in $M$ and $BC$ in $R$. Then by the properties of radiants (Art. 103),

$$O\{DMAQ\} = O\{BRCQ\} = P\{BRCQ\} = P\{AMDQ\};$$

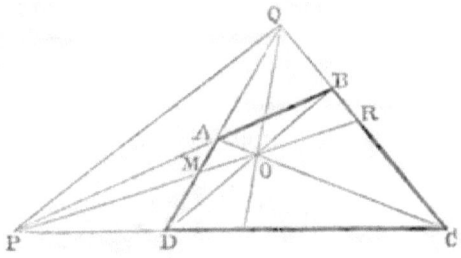

and therefore the range $\{DMAQ\}$, in which the conjugate points $A$ and $D$ have been shewn to be interchangeable, is harmonic.

[Art. 106 (i).

Thus it appears that the two pencils whose vertices are $O$ and $P$ are harmonic, and in like manner that the pencil $Q$ is harmonic; and hence that *each of the three diagonals of the complete quadrilateral is divided harmonically by its remaining two diagonals.*

(ii) Let the sides $AB$, $DC$ of a quadrilateral meet in $P$ and the sides $DA$, $CB$ in $O$. Complete the parallelograms $BODR$ and $AOCQ$, and let $BR$, $PD$ meet in $V$ and $AQ$, $PD$ in $T$.

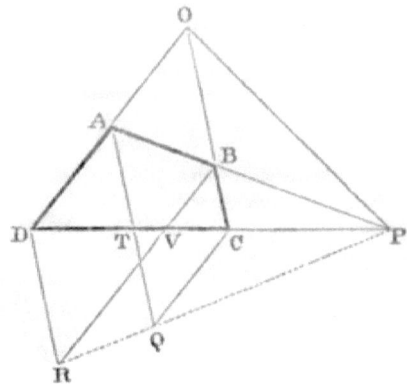

Then by parallels

$$PC : CT = PB : BA = PV : VD,$$

or $\qquad PC : PV = CT' : VD = CQ : VR.$

Hence $PQR$ is a straight line, and therefore the middle points of $OP$, $OQ$, $OR$ lie on a straight line parallel to $PQR$.

But the middle point of $OQ$ is also the middle point of $AC$, since the diagonals of parallelograms bisect one another; and for the same reason the middle point of $OR$ is also the middle point of $BD$.

Therefore *the middle points of the three diagonals $AC$, $BD$, $OP$ of a complete quadrilateral lie on a straight line.*

The straight line bisecting the three diagonals of a complete quadrilateral may be called the DIAMETER of the quadrilateral (Ex. 375, note).

### SCHOLIUM A.

The fundamental anharmonic property of a pencil of four rays constitutes prop. 129 of the seventh book of the Συναγωγή of PAPPUS (vol. II. p. 871, ed. Hultsch), which affirms that *if two straight lines $ABCD$ and $A'B'C'D'$ be drawn across three radiants $OB$, $OC$, $OD$* (fig. Art. 104), *then*

$$AB . DC : AD . CB = A'B' . D'C' : A'D' . B'C'.$$

This property is introduced (pp. 867—919) under the head of Lemmas on EUCLID's three books of Porisms, a subject treated at length by ROB. SIMSON in the work *De Porismatibus* contained in his *Opera quædam reliqua* (Glasguæ, 1776), where some things that were obscure in Pappus are made clear. In the same section of the Συναγωγή (prop. 131) is contained in substance the theorem of Art. 107, that *either diagonal of a quadrilateral is cut harmonically by its other diagonal and the join of the joins* [note, p. 252] *of its opposite sides* (Chasles *Aperçu Historique* p. 36, 1875). And further, it will appear (Schol. C) that a theorem in Pappus implicitly contains the fundamental anharmonic property of four points on a conic.

The ancient theory of transversals was revived and largely developed in the first half of the seventeenth century by DESARGUES, to whom PONCELET, in the introduction to his *Traité des Propriétés Projectives des Figures* (Paris, 1822), applies the title of honour "le MONGE de son siècle" (p. xxxviii). For some further bibliographical information on the subject see Poncelet *loc. cit.* and Chasles *Aperçu Historique*. Of CARNOT's *Géométrie de Position* (Paris, 1803) Poncelet remarks (p. xliv) that in that work "se trouve exposée pour la première fois et dans toute sa généralité cette belle Théorie des transversales dont nous avons déjà si souvent parlé dans ce qui précède, et dont les Anciens n'avaient fait qu'entrevoir les principes et la fécondité." This method has been still further developed by CHASLES "qui a mis en œuvre d'une manière très-heureuse les principes de la Théorie des transversales pour démontrer la plupart des théorèmes de MONGE" (ibid. p. xlv). A good section on **Transversals** was contained in the 12th edition of Hutton's *Course of Mathematics*, vol. II. 214–46 (**London**, 1843).

For the cross **ratio**—or as he called it "Doppelschnittsverhältniss" (*ratio bissectionalis*)*—of four points MOEBIUS introduced the notation (*A, B, C, D*) in his celebrated *Barycentrische Calcul*, § 183, p. 246 Leipzig, 1827); and he used the notation '(*c, d, b, a*) for the cross ratio of the range determined by **four lines** *c, d, b, a* upon any transversal *e* (p. 256). Chasles (*Traité des Sections Coniques* Pt. I. p. 7, 1865), following the example of Möbius, adopts the notation (*a, b, c, d*) **for a** point-tetrad, and uses *O* (*a, b, c, d*), an abbreviation by separation of symbols for (*Oa, Ob, Oc, Od*), to denote the anharmonic ratio of a tetrad of radiants. More generally, whether the letters denote **points** or lines, we may write *O*{*abcd*} for the cross ratio of the "joins" of (*O, a*), (*O, b*), (*O, c*), (*O, d*). The form {*O . abcd*}, used in SALMON's *Conic Sections*, is a convenient substitute for *O* {*abcd*} in cases in which cross ratios are compounded with one another (Möbius, p. 257).

The term *Anharmonic* **was** invented by **Chasles and has been** widely used; but it is an artificial word and wanting in brevity, and perhaps ought rather to mean non-harmonic. The short and significant term *Cross Ratio* (which expresses the result of compounding two simple ratios crosswise) was coined by the late Prof. CLIFFORD, and may be found in his *Elements of Dynamics* (p. 42) published **some** four or five years later (London, 1878). It may be **used** also **with a secondary reference** to the transversal or cross-line on which the ratio is estimated.

## INVOLUTION.

108. If three or more pairs **of points** $A, A'$; $B, B'$; $C, C'$; &c. be taken on a straight line at **such distances from a point** $O$ thereon that

$$OA.OA' = OB.OB' = OC.OC' = \&c.,$$

they are said **to constitute** a system in *Involution*. The point $O$ is called the *Centre* and the points $(A, A')$, $(B, B')$, $(C, C')$, &c. are called *Conjugate Points* or *Couples* of the involution. The centre is evidently conjugate to the point at infinity **upon the axis.**

---

* STEINER called it "Doppelverhältniss."

If the several points of an involution lie on the same side of the centre as their conjugates respectively, the products $OA.OA'$, &c. are to be regarded as positive, and the involution is said to be *positive*; but if conjugate points lie on opposite sides of the centre the involution is said to be *negative*. In a positive involution there are two points $F$ and $F'$ (on opposite sides of the centre) each of which coincides with its own conjugate: these are called the *Double Points* or *Foci* of the involution. In a negative involution the foci are imaginary.

When two pairs of conjugate points $A, B$ and $A', B'$ are given the involution is completely determined. For if two pairs of parallels be drawn through $A, A'$ and $B, B'$ respectively, and if they intersect in $P$ and $Q$, the line $PQ$ evidently meets the axis of the involution in a fixed point $O$ such that

$$OA : OA' = OP : OQ = OB' : OB,$$

or
$$OA.OB = OA'.OB'.$$

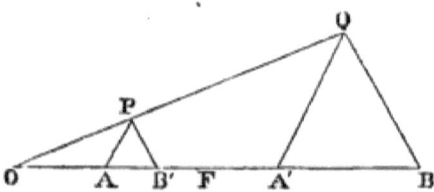

Thus the centre $O$ and the constant of the involution are known, and the system is completely determined.*

A *Pencil* of lines in involution is a pencil which determines a row of points in involution on any transversal: every such pencil has one pair of *Double Rays*, on which lie the double points of all its transversals. [Art. 112.

109. *A system of circles having a common radical axis determine pairs of points in involution on any transversal.*

For if a system of circles be drawn through two points $P$

---

\* The six lines joining any four points $PQP'Q'$ determine an involution on any transversal: the construction in the text results from taking $P'$ and $Q'$ at infinity.

and $Q$, and if a transversal meet $PQ$ in $O$ and meet any two of the circles in $A, A'$ and $B, B'$, it is evident that

$$OA.OA' = OB.OB' = OP.OQ.$$

Conversely, if circles be drawn through the several pairs of conjugate points of an involution and through a common external point $P$, they will all meet at a second fixed point $Q$, such that $PQ$ passes through the centre of the involution. The *Foci* of the involution are the points of contact of the two circles which can be drawn through $P$ and $Q$ so as to touch the transversal: these points are *imaginary* when $P$ and $Q$ lie on opposite sides of the transversal.

**110.** *A pencil of rays in involution has in general two conjugate rays only at right angles; but if two pairs of conjugate rays be at right angles every two conjugate rays are at right angles.*

Let $P$ be the vertex of a pencil in involution, and let a transversal meet two pairs of conjugate rays in $A, A'$ and $B, B'$. Then the circles $APA'$ and $BPB'$ determine by their intersection a second fixed point $Q$, and every circle through $P$ and $Q$ meets the transversal in points $C$ and $C'$ such that $PC, PC'$ are conjugate rays. [Art. 109.

Now in order that the angle $CPC'$ may be a right angle the centre of the circle $CPC'$ must lie on the axis of involution. In general one circle only can be drawn through $P$ and $Q$ so as to have its centre upon a given transversal; but if two circles can be so drawn the transversal must bisect $PQ$ at right angles, and must therefore be the locus of the centres of all the circles that can be drawn through $P$ and $Q$.

It is hence evident that a right angle turning about its vertex generates a pencil of rays in *negative* involution: for example, *the conjugate diameters of a circle constitute a negative involution.*

It is further evident that there are two points (one on each side of the axis) at which all the segments of a range in negative involution subtend right angles.

**111.** *If a row of points be in involution any four of them and their four conjugates are equicross, and conversely.*

(i) For if $(A, A')$, $(B, B')$, $(C, C')$, $(D, D')$ be conjugate points in an involution of which $O$ is the centre, then since by definition,
$$OA' : OB' = OB : OA,$$
therefore $\quad A'B' : OB' = AB : OA;$
and similarly $\quad C'D' : OD' = CD : OC.$

Hence $A'B'.C'D' : OB'.OD' = AB.CD : OA.OC;$
and in like manner (or by merely interchanging $B$ and $D$, and $B'$ and $D'$),
$$A'D'.C'B' : OB'.OD' = AD.CB : OA.OC;$$
and therefore $\{A'B'C'D'\}$ is equal to $\{ABCD\}$, or every four points and their conjugates are equicross, as was to be proved.

From the general relation $\{ABCD\} = \{A'B'C'D'\}$, where the several couples may be taken arbitrarily, we deduce as one form of the relation between three couples in involution (by substituting $A'$ for $D$ and $A$ for $D'$), that
$$\{ABCA'\} = \{A'B'C'A\},$$
which readily reduces to
$$AB.B'C'.CA' = A'B'.BC.C'A.$$

Another form of the relation between the three couples is
$$\{ABA'C\} = \{A'B'AC'\},$$
or $\quad \dfrac{AB.A'C}{AC.A'B} = \dfrac{A'B'.AC'}{A'C'.AB'},$

or $\quad AB.AB' : AC.AC' = A'B.A'B' : A'C.A'C'.$

Conversely, when one of these relations is established the three couples $AA'$, $BB'$, $CC'$ are proved to be in involution.

(ii) Otherwise thus. Let circles be drawn each through two conjugate points of an involution so as to cointersect at points $P$ and $Q$ (Art. 109). Join $P$ to any four of the points and join $Q$ to their conjugates. Then it may be easily shewn that the two pencils thus formed have their angles equal (or supplementary) each to each,* and are therefore equianharmonic.

---

* For this proof see McDowell's *Exercises on Euclid and in Modern Geometry*, Arts. 206, 7 (Cambridge, 1878).

**112.** *The foci and any two conjugate points of an involution constitute a harmonic range.*

This is evident from Art. 35, Cor. 1, where $A$ and $A'$ may be taken as the foci of an involution of which $S$ and $X$ are any two conjugate points.

Conversely, any two points $S$ and $X$ which form a harmonic range with a pair of fixed points $A$ and $A'$ are conjugate points in an involution whereof $A$ and $A'$ are the foci.

### SCHOLIUM B.

The theory of Involution was laid down by DESARGUES (1593—1662), the friend of Descartes and teacher of Pascal, in his BROUILLON PROIECT *d'une atteinte aux événemens des rencontres d'un Cone auec un Plan*, discovered in manuscript by M. Chasles in 1845, and printed in Poudra's two volume edition of the *Œuvres de Desargues*, vol. I. pp. 97—230 (Paris, 1864); an analysis of the work being also given, in which its strange and embarrassing terminology is replaced by the expressions now in use.

The germ of the theory is contained in lib. VII. prop. 130 of the *Collectio* of PAPPUS (p. 873, ed. Hultsch), which may be stated conversely as follows:

*If the sides of a triangle PQR meet a transversal in A, B, C, and if the three radiants from any point O to its opposite vertices meet the transversal in F, E, D respectively, then*

$$AF \cdot CB : AB \cdot CF = FA \cdot DE : FE \cdot DA.$$

That is to say, the opposite sides and the two diagonals of any quadrilateral $OPQR$ meet any transversal in pairs of points $(A, F)$, $(B, E)$, $(C, D)$ in involution, the cross ratio $(AFCB)$ being equal to $(FADE)$.

DESARGUES, having defined and established some properties of his *Involution de six Points*, and having enunciated the so-called theorem of Ptolemy in a new form, next shews that the pencil subtended at any vertex by six points in involution is cut in points in involution by any transversal (Poudra, pp. 247, 256—8); he fully establishes the theory of poles and polars (sometimes wrongly attributed to de La Hire), shewing *inter alia* that the three pairs of chords joining four points on a conic determine a self-conjugate triangle, and not omitting to notice also the case of polar planes (pp. 263, 72, 90); and he proves that any conic and the sides of an inscribed quadrilateral determine points in involution on any transversal (p. 268). It is to be observed that he regards six points as constituting a complete involution (which however does not really detract from the generality of his conclusions), and that he uses the term *Involution de quatre Points* to denote the two double points or foci together with a single pair of conjugate points; that is to say, he regards a harmonic range as an involution of four points.

It may be well to quote from Poudra's vocabulary of terms used by Desargues (*Œuvres*, I. 101) the explanation of the still surviving term *Involution*, which signifies "Trois couples de points tels que $oa \times oa' = ob \times ob' = oc \times oc'$ et que tous les points conjugués sont tous mêlés ou démêlés entr' eux."

## ANHARMONIC PROPERTIES OF CONICS.

### PROPOSITION I.

**113.** *The cross ratios of four fixed points on a conic and of the tangents thereat are constant and equal to one another.*\*

(i) From four fixed points $A$, $B$, $C$, $D$ on a conic draw chords to any point $P$ on the curve, and produce them to meet the $S$-directrix in $abcd$ respectively, so that

$$P\{ABCD\} = P\{abcd\} = S\{abcd\}.$$

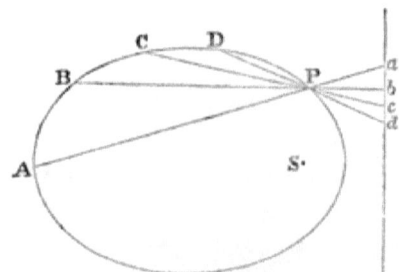

Then since the angles $aSb$, $bSc$, $cSd$, being equal or supplementary to the halves of the constant angles $ASB$, $BSC$, $CSD$ respectively (Art. 13), are themselves constant, it follows that $S\{abcd\}$ and $P\{ABCD\}$ are constant, whatever be the position of $P$ on the curve. Conversely, a conic can in general be drawn through six points† $PP'ABCD$ so related

---

\* The reciprocal properties (i) and (ii) were stated in their direct and converse forms in **Steiner's** *Systematische Entwickelung der Abhängigkeit geometrischer Gestalten von einander*, pp. 156, 7 (Berlin, 1832), where it is remarked that: "Die Sätze links [= (ii)] sind, unter anderer Form abgefasst, bekannt." The property (iii) was given (for the circle) in **Chasles'** *Géométrie Supérieure*, § 663 (Paris, 1852).

† The six points in Art. 105 may be regarded as lying on a conic which has degenerated into a pair of straight lines.

that $P\{ABCD\} = P'\{ABCD\}$; or if $ABCD$ be fixed points, and $P$ a variable point such that $P\{ABCD\}$ is constant, **the locus** of $P$ is a conic through $ABCD$.

(ii) Next let the tangents to **a conic** at four fixed **points** $ABCD$ meet the **tangent at any** fifth point in the range $abcd$.

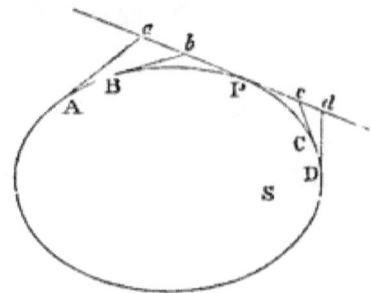

Then since the angles $aSb$, $bSc$, $cSd$, being equal or supplementary to the halves of the constant angles $ASB$, $BSC$, $CSD$ respectively (Art. 13, Cor.), are themselves constant, it follows that $S\{abcd\}$ or $\{abcd\}$ is constant. Conversely, a conic can in general be described touching six straight lines so related that four of them determine ranges of equal **cross** ratios $\{abcd\}$ and $\{a'b'c'd'\}$ on the remaining two; or **if four fixed straight lines meet** a variable straight **line in a range of constant** cross ratio, the variable line envelopes a conic touching the four fixed lines.

(iii) Lastly let $P$ be a fixed point on a conic, $PQ$ a variable chord meeting the $S$-directrix in $R$, and let $T$ be the point of concourse of the tangents at $P$ and $Q$.

Then since the angle $RST$ is always a right angle (Art. 9, Cor. 1),
$$P\{Q\} = P\{R\} = S\{R\} = S\{T\} = \{T\},$$
or the pencil **subtended at** $P$ **by** any four positions of $Q$ on the curve is equal to the **range in** which the tangents at the $Q$'s intersect the tangent at $P$.

*Corollary* 1.

*If two angles* **MBD**, *MCD of given magnitudes turn about* $B$, $C$ *as poles in such a manner that the intersection $M$ of two* *of their arms describes* **a straight line, the intersection** $D$ *of their remaining arms describes* (**in** *general*) *a* **conic passing** *through* $B$ *and* $C$;* for since

$$B\{D\} = B\{M\} = C\{M\} = C\{D\},$$

a conic can be drawn through $B$, $C$ and any four positions of $D$; and three assumed positions of $D$ together with $B$ and $C$ determine a single conic on which every other position of $D$ must lie. But if the straight line described by $M$ passes through $B$ or $C$, or if the pencils $B\{D\}$ and $C\{D\}$ have a common ray, the locus of $D$ is a straight line. The general case affords a ready means of drawing a conic through five given points. [Ex. 367.

### Corollary 2.

*If three straight lines* $md$, $dr$, **rm** *turn* **about given poles** $B$, $C$, $D$, *whilst* $m$ *and* $r$ *move along fixed* **straight lines** $PG$ *and* $PQ$, *the point* $d$ *describes* **a conic passing through** $B$ *and* $C$; for it is **evident that**

$$B\{d\} = \{m\} = \{r\} = C\{d\}.$$

**It appears (by taking** special cases) that **the** point $P$ and **the** intersections of $BD$, $PQ$ and $CD$, $PG$ likewise belong to the locus: and **conversely, if** these three points and $B$ and $C$ be given, the lines $PG$ and $PQ$ can be drawn, and the locus of $d$, which is the conic through the five given points, can be traced.†

---

\* This is Newton's method of generating conic sections. The theorem is proved in the *Principia*, Lib. I. Sect. v. lemma 21, where it is deduced from lemma 20 [= Ex. 80], and this again from the theorem *Ad quatuor lineas* (Scholium C). It is also stated, with generalisations and limitations, in his *Enumeratio Linearum Tertii Ordinis*, Cap. VII. (see p. 26 in Talbot's translation, London 1861; or Newtoni *Opera quæ exstant omnia*, vol. I. 556, ed. Horsley, London 1779), under the head: *De Currarum Descriptione Organica.*

† This method of drawing a conic through five given points was discovered by Maclaurin in 1722 although not published by him till 1735. In the *Philosophical Transactions of the Royal Society of London* (vol. VIII. p. 50, 1805) he shews how to deduce it by elementary geometry from the above mentioned lemma 20 of the *Principia*, "which itself is a case of this description" (p. 43). Braikenridge, who had already published Maclaurin's construction, is said to have been in communication with him and to have been made acquainted with his theorems in 1727 (*ibid.* pp. 6, 43).

## PROPOSITION II.

**114.** *The diameters of a conic form a pencil in involution in which conjugate diameters are conjugate rays.*

(i) For since any two conjugate diameters meet the $S$-directrix in points $V$ and $V'$ such that

$$XV \cdot XV' = CX \cdot SX = \text{a constant,} \quad [\text{Art. 14, Cor. 4.}]$$

therefore $CV$ and $CV'$ are conjugate rays in a pencil in involution. This involution is *negative* in the ellipse, and *positive* in the hyperbola: in the latter case the asymptotes are the double rays. [Art. 53.

(ii) The centre of the involution determined by the pencil of diameters on any tangent is at its point of contact, since this is evidently conjugate to the point at infinity upon the tangent. This is in accordance with Art. 47, from which a second proof of the proposition may be derived.

### Corollary 1.

From a fixed point $O$ draw the perpendicular $OP$ to a variable diameter $CP$, and produce it to meet the conjugate diameter in $D$: then since

$$C\{D\} = C\{P\} = O\{D\},$$

the locus of $D$ is a conic through $O$ and $C$, and it evidently has real points at infinity on the axes of the original conic. At the four points of concourse of the conics the positions of $OD$ are normal at $D$ to the original conic. Hence we infer that *there are four points on a given conic such that the normals thereat cointersect at a given point $O$, and the four points lie on an* **equilateral** *hyperbola, which passes through $O$ and through the* **centre of** *the given conic, and has its asymptotes parallel to the axes of the latter.*[*]

---

[*] This method of drawing a normal $OD$ to a conic from a given point $O$, with the help of an equilateral hyperbola, is given by APOLLONIUS in his *Conics*, Lib. **v.** props. 58–63, where he regards the normal as a line drawn from $O$ so that the intercept upon it between $D$ and the axis of the given conic is a minimum. For another treatment of normals see Prof. CREMONA's article *On Normals to Conics*, in the *Oxf. Camb. and Dublin Messenger of Mathematics*, vol. III. p. 88. See also Ex. 286,

## Corollary 2.

By drawing $OP$ to meet $CP$ (Cor. 1) at any other *given angle* and in a given "sense" or direction of rotation, and proceeding as above, we determine a conic passing through $O$ and $C$ and having its asymptotes (real or imaginary) parallel to those conjugate diameters of the given conic which contain an angle equal to the given angle; and the two conics intersect at four points $D$ such that $OD$ meets the given conic at the given angle. If the sense of rotation be not given, four other positions of $OD$ (making in all eight), which meet the given conic at the given angle, can be in like manner determined.

### SCHOLIUM C.

The celebrated problem of the "*Locus ad quatuor lineas*" (τόπος ἐπὶ δ' γραμμάς)—handed down by PAPPUS from his predecessors without solution (*Collectio* Lib. VII, vol. II. pp. 677—9, ed. Hultsch), solved only by his new method of coordinates by DESCARTES (*Geometria* Libb. I. II. 7—16, 24—34, ed. Schooten, 1659), and at length completely solved by NEWTON (*Principia* Lib. I. Sect. V. lemm. 17—19) by the elementary geometry of Apollonius—implicitly contains the fundamental anharmonic property of four points on a conic (Art. 113). The problem and its Newtonian solution are as follows.

(i) *If $P$ be any point of a conic and $ABDC$ a given inscribed trapezium, and if straight lines $PQ$, $PR$, $PS$, $PT$ meet the sides $AB$, $CD$, $AC$, $DB$ respectively at given angles: the rectangle $PQ \times PR$ is to the rectangle $PS \times PT$ in a given ratio.*

*a.* First let $PQ$ and $PR$ be parallel to $AC$, and $PS$ and $PT$ parallel to $AB$; and let the side $BD$ be also parallel to $AC$. Then since $PQ.QK$ varies as $AQ.QB$, "per prop. 17, 19, 21 & 23 lib. III. conicorum *Apollonii*" (Art. 16, Cor. 1), if $K$ be the point in which $PQ$ meets the conic again; and since the diameter of the chords $AC$, $BD$, $KP$ bisects also the intercept $QR$, so that $QK = PR$; it follows that $PQ.PR$ varies as $AQ.QB$ or $PS.PT$, as was to be proved.

*b.* Next let $BD$ be not parallel to $AC$.* Draw $Bd$ parallel to $AC$ meeting $ST$ in $T$ and the conic in $d$. Join $Cd$ cutting $PQ$ in $r$,

---

note, where if $T$ and $U$ be given $T'$ and $U'$ can be at once determined; and thus from the intersection of any two normals to a conic two other normals can be drawn. Corollaries 1 and 2, as they stand, are taken from CHASLES' *Traité des Sections Coniques*, chap. VII. pp. 142—4. On the parabola see Ex. 617, note.

* See the lithographed figure No. 3.

and draw $DM$ parallel to $PQ$ cutting $Cd$ in $M$ and $AB$ in $N$. Then by similar triangles, and by parallels, $Bt$ (or $PQ$) is to $Tt$ as $DN$ to $NB$; and $Rr$ is to $AQ$ (or $PS$) as $DM$ to $AN$; and therefore

$$PQ.Rr : PS.Tt = \text{rect. } NDM : \text{rect. } ANB,*$$
$$= PQ.Pr : PS.Pt, \text{ by case } a,$$
$$= PQ.PR : PS.PT, \text{ dividendo.}$$

*c.* Having **thus shewn** that this last ratio has the constant value $DN.DM : AN.NB$, we see at once that $PQ.PR$ will still vary as $PS.PT$ if $PQ$, $PR$, $PS$, $PT$ be drawn each at its own constant inclination to $AB$, $CD$, $AC$, $DB$ respectively. It is further evident that if **$PX$** and **$PY$** be drawn to meet the diagonals $AD$ and $BC$ at constant angles, each of the ratios $PX.PY : PQ.PR : PS.PT$ is constant. Conversely (lemma 18), if **$PQ.PR$** varies as $PS.PT$, the locus of $P$ is a conic circumscribing **$ABCD$**. In lemma 19 and its two corollaries NEWTON completes the solution of the problem by shewing how to determine the actual locus of $P$ for a given value of the ratio $PQ.PR : PS.PT$; and he concludes by remarking with evident satisfaction: "Atque ita problematis veterum de quatuor lineis ab *Euclide* incœpti & ab *Apollonio* continuati non calculus, sed compositio geometrica, **qualem veteres quærebant**, in hoc corollario exhibetur."

(ii) The anharmonic property of four points on a conic follows immediately from the above theorem *Ad quatuor lineas*. For if $PQ$, $PR$, $PS$, $PT$ be perpendiculars to $AB$, $CD$, $AC$, $BD$ respectively (cf. Art. 94), so that $PQ.AB = PA.PB \sin APB$, &c., and thus

$$\frac{PQ.PR.AB.CD}{PS.PT.AC.BD} = \frac{\sin APB}{\sin APC} \cdot \frac{\sin CPD}{\sin BPD} = P\{ABCD\};$$

and if the **ratio** $PQ.PR : PS.PT$ and the trapezium $ABDC$ be given; the cross ratio $P\{ABCD\}$ will be constant.

(iii) CHASLES and others have proved **the constancy of the** cross ratios of four given points on or tangents **to a conic by** projection **from** the circle, and have taken the properties thus **proved** as the foundation of their **treatises on** conics: **but the most** elementary **proofs of** the properties **in** question are those which **we have** adopted **in** Art. 113 from the *Geometrical Demonstration of the Anharmonic Properties of a* **Conic** contributed by Mr. B. W. HORNE, Fellow of St. John's College, to the *Quarterly Journal of Mathematics*, vol. IV. 278 (1861): his proofs assume only those simple angle-properties of the focus of a conic which reduce, when the directrix is removed to infinity, to **the** fundamental angle-properties of the circle (Scholium A, p. 22).

---

\* This notation was formerly in use for the **rectangle** $AN.NB$.

## RECIPROCAL POLARS.

### PROPOSITION III.

**115.** *The polar of any point with respect to a conic is a straight line, and conversely.*\*

(i) Let a variable chord be drawn to a conic through a given point $O$, and let the tangents at its extremities intersect in $T$; then will the locus of $T$ be a straight line.

For if the variable chord and its diameter $CT$ meet the

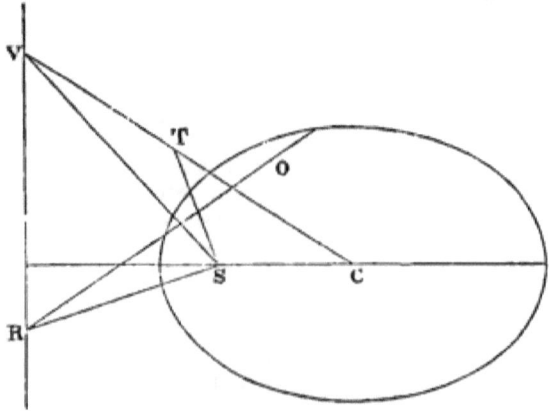

directrix in $R$ and $V$ respectively, then since $SR$, $ST$ and $OR$, $SV$ are at right angles (Arts. 9, 14), therefore

$$S\{T\} = S\{R\} = O\{R\} = S\{V\} = C\{T\}.$$

And since the homographic pencils $S\{T\}$ and $C\{T\}$ have a common ray $CS$ (the point $T$ lying on the axis when $OR$ is a principal ordinate), the locus of $T$ is a straight line.

[Art. 105.

Conversely, if $T$ be any point on a straight line, it may be shewn in like manner that its chord of contact determines homographic ranges $\{R\}$ and $\{\infty\}$ on the directrix and the straight line at infinity, and hence that it passes through a fixed point. [Art. 104.

(ii) Otherwise thus. Let $PL$ and $PN$ be given tangents to a conic, $PMR$ any chord through $P$, and $O$ its intersection

---

\* For other proofs of the properties of polars see Arts. 17, 18, 41, 93.

CROSS RATIO AND INVOLUTION.

with $LN$; and let $A$ and $B$ be the intersections of the tangents at $N$, $M$ and $N$, $R$.

Then since the cross ratios of the tangents at $L$, $M$, $N$, $R$

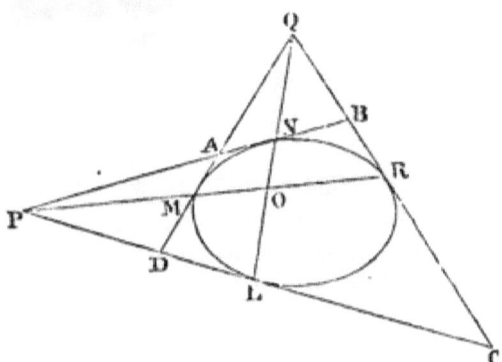

and of their points of contact are equal (Art. 113), therefore (estimating the former on the tangent at $N$ and the latter on the transversal $MR$) we have

$$\{PANB\} = L\{LMNR\} = \{PMOR\};$$

whence it follows that the tangents $AM$ and $BR$ intersect on the fixed straight line $NO$, and conversely.  [Art. 104.

### PROPOSITION IV.

**116.** *A row of points on any axis and their polars with respect to a conic are homographic, and they determine an involution on that axis.*

(i) It follows at once from the construction of Art. 115 (i), where $RST$ is a right angle, that if $T$ be taken on the polar of a given point $O$, then

$$\{T\} = S\{R\} = O\{R\},$$

or the points $T$ and their polars $OR$ are homographic.

(ii) Otherwise thus. In the preceding figure, if the equal cross ratios $L\{LMNR\}$ and $N\{LMNR\}$ be estimated on the transversal $MR$, then

$$\{PMOR\} = \{OMPR\},$$

or $O$ and $P$ are harmonic conjugates with respect to $M$ and $R$.

[Art. 106.

Hence, as $P$ moves along $MR$, the points $O$ and $P$ are couples in the involution of which $M$ and $R$ are the foci (Art. 112); that is to say, if $P$ be a variable point on the polar of any given point $Q$, and $QO$ be the polar of $P$, the range $\{P\}$ is homographic with the pencil $Q\{O\}$, and the two together determine an involution on the polar of $Q$.

*Corollary* 1.

If a pair of points divide a chord of a conic harmonically, each point lies on the polar of the other, and the two are said to be *conjugate with respect to the conic*. From the reciprocity of the relation between such points it is easy to deduce the theorem of Art. 17, Cor. 1, that *the intersection of any two straight lines is the pole of the line which joins their poles.* It is evident that *a system of conics having a common self-polar triangle* (note, p. 272) *determine an involution on any transversal drawn through a vertex of the triangle.*

*Corollary* 2.

Two straight lines are said to be *conjugate with respect to a conic when they pass each through the polar of the other*, or in other words, when they are harmonic conjugates with respect to the two tangents (real or imaginary) that can be drawn to the conic from their point of concourse. From a given point there can in general be drawn one pair only of straight lines *at right angles* to one another and conjugate to a given conic (Art. 110); but *if the given point be a focus, every two conjugate lines drawn through it are at right angles* (Art. 7); and conversely it may be shewn that no other real point is so related to the conic. Notice that conjugate diameters are also conjugate lines in the sense of this corollary.

PROPOSITION V.

117. *If the locus of a point be a conic the envelope of its polar with respect to a conic is a conic, and conversely.*

Take four fixed points $ABCD$ and their polars with respect to a conic: and take a variable point $P$ and its polar with respect to the same conic: and let the fixed polars meet the polar of $P$ in the points $a$, $b$, $c$, $d$ respectively.

Then **by** Art. 116, the points $a$, $b$, **c**, $d$ are the poles **of** $PA$, $PB$, $PC$, $PD$; and $P\{ABCD\}$ is equal to $\{abcd\}$; and **if** the one be constant the other is constant.

Hence, if **the locus of** $P$ **be a conic** passing through the points $ABCD$, **the envelope of the polar of** $P$ will be a conic touching the **polars of** $ABCD$, **and conversely.** [Prop. I.

*Corollary.*

**To any** figure generated by **points or** *poles* corresponds a reciprocal figure generated by their *polars* with respect **to any** assumed auxiliary conic; and **any** property of the **one** figure implies a reciprocal property of **the other.** The method of deducing reciprocal properties **from one another will form the** subject **of** Chapter XII. Notice, **as a** special case of **the** proposition, **that any conic may be regarded as its own** reciprocal, its several points being the poles of **the tangents** thereat. Also see Scholium E at the end of this chapter.

## THE TRIANGLE.

#### PROPOSITION VI.

**118.** *If two triangles circumscribe a conic their six vertices lie on a conic, and conversely.*

Let $ABC$ and $DEF$ be two triangles whose six sides touch the same conic: let $BC$ meet $DF$ in $d$ and $EF$ in $e$: and let $DE$ meet $AB$ in $b$ and $AC$ in $c$.

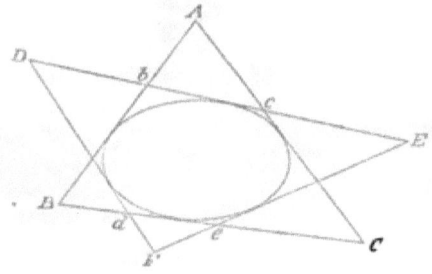

Then by Prop. I, since the four tangents $AB$, $AC$, $FD$, $FE$ are homographic with respect to the tangents $BC$ and $ED$, therefore

$$\{BCde\} = \{bcDE\},$$
or $$F\{BCDE\} = A\{BCDE\};$$

and therefore the six points $ABCDEF$ lie on one conic.

Conversely, if the six vertices of two triangles lie on a conic, it may be shewn in like manner that their six sides touch a conic.

### Corollary 1.

If the enveloped conic be a parabola, and if $D$ be taken at its focus and $E$ and $F$ at the circular points at infinity (Chap. XI), the conic through $ABCDEF$ becomes a circle. Hence *the circumscribed circle of any triangle $ABC$ whose sides touch a parabola passes through the focus* (Art. 29).

### Corollary 2.

*If two conics be so related that a single triangle can be inscribed in the one and circumscribed to the other, an infinity of triangles can be so circuminscribed to the two conics.* For let $ABC$ be the first triangle, and $ab$ any chord of the outer conic which touches the inner: complete the triangle $abc$ by drawing the second tangents from $a$ and $b$ to the inner conic: then the point $c$ must always lie on the same fixed conic with the points $ABCab$.

### PROPOSITION VII.

**119.** *If two triangles be self-polar\* with respect to a conic, their six vertices lie on a conic, and their six sides touch a conic.*

Let $ABC$ and $DEF$ be two triangles self-polar with respect to a conic:† then evidently the join of any two of their six sides is the pole of the join of the opposite vertices. [Art. 116.

---

\* A triangle may be called *self-polar* with respect to a conic— cf. the French term "*autopolaire*"—when each vertex is the pole of the opposite side. Such triangles are usually termed *self-conjugate*, and some writers call them *self-reciprocal*. The vertices of a self-conjugate triangle constitute what is called a *Conjugate* (or self-conjugate) *Triad* of points.

† See the lithographed figure No. 4.

Let $BC$ meet $AD$ in $d$ and $EF$ in $d'$: then $d'$ is the pole of $AD$, and $d$ and $d'$ are therefore conjugate points (Art. 116) with respect to the conic.

Also let $BC$ meet $AE$ in $e$ and $DF$ in $e'$: then $e$ and $e'$ are likewise conjugate points with respect to the conic: and it is evident that $B$ and $C$ are conjugate with respect to it.

It follows that the points $BC$, $dd'$, $ee'$ are couples in involution, and hence (Arts. 111, 102) that

$$\{BCde\} = \{CBd'e'\} = \{BCe'd'\},$$

or
$$A\{BCDE\} = F\{BCDE\}.$$

Therefore a conic goes through the six points $ABCDEF$; and therefore another conic (Prop. VI) touches the six sides of the two triangles.

### Corollary 1.

*If upon a given conic one triad of points self-conjugate with respect to a second given conic can be determined, an infinity of such triads can be determined upon it.* For if $ABC$ be the first triad, $R$ any other point on the first conic, $S$ one of the points in which the polar of $R$ with respect to the second conic meets the first, and $T$ the pole of $RS$ with respect to the second conic; the point $T$ must always lie on the same conic with $ABCRS$. By taking $R$ at the centre of an equilateral hyperbola, and $S$ and $T$ at the circular points at infinity (Chap. XI.), we deduce that *the circle through any conjugate triad with respect to an equilateral hyperbola passes through its centre.*

[Art. 64, Cor. 4.

### Corollary 2.

Let the first conic become a circle, and let $Q$ be one of its points of concourse with the second conic. Next let the points $R$ and $S$ coalesce at $Q$, so that the inscribed self-polar triangle $RST$ degenerates into the vanishing triangle $QQT$: then $T$ becomes *the pole with respect to the conic of the tangent to the circle at $Q$.* Conversely, if $QQ'$ be any chord of a conic and $T$ its pole, the circle drawn through $T$ so as to touch $QQ$ at $Q$ (or $Q'$) is such that an infinity of triangles self-polar with respect to the conic can be inscribed in it.

### Corollary 3.*

Let the circle described as above meet the diameter $CT$ of the conic in a second point $t$, and let $V$ be the middle point of $QQ'$, and $CP$, $CD$ the semi-diameters conjugate and parallel to $QQ'$. Then since

$$VT \cdot Vt = QV^2 = \frac{CD^2}{CP^2}(CP^2 \sim CV^2) = CD^2 \cdot \frac{CT \sim CV}{CT}$$

it follows that $CT \cdot Vt$ is equal to $CD^2$, and hence that

$$CT \cdot Ct = CP^2 \pm CD^2 = CA^2 \pm CB^2;$$

and hence that *the circumscribed circle of any triangle self-polar with respect to a conic is orthogonal to its director circle*. In the case of the parabola every such circle is orthogonal to the directrix, or *has its centre upon the directrix*. Otherwise thus: let any circle in which self-polar triangles with respect to a conic can be inscribed meet the director circle in $O$ and the chord of contact of the tangents $OQ$, $OQ'$ to the conic in $P$ and $P'$; and let $V$ be the middle point of $QQ'$. Then since the points $P$ and $P'$ are conjugate, and since the angle $QOQ'$ is a right angle,

$$VP \cdot VP' = VQ^2 = VO^2;$$

and therefore the line $VO$, which is normal to the director circle at $O$, is the tangent at $O$ to the circle $OPP'$.

## THE QUADRILATERAL.

### PROPOSITION VIII.

120. *The intersections of the opposite sides and of the diagonals of a quadrilateral are a conjugate triad with respect to every conic circumscribing the quadrilateral.*

---

* The former of the two proofs of GASKIN's theorem (note, p. 186) in Cor. 3 is due to M. Paul Serret (*Nouvelles Annales* xx. 79, 1861). The theorem may also be proved independently of Prop. VII., as follows: let $QQ'$ be any chord of an ellipse, $V$ and $T$ its middle point and pole, and $P$, $R$ any two conjugate points upon it, so that $VP \cdot VR = QV^2$. Let the circle round $TPR$ meet $VT$ again in $O$: then it may be shown that $VO \cdot VT = QV^2$, and hence that $CO \cdot CT = CA^2 + CB^2$ (*Nouv. Ann.* xx. 25). See also the *Quarterly Journal of Mathematics* x. 129.

If $ABCD$ be any four points on a conic, and $OPQ$ the intersections of $(AC, BD)$, $(AB, CD)$, $(AD, BC)$; then since

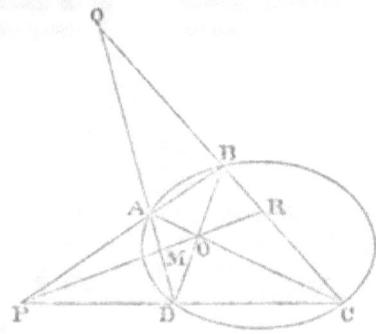

the line $OP$ and the point $Q$ divide both $AD$ and $BC$ harmonically (Art. 107), therefore $OP$ is the polar of $Q$.

In like manner $OQ$ is the polar of $P$: and therefore $O$ is the pole of $PQ$, and the points $OPQ$ are a conjugate triad, as was to be proved.

*Corollary* 1.

To draw tangents to a conic from a given point $Q$ *with the ruler only*, draw any two chords $QAD$ and $QBC$ from the given point: let the line $PO$ (the join of the joins of $AB$, $CD$ and $AC$, $BD$) cut the conic in $T$ and $T'$: draw $QT$ and $QT'$, which will be the tangents required.

*Corollary* 2.

From a given point $P$ draw a fixed chord $PAB$ and a variable chord $PDC$ to a conic. Then since $AC$, $BD$ and $AD$, $BC$ meet on the polar of $P$, therefore

$$A\{C\} = B\{D\} = A\{D\}, \qquad \text{[Prop. I.}$$

where $C$ (or $D$) may be *either* extremity of the variable chord. Hence, taking any three positions of $CD$, we have

$$A\{CC'C''D\} = A\{DD'D''C\};$$

and therefore *any variable chord* $CD$ *drawn through a fixed* *point determines an involution at any point $A$ on the conic,*\* *and conversely.* For example, a chord of a conic which subtends a right angle (Art. 110) at a given point on the curve passes through a fixed point on the normal thereat.†

### PROPOSITION IX.

**121.** *The three diagonals of a complete quadrilateral determine a triangle which is self-polar with respect to every conic inscribed in the quadrilateral.*

Let $a$, $b$, $c$, $d$ denote the tangents at any four points $A$, $B$, $C$, $D$ on a conic, and $ab$ the join of any two of them $a$ and $b$.

Then (in the preceding figure), since $AB$ and $CD$ pass through $P$, their poles $ab$ and $cd$ lie on $OQ$ the polar of $P$. And in like manner $ad$ and $bc$ lie on $OP$, and $ac$ and $bd$ on $PQ$.‡

That is to say, the three diagonals of the circumscribed quadrilateral $abcd$ lie upon the sides of the self-polar triangle $OPQ$.

### Corollary.

In the reciprocal quadrilaterals $abcd$ and $ABCD$ determined by any four tangents to a conic and their points of contact respectively, two pairs of diagonals cointersect and form *a harmonic pencil* $O\{PAQB\}$ (Art. 107); and the third diagonals lie in one straight line, and their extremities form *a harmonic range* $\{ac, P, bd, Q\}$. [Prop. IV.

---

\* Otherwise thus: if $PE$ and $PF$ be the tangents to the conic from the given point $P$, then $E\{ECFD\}$ is harmonic (Art. 18), and therefore $A\{ECFD\}$ is harmonic, or $AC$ and $AD$ are conjugate rays in the involution of which $AE$ and $AF$ are the double rays. Note that four *points on* a conic are said to be *harmonic* when they subtend a harmonic pencil at every fifth (Prop. I.); and the tangents at four such points are said to be harmonic.

† This theorem is due to Frégier (Gergonne's *Annales* VI. 231, 1816).

‡ This was proved by MACLAURIN in Sect. II. §§ 35, 36 of the *Appendix* to his work on algebra above referred to (Ex. 324, note), in which he applied COTES' theorem of harmonic means to curves of the second order. He thus virtually reciprocated a theorem of Desargues (Prop. VIII.), although reciprocation, as a method, was only discovered in the century following.

## PROPOSITION X.

**122.** *Any conic and the three pairs of* **chords joining any** *four points upon it meet* **every** *transversal in four pairs* **of points** *in involution.*\*

(i) For if any **transversal meet a conic** in the points $AA'$, and any two **of the three pairs of chords joining** four points on the curve in the **points** $BB'$ **and** $CC'$, then by the theorem *Ad quatuor lineas* (Scholium C),

$$AB.AB' : AC.AC' = A'B.A'B' : A'C.A'C';$$

and therefore $AA'$, $BB'$, $CC'$ are couples in an involution (Art. 111); and the third pair of connecting chords determine a fourth couple in the same involution.

(ii) Otherwise thus. Let any transversal meet a conic in the points $AA'$, and any two chords $ab$ and $cd$ in $BB'$, and the chords $ad$ and $bc$ in $CC'$.

Then since the points $AdbA'$ on the conic are equicross with respect to $a$ and $c$, therefore (estimating $a\{AdbA'\}$ and $c\{AdbA'\}$ on the transversal) we have

$$\{ACBA'\} = \{AB'C'A'\} = \{A'C'B'A\};$$

and therefore $AA'$, $BB'$, $CC'$ are couples in an involution

(iii) The four sides of a given trapezium suffice to determine an involution on any transversal (Art. 108), and every conic circumscribing the trapezium passes through an additional couple in such involution; and no conic which does not circumscribe the trapezium can pass through a couple in such involution for every position of the transversal. For if the transversal, as it turns about any point $A$ of a conic, meets it again always

---

\* This is one of the fundamental theorems of DESARGUES. Having first proved it for the circle he extended it to the general conic by projection, leaving to others to devise some direct proof applicable to the general case (Poudra's *Œuvres de Desargues* I. 174, 193). The proof given above in Art. 122 (i) shews that the theorem is an immediate corollary from the ancient theorem *Ad quatuor lineas*. For the second proof (with a diagram) see Salmon's *Conic Sections*, Art. 344. The theorem seems to have been first stated for the case of *three conics*, instead of one conic and an inscribed quadrilateral, by Sturm (Gergonne's *Annales* XVII. 180). At the end of the same memoir (p. 198) Sturm alludes to the reciprocal theorem of Prop. XI. See also Poncelet *Traité des Propriétés Projectives* II. 149.

in the conjugate **point** $A'$, **this conic must have an** infinity of points $A'$ **in common** with the conic through $A$ and the four summits of **the trapezium.** Hence the proposition may be stated **as** follows:

*A* **system of conics** *through four* **common points** *(with* **their three** *pairs of common chords)*\* *meet every transversal* **in pairs** *of points in involution, and conversely.*

Each of the two double points belonging to any transversal **must be at** its point of contact with one of the conics of the system, or at an intersection of a pair of their common chords. There can therefore in general be drawn *two* **conics** through **four** given points to touch **a given** straight line.

### Corollary 1.

The foci of any transversal† with **respect to a** quadrilateral are evidently *conjugate points* with respect to every conic circumscribing it. Hence *the polar of a given point F with respect to a system of conics through four given points ABCD passes through a fixed* **point** $F'$,‡ which **is** determined **as the second** focus of the transversal which touches the conic $ABCDF$ at $F$.

### Corollary 2.

Through three given points $ABC$ draw **two conics** touching **a given line at its extremities** $F$ and $F'$ respectively. These **conics** intersect at **a** fourth point $D$, through which must pass every conic through $ABC$ which has $F$, $F'$ for a pair of conjugate points.§ Hence we are led to infer generally that, **in describing** conics subject to given conditions, *the* **condition** *that* **two specified** *points F and F'' should be conjugate* **with respect to a** *conic is equivalent to having given one* **point** $D$ **on the** *curve.*

---

\* Each pair may **be regarded as a** degenerate hyperbola of the system.

† **We** use this expression as an abbreviation for "the foci of the involution determined **upon any** transversal by the **sides of the** quadrilateral, taken in **opposite pairs."**

‡ It **was** proved analytically **by** Lamé (*Examen des différentes méthodes employées pour résoudre les problèmes de Géométrie* pp. 34—38, Paris, **1818**) that *if a system of conics* (or quadrics) *have the same points of intersection, their diameters* (or diametral planes) *severally conjugate to a* **system** *of parallel diameters meet in a point.*

§ For example, if $F$ and $F'$ be the circular points at infinity, the conics are equilateral hyperbolas, and they pass through the orthocentre of the triangle $ABC$.

### Corollary 3.

Each common tangent $FF'$ to two conics is cut harmonically by every other conic (or pair of chords) through their four common points. By supposing three of the four common points to coalesce, and one of the conics to become a circle, we deduce that the common chord of a conic and its *circle of curvature* at any point and their common tangent thereat divide their further common tangent harmonically. [Ex. 647.

### Corollary 4.

A system of conics having *double contact* cut every transversal in pairs of points in an involution, having one focus upon their common chord of contact $CC'$. In particular, if a transversal meet one of the conics in $AA'$, and their fixed common tangents (which themselves constitute a degenerate conic of the system) in $BB'$, then $CC'$ passes through a focus of the involution $AA'$, $BB'$.* Hence, if two points $AA'$ on a conic and also two tangents to it be given, their chord of contact passes through one or other of two fixed points on the line $AA'$; and if a third point $A''$ on the conic be given, the same chord of contact passes also through one of two fixed points on the line $AA''$, and may therefore have any one of *four* positions. There are therefore four solutions of the problem, *to draw a conic through three given points so as to touch two given lines*.

### Corollary 5.

If one focus of an involution be at infinity its other focus bisects every segment of the involution (Art. 112). Hence and from Cor. 4, *any two conics having* double contact *make equal and opposite intercepts on every transversal parallel to their common chord*, and therefore on *every* transversal without exception in the case in which their common chord is the straight line at infinity (Art. 57). Conversely, we are led to infer from Ex. 50 that *every two similar and coaxal conics are to be regarded as having double contact with one another upon the line at infinity*.

---

* Notice the special case of Ex. 69.

## PROPOSITION XI.

123. *The pairs of tangents from **any** point to a system of conics inscribed in the same quadrilateral form a pencil in involution, and conversely.*

**Let four** tangents to a conic intersect **in the three pairs of** points $aa'$, $bb'$, $cc'$; and let any transversal meet their polars in $AA'$, $BB'$, $CC'$ and the conic in $DD'$; and **let $O$ be the point** of which the transversal is the polar.

Then **each ray of** the pencil $O\{aa'bb'cc'DD'\}$ is the polar **of the corresponding** point in the range $\{AA'BB'CC'DD'\}$, and the two systems are therefore homographic (Prop. IV); and **since the** latter is in involution **(Prop. X)** the former **is also in involution.**

If now we suppose **the four tangents to remain** fixed whilst the conic varies, the pairs of **tangents $OD$, $OD'$ from any** assumed point $O$ to every conic **of the system are conjugate** rays in the involution $O\{aa'bb'cc'\}$, **as was to be proved.**

Conversely **it may be** shewn (Art. 122 § iii) that if the tangents from **every point $O$ to a conic belong to the** involution $O\{aa'bb'cc'\}$, **the conic must be one** of the **system inscribed** in the quadrilateral whose summits are $aa'$, $bb'$, $cc'$.

### Corollary 1.

*The director circles of **all** conics inscribed in the same quadrilateral are coaxal.*\* For if $O$ be taken at either point of concourse **of any two of** these circles, the tangents from it to their two conics will be at right angles, **and therefore the** tangents from it to every conic of the **system will be at right**

---

\* This is one of GASKIN's theorems, for the reciprocal of which see Art. 69. **It** may also be deduced from Art. 119, **Cor. 3. combined** with Prop. XII., which require that the *limiting points* of every **two of the director** circles should lie upon a fixed straight line, and also upon the **circle through the** intersections of the three diagonals of the quadrilateral, and should **therefore** be two fixed points. **Prof. TOWNSEND has established the** analogous theorem in three dimensions, **that *the director spheres of the system of quadrics touching eight fixed planes (and therefore inscribed in the same developable surface)* have a common radical plane. See the** *Quarterly Journal of Mathematics* vol. VIII. 10–14. The same theorem **appears to** have been proved independently **by M. PICQUET** (Chasles *Rapport sur les progrès de la Géométrie* p. 370, Paris, 1870). It occurs to me **that the** director circle and sphere might **have been** called the ORTHOCYCLE and ORTHOSPHERE.

angles (Art. 110). To the same coaxal system belong the circles on the three diagonals of the quadrilateral as diameters.

[Art. 33, Cor. 3.

### Corollary 2.

If one side of the quadrilateral be at infinity, its three diagonals become the parallels through the vertices to the opposite sides of a triangle; and the circles upon these diagonals become the perpendiculars of the triangle, whose intersection must therefore be a point on the directrix of every *parabola inscribed in the triangle*. [Art. 29, Cor. 1.

### Corollary 3.

It may be shown by **reciprocation*** from Art. 122 (or directly, by the kind of reasoning there employed), that *the pole of a given straight line with respect to a system of conics inscribed in a quadrilateral lies upon a fixed straight line;* and that to have given that a specified pair of straight lines are conjugate with respect to a conic is equivalent to having given *one tangent* to the curve; and that *two* conics can in general be drawn through one given point so as to touch four given lines; and that *four* conics can be drawn through two given points so as to touch three given lines.

### Corollary 4.

Every pair of tangents $TP$, $TQ$ to a conic whose foci are $S$ and $H$ are harmonic conjugates with respect to the bisectors of the angle $STH$ (Art. 50), as are also the lines from $T$ to the circular points at infinity $\phi$ and $\phi'$ (Chap. XI). The tangents $TP$, $TQ$ are therefore a couple in the same involution with $T\{SH\phi\phi'\}$, and *every conic which has $S$ and $H$ for foci may accordingly be regarded as inscribed in the trapezium $S\phi H\phi'$*. On account of this affinity of the points $\phi$, $\phi'$ to the foci of every conic in their plane we shall sometimes speak of them as the Focoids, comparing the use of the term *centroid*, or quasi-centre.

---

* Notice that the proof of Prop. **XI.** is itself an example of reciprocation.

## PROPOSITION XII.

**124.** *The locus of the centres of all conics inscribed in a given quadrilateral is a straight line,*[*] *which also bisects its three diagonals.*

(i) Since the director circles of the system of inscribed conics are coaxal (Art. 123, Cor. 1), their centres, which are also the centres of their conics, lie on a straight line. This line is evidently the diameter of the quadrilateral (Art. 107), since the middle points of the three diagonals (regarded as flat conics inscribed in the quadrilateral) belong to the locus of centres.

The proposition also follows as a special case from Art. 123, Cor. 3 by regarding the centre of every conic as the *pole of the line at infinity* with respect to it.

(ii) In the following proof the parallelogram of forces is assumed.

Let any conic touch four fixed lines $AB$, $BC$, $CD$, $DA$ in $N$, $R$, $L$, $M$ respectively. Then the resultant of $AM$ and $AN$, regarded as representing forces, bisects the chord of contact $MN$, and therefore passes through the centre of the conic.

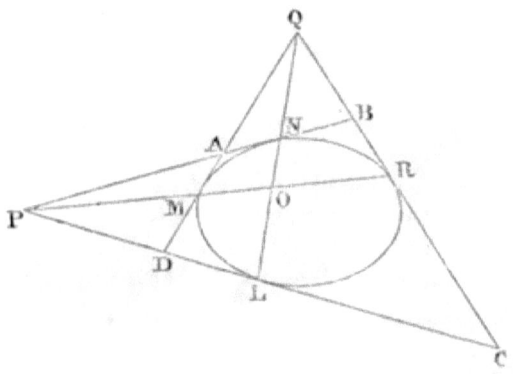

---

[*] This important theorem of NEWTON (Ex. 505, note), which was originally proved by the elementary method of Exx. 370-2, served as a starting point for later researches into the properties of systems of conics subject to less than five conditions. Notice the use made of it by BRIANCHON and PONCELET in Gergonne's *Annales* XI. 219. It might have been deduced from it by projection—a method not unknown to NEWTON—that there are an infinity of pairs of straight lines conjugate to the entire system of conics touching four given lines, which is the equivalent of Prop. XI. on *involution*.

Thus the centre is a point on the resultant of each of the pairs $(AM, AN)$, $(NB, RB)$, $(CR, CL)$, $(LD, MD)$; that is to say, it is a point on the resultant of $AB$, $CB$, $AD$, $CD$, and its locus is therefore a straight line.* It is evident that the resultant of these forces bisects the two diagonals $AC$ and $BD$; and by resolving them severally into

$$PB + AP, \quad CQ + QB, \quad QD + AQ, \quad CP + PD,$$

we see that it bisects the third diagonal $PQ$ also.

### Corollary.

One conic and one only can be inscribed in a given quadrilateral so as to have its centre upon any given straight line, since this line by its intersection with the diameter of the quadrilateral determines a single position of the centre of the conic. Hence we are led to infer that *to have given a diameter of a conic is equivalent to having given a tangent to it*, since either datum alike (when four other tangents are given) determines one conic and one only. This is in accordance with Art. 123, Cor. 3, since every diameter of a conic is conjugate to the line at infinity.

### PROPOSITION XIII.

**125.** *The centres of* all the conics which circumscribe a given quadrilateral lie upon its eleven-point *conic*.

(i) Through four given points *two* conics can be drawn so as to have their centres (real or imaginary) upon any given straight line. [Prop. XII. Cor.

The locus of the centres of all the conics through four given points $ABCD$ is therefore of the second order, since every straight line meets it in two points and two only.

The join $AB$ of any two of the four points meets this conic of centres in two points, which must evidently be the middle point of $AB$ and its intersection with $CD$.

(ii) Otherwise thus. If $O$ be the centre of any conic through $ABCD$, the radiants from $O$ parallel to the six joins of the

---

* *Nouvelles Annales* I. 24 (1862). For a proof depending upon the dynamical principle of moments see the *Quarterly Journal of Mathematics* VI. 215.

points $ABCD$ are homographic with the radiants from $O$ to their middle points (**Prop. 11**); and **the locus** of $O$ is therefore a conic through those **middle points**. And it is obvious that the three intersections $(AB, CD)$, $(AC, BD)$, $(AD, BC)$ are also points on the locus.

(iii) The conic of centres will have *two* real points **at infinity** or *one*, or *none* according as two real parabolas or one or **none can be** circumscribed to the quadilateral $ABCD$. **The locus will** therefore be in general **a hyperbola if** this quadrilateral be **convex** (Ex. 184), **or a parabola if** two of its sides be parallel, **and an** ellipse **if the quadrilateral** be reentrant.

(iv) **Let** either **of the two parabolas (real** or imaginary) which pass through $ABCD$ touch the line at infinity in $K$, which will accordingly be the pole of that line with respect to the parabola, and therefore the *centre* of the parabola.

The conic of centres therefore passes through the two **points on the line at infinity which are** conjugate with respect **to all** the conics through $ABCD$ (Art. 122, Cor. 1), as well as through **the six middle points and the** three intersections of **their three pairs of common chords; and we have therefore called it the** *Eleven-Point Conic* **of the quadrilateral** $ABCD$. **Its centre** *is at the centroid of the points* $ABCD$, since **at that point the joins of** the middle points of $(AB, CD)$, $(AC, BD)$, $(AD, BC)$ **meet** and bisect one another.*

**It is** evident that the *polars of any point on the eleven-point conic of $ABCD$ with respect to all the conics round $ABCD$ are parallel*, since they all meet in a point **(Art. 122, Cor. 1)**, and one of them is the line at infinity.

*Corollary* 1.

Since the eleven-point conic $E$ contains **a** conjugate triad **(Art. 120) with respect to every** conic $F$ through $ABCD$,

---

* This is at once evident, since four equal particles at $ABCD$ balance two and two about the middle points of any **pair of** the above lines; and therefore the centroid of the four particles is **at the middle** point of the line joining any such pair of middle points. In the *Quarterly Journal of Mathematics* VI. 127 I have shewn how to verify a simple construction **for the** centroid of the *area of any quadrilateral* by an extension of the barycentric principle.

therefore if $O$ be an intersection of $E$ and $F$, the tangent to $E$ at $O$ has its pole with respect to $F$ on $E$.   [Art. **119**, Cor. 3.

### Corollary 2.

The eleven-point conic of any quadrilateral $ABCD$ inscribed in a conic touches the diameter of the quadrilateral formed by the tangents to it at $ABCD$, since the complete locus of centres of the system of conics inscribed in the latter quadrilateral is its diameter (Prop. XII.), and the locus of centres of all conics round $ABCD$ is its eleven-point conic, and *one* conic only can be both inscribed in the one quadrilateral and circumscribed to the other.

### Corollary 3.

When the two points at infinity which are conjugate to all the conics round $ABCD$ are the *circular points* the eleven-point conic becomes the nine-point circle, and the points $ABCD$ become a triad and their orthocentre.  The nine-point circle really belongs to this form of tetrastigm, and not specially to any one of the four triangles determined by its vertices; in the same way that the system of equilateral hyperbolas circumscribing any one of these four triangles is a system of conics circumscribing the tetrastigm.

### SCHOLIUM D.

We have seen that an ellipse or a hyperbola may degenerate into a straight line $AA'$ or its complement (Art. 33, Cor. 3). For example, the diagonals of a quadrilateral may be regarded as flat conics inscribed in it, and accordingly their middle points belong to the locus of centres of all conics inscribed in it (Prop. XII).  This agrees with the bifocal definition $SP \pm HP = a$ constant, in accordance with which the point $P$ may in the limit lie anywhere upon the line $SH$; or upon the complement of $SH$, if the lower sign be taken.

Again, if $TP$, $TQ$ and $TP'$, $TQ'$ be the tangents from any point $T$ to two ellipses whose common foci are $S$ and $H$, the angles $PTP'$ and $QTQ'$ are always equal; and hence when the inner ellipse assumes the line-form $SH$ the angles $STP$ and $HTQ$ are equal. But since this is also the case when the second ellipse is left out of consideration, and the lines $TS$ and $TH$ are simply drawn through the fixed points $S$ and $H$, the point-pair $S$ and $H$ are so far indistinguishable from the flat conic $SH$.

Again, let the ellipse be regarded as the envelope of a straight line subject to the relation $\lambda\mu = b^2$, where $\lambda$ and $\mu$ are the perpendiculars upon it from $S$ and $H$. When $b$ vanishes the ellipse again appears to coincide with the points $S$ and $H$, which are represented by $\lambda = 0$ and $\mu = 0$ taken *separately*; but by taking $\lambda = 0$ and $\mu = 0$ simultaneously we find that (besides the tangents whose points of contact are at $S$ and $H$) the limiting conic has an infinity of tangents which ultimately coincide with the line joining $S$, $H$ and have their points of contact *distributed along* $SH$.* We may therefore say (1) that an ellipse degenerates into the line $SH$ joining its foci when its minor axis vanishes, meaning that $SH$ is an actual limiting form of the curve;† or (2) we may say that it "degenerates into" the point-pair $S$, $H$, understanding that at the instant at which it does so degenerate it ceases to belong properly speaking to the class conic, although the point-pair $S$, $H$ and the line joining them may be, as regards some properties, indistinguishable.

In like manner the hyperbola may be said to degenerate into its asymptotes $ECe$ and $E'Ce'$ (Art. 54); but strictly speaking it becomes the pair of vertically opposite angles $ECE'$ and $eCe'$, and then has for its tangents at $C$ those lines only through $C$ which lie *within* the said angles. The conjugate hyperbola at the same time becomes coincident with the two supplementary angles, and has for its tangents at $C$ all the lines through $C$ which fall within those angles. It is therefore practically sufficient to say that either of the two conjugate hyperbolas "degenerates into" the line-pair $ECe$, $E'Ce'$ and has for tangents every straight line through $C$; but the theoretical difference between these two views of the limit becomes apparent when we observe that the one makes the curvature at $C$ zero whilst the other makes it *infinite*.

For some further discussion of these matters see the *Quarterly Journal of Mathematics* VIII. 126, 235, 343. X. 93; *Oxf. Camb. Dubl. Messenger of Mathematics* IV. 86, 129, 140, 148; Chasles *Sections Coniques* pp. 30—33; Salmon's *Higher Plane Curves* pp. 377, 383 (ed. 3, 1879).

## HEXAGRAMMUM MYSTICUM.
### PROPOSITION XIV.

**126.** *The three pairs of opposite sides of any hexagon inscribed in a conic have their intersections in one straight line.*

(i) Let $ABCDEF$ be any six points on a conic, and let $O$, $P$, $Q$ be the intersections of $(AB, DE)$, $(BC, EF)$, $(CD, AF)$.

---

* This appears also by projecting the conic upon any plane from any vertex *in its own plane*.

† If $\lambda$, $\mu$, $\nu$ be the perpendiculars from three points upon a straight line, the envelope of a line subject to the relation $\lambda\mu\nu = c^3$ assumes a corresponding line-form when $c$ vanishes. By supposing each coordinate to become equal to a perpendicular of the triangle of reference whilst the product of the remaining two coordinates vanishes, we see that the limit of the curve is made up of three parts each of which constitutes a side of the triangle of reference or its complement. See also *Mathematical Questions from the* EDUCATIONAL TIMES, vol. XVI. 43.

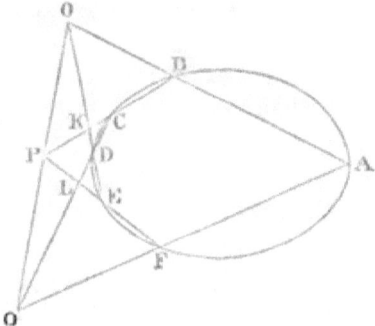

Then since the four points $ACDE$ are equicross with respect to $B$ and $F$, therefore

$$\{OKDE\} = B\{ACDE\} = F\{ACDE\} = \{QCDL\},$$

if $K$ be the intersection of $BP$, $DO$ and $L$ the intersection of $FP$, $DQ$.

And since the ranges $\{OKDE\}$ and $\{QCDL\}$ are thus equal and have a common point $D$, therefore the lines $OQ$, $EL$, $KC$ meet in a point, or the points $OPQ$ lie in a straight line,* as was to be proved.

(ii) *Otherwise thus.* Let $O$ and $Q$ be the intersections of $AB$, $DE$ and $AF$, $CD$ respectively. And let $OQ$ meet $AD$ in $R$, and $BC$ in $P$, and $EF$ in $P'$. Then will $P'$ coincide with $P$.

For the points $O$, $Q$ and the conic determine upon the transversal $OQ$ an involution to which, by a property of the inscribed quadrilateral $ABCD$, the couple $PR$ belong (Prop. x); and by a property of the quadrilateral $ADEF$ the couple $P'R$ belong to the same involution,† and therefore $P'$ coalesces with $P$.

(iii) *Otherwise thus.*‡ Consider the surface generated by

---

\* This line changes its position when the points $ABCDEF$ are taken in a different order. On the various Pascal-lines $OPQ$ see the note on Pascal's theorem at the end of Salmon's *Conic Sections*; and see Townsend's *Modern Geometry* chap. 17.

† The proposition is thus virtually a corollary (Art 122 §i) from the theorem *Ad quatuor lineas*. See also Salmon's *Conic Sections*, Art. 267.

‡ This proof, as it stands, is taken from *Math. Questions from the* EDUCATIONAL TIMES XVIII. 83 (1873). For the corresponding proof of Prop. XV. see vol. XIX. 65. Both theorems had been treated in this way by DANDELIN in vol. III. of the *Nouveaux Mémoires de l'Académie &c. de Bruxelles* (1826).

a straight line which always meets three fixed non-intersecting straight lines in space.

Let 1, 2, 3, be the fixed lines and 1', 2', 3' any three positions of the moving line. Then the common section of the two planes through 3', 1 and 2, 2' respectively passes through the points (3', 2) and (1, 2').

In like manner the common section of the planes through 1', 2 and 3, 3' passes through the points (1', 3) and (2, 3'); and the common section of the planes through (2', 3) and (1, 1') passes through the points (2', 1) and (3, 1'). The three common sections therefore form a triangle, and consequently *lie in one plane*.

Now let the surface be cut by any arbitrary plane. This plane will be met by the planes through 3', 1; 2, 2'; &c. in a hexagon 1, 1', 2, 2', 3, 3'; and by the three common sections (since they are co-planar) in three *collinear* points $PQR$, which are also the intersections of the opposite sides of the hexagon. The proposition is thus true for any plane section of the ruled quadric, and therefore for any conic.

### *Corollary* 1.

Five points $BCDEF$ on a conic being given, we may now find any number of sixth points $A$ on the curve, viz. by drawing arbitrary lines $BO$ through one of the given points $B$, and then determining successively the points $OPQ$, and the line $QF$, and its intersection with $BO$. Notice that $A$ is a vertex of the variable triangle $AOQ$, the extremities of whose side $OQ$ slide along fixed lines $ED$ and $CD$, whilst its three sides pass through three fixed points $PBF$ respectively.

[Prop. I. Cor. 2.

### *Corollary* 2.

If $ABCEF$ be five given points on a conic, the *tangent* at any one of them $C$ may be constructed by this proposition; for we have only to make $D$ coincide with $C$, in which case the line $CQ$ becomes the tangent at $C$. Again, by supposing $C$ to coincide with $B$ and $E$ with $F$, we deduce that the tangents at the vertices $B$ and $F$ of a *quadrilateral ABDF* inscribed in a conic intersect upon the straight line which joins the points of concourse of its sides $AB$, $DF$ and $AF$, $BD$.

### PROPOSITION XV.

**127.** *The joins of the three pairs of opposite vertices of any hexagon circumscribing a conic meet in a point.*

(i) Let the tangents at $A$, $B$, &c. in the preceding figure be $a$, $b$, &c.; and let $ab$ denote the intersection of any two of them $a$ and $b$.

Then the join of $ab$ and $de$ is the polar of $O$; the join of $bc$ and $ef$ is the polar of $P$; and the join of $cd$ and $fa$ is the polar of $Q$. And these three joins meet in a point, since their poles $OPQ$ are in one straight line.

(ii) Otherwise thus. Let $AA'$, $BB'$, $CC'$ be the opposite vertices of any hexagon circumscribing a conic; and let the four tangents $AB$, $BC$, $A'B'$, $C'A$ determine the range $\{ECA'F\}$ on the tangent $CA'$, and the range $\{GKB'C'\}$ on the tangent $B'C'$.

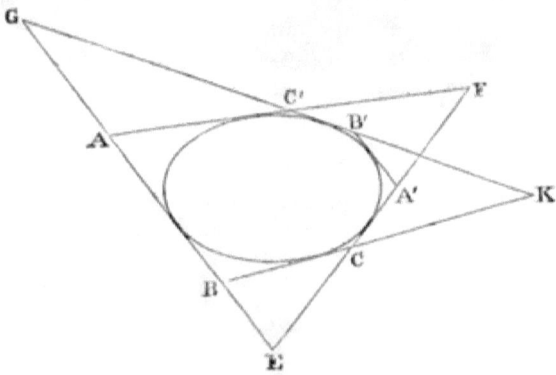

These ranges being equal (Prop. I), we have
$$A\{BCA'C'\} = \{ECA'F\} = \{GKB'C'\} = B\{ACB'C'\};$$
and therefore, $AB$ being common to the two pencils, their rays $(AC, BC)$, $(AA', BB')$, $(AC', BC')$ meet on a straight line (Art. 105), or the diagonals $AA'$, $BB'$, $CC'$ of the hexagon meet in one point.

### Corollary 1.

Having given five tangents to a conic we may determine their points of contact by this proposition; for if the summit $C'$ of the circumscribing hexagon be *on the curve*, the tangents $AC'$ and $B'C'$ being supposed to coalesce, then the line joining the opposite summit $C$ to the intersection of $AA'$ and $BB'$ determines by its intersection with $AB'$ the required point of contact $C'$. We may also determine an infinity of other tangents to a conic when five tangents $AB$, $BC$, $CA'$, $A'B'$, $B'C'$ are given; for if we draw any line through the given point $A'$ to meet $AB$ in $A$, the point $C'$ may be determined as above.

### Corollary 2.

*The orthocentre of any triangle is a point on the directrix of every parabola inscribed in it.* For if $abc$ be any three tangents to a parabola, $a'c'$ the tangents at right angles to $a$ and $c$ respectively, and $\infty$ the line at infinity, which together make up a hexagon $abcc'\infty a'$ circumscribing the parabola, then the joins of $ab$, $c'\infty$ and $bc$, $a'\infty$ are two of the perpendiculars of the triangle $abc$; and the join of the joins of the orthogonal tangents $aa'$ and $cc'$ is the directrix; and, by the proposition, these three joining lines cointersect.*

#### SCHOLIUM E.

PASCAL's theorem (Prop. XIV)—elsewhere called by him the theorem of the *Mystic Hexagram*—was enunciated without proof in his *Essais pour les Coniques* (1640) as a property of the circle, which might be generalised by projection, and then used as the foundation of a complete treatise on conics. See *Œuvres de Blaise Pascal*, IV. 1—6 (nouv ed. Paris, 1819); Chasles' *Aperçu Historique*, pp. 68—74.

BRIANCHON's theorem (Prop. XV) was deduced from Pascal's by means of Desargues' properties of what are now called polars (Scholium B). The author's proof of his theorem, given in his *Mémoire sur les Surfaces courbes du second Degré* (*Journal de l' Ecole polytechnique*, tome VI. 297—311, 1806), was as follows.

---

* This proof is given, as by Mr. John C. Moore, in Salmon's *Conic Sections* (Art. 268, Ex. 3, sixth ed. 1879). See also Scholium p. 57. Brianchon and Poncelet had deduced the reciprocal theorem of Art. 69 from Pascal's hexagram.

Take any three concurrent straight lines $PAA'$, $PBB'$, $PCC'$ in space, and let $LNMlmn$ denote the six intersections

$$(AB, A'B'), \quad (BC, B'C'), \quad (CA, C'A'),$$
$$(AB', A'B), \quad (BC', B'C), \quad (CA', C'A)$$

respectively. Then the four triads of points $LMN$, $Lmn$, $lMn$, $lmN$ are evidently collinear, since they lie severally upon the common sections of the four pairs of planes

$$(ABC, A'B'C'), \quad (AB C', A'B'C), \quad (AB'C, A'BC'), \quad (A'BC, AB'C').$$

And since every two of these triads have one point in common the four common sections and therefore the six points $LMNlmn$ lie in one plane, which also together with point $P$ divides each of the segments $AA'$, $BB'$, $CC'$ harmonically. [Art. 107.

If the whole figure be now projected orthogonally upon any plane, then (with the same notation) the six points $LMNlmn$ will in general still lie by threes upon four separate straight lines, in the order above-mentioned; but if any *other* three of them as $LmN$ be also collinear the *six* points will then lie in one straight line, since the plane of projection must be at right angles to the plane of the original six points; and this line together with the point $P$ will divide the segments $AA'$, $BB'$, $CC'$ harmonically.

This is the case when $AA'$, $BB'$, $CC'$ are concurrent chords of a conic, since their extremities may be taken in any order to form an inscribed hexagon (Prop. xiv). For example, the hexagon $ABC A'B'C$ has for its Pascal-line $LMn$, on which the remaining three points $lmN$ must also lie. Brianchon then observes that two of the three concurrent chords suffice to determine this line, whilst the third $CC'$ may be supposed to turn about $P$, and to coincide with either of the former, or to become itself a tangent (if $P$ be an external point). Having thus virtually given a fresh proof of the properties of polars,* he at once deduces his own theorem (Prop. xv.) from the reciprocal theorem of Pascal, which he takes from the *Géométrie de position* (Carnot), probably not knowing to whom it was due. See also Gergonne's *Annales* iv. 196, 379 (1813—14).

This brilliant application of Desargues' theory of polars, in conjunction with the property that *the polar planes of all points on one quadric with respect to a second envelope a third*, which Brianchon proved in the same article (as an extension from the case of similar and coaxal quadrics), served as a basis for the method of Reciprocal Polars, the full development of which was so largely due to PONCELET (Crelle's *Journal* iv. 1—71, 1829).

---

* Pascal himself also had doubtless deduced the properties of polars (which he would have learned from Desargues) from his hexagram.

## EXAMPLES.

696. If $ABC$, $DEF$ be two triangles such that $AD$, $BE$, $CF$ meet in a point, the intersections of $(BC, EF)$, $(CA, FD)$, $(AB, DE)$ lie in one straight line, and conversely*; and every tetrad of radiants or collinear points in the figure is harmonic.

697. If the vertices of a triangle slide severally on three fixed radiants, and if two of its sides pass through fixed points, the third side passes through a third fixed point in a line with the former two, and conversely.

698. If one quadrilateral be divided into two others by any straight line, the diagonals of the three intersect in three collinear points.

699. Prove for the case of the circle that any four points on the curve and the tangents thereat are equicross; and that the cross ratio of any four points $ABCD$ on the curve is equal to $\dfrac{AB \cdot CD}{AD \cdot BC}$.

700. Prove that the sides and diagonals of a quadrilateral determine an involution on any transversal; and that its six summits subtend a pencil in involution at any point in its plane.

701. The circles on the three diagonals of a complete quadrilateral as diameters are coaxal; and they are orthogonal to the circle through the three intersections of its diagonals; and they determine an involution on any transversal.

702. Any two triangles which are reciprocal polars with respect to a circle† are in homology.

703. Find the locus of intersection of tangents to two given circles whose lengths are in a constant ratio.

---

\* Two such triangles are said to be *in perspective* or *in homology*. Solutions of Exx. 696-702, 705 are given in McDowell's *Exercises in Euclid and in Modern Geometry*, pp. 134-187, 227, 239 (new edit. 1878).

† The same may be proved for any conic, as (for example) in Cremona's *Eléments de Géométrie Projective*, p. 227 (1875).

704. The pairs of radiants from any point to the vertices of a triangle and parallel to its opposite sides respectively form a pencil in involution.

705. Deduce by reciprocation from the property of the orthocentre that, if from any point radiants be drawn to the summits of a triangle, the radiants at right angles to them meet the opposite sides of the triangle in three collinear points.

706. The nine-point circle $N$ of the tetragon determined by a triad of points and their orthocentre (Art. 125, Cor. 3) touches the sixteen circles inscribed or escribed by fours to the four triangles determined by the summits of the tetragon (note, p. 191). If $ABC$ and $O$ be the points of contact of any of these sixteen circles with its triangle and with $N$ respectively, the sixteen sets of lines $OA$, $OB$, $OC$, making in all forty-eight lines, pass by fours through the extremities of the six diameters of $N$ parallel to the sides and diagonals of the tetragon; and every two tetrads which pass through opposite extremities of the same diameter have equal cross ratios.*

707. Prove by reciprocation from the theorem *Ad quatuor lineas* (or otherwise), that if a quadrilateral be circumscribed to a circle, the ratios of the products of the distances of its three pairs of opposite summits from any fifth tangent are invariable.†

708. From the anharmonic point-property of a conic deduce the theorem *Ad quatuor lineas*; and thence deduce the theorems of Art. 16, and the property of any principal or oblique ordinate. Shew also how to deduce the anharmonic property of four tangents from that of four points‡

---

\* See Dr. Casey's article in the *Quarterly Journal of Mathematics*, IV. 245.

† See Mulcahy's *Principles of Modern Geometry*, p. 43 (ed. 2, 1862).

‡ All the chords $PQ$ drawn to a conic from a given point $P$ upon it are bisected by a similar conic touching the former at $P$ and passing through its centre $O$. Let the tangent at $Q$ meet that at $P$ in $R$. Then $OR$ meets $PQ$ in a point $q$ lying on the inner conic; and by the point-property of the latter,
$$P\{Q\} = P\{q\} = O\{q\} = O\{R\} = \{R\}.$$
This proof is from GASKIN's *Geometrical Construction of a Conic Section* p. 26 (1852).

709. From the anharmonic tangent-property of a conic deduce the relation between the intercepts made upon a pair of parallel tangents by any third tangent.*

710. From the point-property of a conic, $A \{ABCD\} = B \{ABCD\}$, deduce that if from any point $K$ on the chord $AB$ a transversal be drawn meeting the tangents at $A$, $B$ in $T$ and $T'$, and the conic in $C$ and $D$, then
$$KC.KT'.TD = KD.TK.T'C.$$

711. Deduce elementary properties of the hyperbola from the relation.
$$\infty \{AB\infty\infty'\} = \infty' \{AB\infty\infty'\},$$
where $\infty$ and $\infty'$ are its two points at infinity, and $AB$ any other two points on the curve.

712. Deduce Art. 23, Cor. 3 from the relation
$$P\{PEQ\infty\} = \infty \{PEQ\infty\},$$
where $\infty$ is the point at infinity on the parabola.

713. Shew also by cross ratio that three fixed tangents to a parabola divide any fourth in a constant ratio.

714. Deduce from Prop. VI. that, if a conic touches the sides of a triangle and passes through the centre of its circumscribed circle, this circle touches the orthocycle (note, p. 280) of the conic.

715. Deduce from Prop. IX. that the nine-point circle of every triangle self-polar with respect to a parabola passes through the focus; and construct a triangle self-polar to every parabola inscribed in a given triangle.

716. If $OP$ and $OQ$ be tangents to a conic, the circle through $P$ which touches $OQ$ in $Q$ is such that triangles self-polar with respect to the conic can be circumscribed to it.

---

* CHASLES has founded his *Traité des Sections Coniques* upon the anharmonic properties of conics (cf. *Aperçu Historique*, pp. 39, 334–344). The properties of diameters and of the foci are deduced in chaps. VI. and X. The same general method is followed by Cremona; and it is given as an alternative by Rouché and De Comberousse (*Traité de Géométrie* § 1125, 4me ed. Paris 1879).

717. Shew how to inscribe in a given conic a **triangle** (or *n*-gon) whose sides pass severally through given points.*

718. Having proved the properties of polars by **cross ratio**, deduce the fundamental property of a diameter of a conic.

719. Prove Prop. II. by the same method, and deduce the elementary properties by **which** it was proved in **the** text.

720. **From the** properties **of** quadrilaterals inscribed **or circumscribed to a conic, deduce that** the diagonals of every **inscribed parallelogram are diameters** of the conic; and that supplemental chords are parallel to conjugate diameters; and that the diagonals of every circumscribed quadrilateral are conjugate diameters.

721. If a variable tangent **to a conic meet the tangents from a** given point $L$ in $F$ and $H$, and if $M$ and $K$ be a certain pair **of fixed** points on the fixed **tangents**; shew that $MF.KH$ is constant,† and deduce that a variable tangent **to a conic** divides any two fixed tangents homographically.

722. If $ABCDE$ be a pentagon circumscribing a parabola, the parallels from $B$ to $CD$ and from $A$ to $DE$ intersect upon $CE$.

723. If $ABCD$ be a quadrilateral circumscribing a parabola, the parallels from $A$ to $CD$ and from $C$ to $AD$ intersect on the diameter through $B$; **and every** other tangent divides $AD$ and $BC$ (or $AB$ and $CD$) proportionally.‡ Consider also the limiting case in which $ABC$ is a straight line.

---

\* On Exx. 717 &c. see Salmon's *Conic Sections* §§ 297, 326–8, **Exx**, where **Townsend's solution** is given; Rouché **et De Comberousse** *Traité de Géométrie* § 1134. The problem—for a simple case of which see Pappus *Collect.* lib. VII, prop. 117—**was** solved by PONCELET, and analytically by GASKIN. See *Historical Notices respecting an Ancient Problem* in The Mathematician vol. III. pp. 75, 140, 225, 311, 42 (suppl.).

† See NEWTON's theorem **Ex. 371**, with Ex. 364, note; and compare Ex. 726, note. See also CHASLES *Géométrie Supérieure* § 120, *Sections Coniques* § 56.

‡ Exx. 722–3 having been deduced from Brianchon's hexagon in Quetelet's *Correspondance mathématique et physique* IV. 155, Chasles was led (ibid. IV. 364, v. 289) from Ex. 723 to Ex. 724 (which is equivalent to the *anharmonic property* of **four tangents to** a conic), apparently without being aware that an equivalent theorem (Ex. 721, note) had been proved by NEWTON. See also *Aperçu Historique* pp. 341–4 (Note XVI.).

724. In a quadrilateral $ABCD$ circumscribed to a conic, the ratio of the ratios in which any fifth tangent divides either pair of its opposite sides is constant.

725. If the fifth tangent meet $AB$, $CD$ in $M$ and $N$, and if a sixth meet $AD$, $BC$ in $P$ and $Q$, then
$$AM.BQ.CN.DP = AP.DN.CQ.DM.$$

726. *Quatuor rectis $BL$, $BI$, $DK$, $DH$ positione datis, ducere quintam $LH$ talem, ut partes abscissæ $HI$, $IK$, $KL$ sint in ratione data.**

727. If a fixed conic $S$ and a variable conic $S'$ be inscribed in the same quadrilateral, the four points in which $S'$ intersects $S$ subtend at any point on $S'$ a pencil whose cross ratio is constant, being equal to that of the range in which the sides of the quadrilateral meet any fifth tangent to $S$.†

728. If the tangent at $O$ to a conic meet any other three tangents in the points $abc$, and meet their three chords of contact in $a'b'c'$, prove that $\{Oabc\} = \{Oa'b'c'\}$.

729. If $AB$ be a given chord of a conic, and $PQ$ a variable chord such $\{APQB\}$ is constant, the envelope of $PQ$ is a conic touching the former at $A$ and $B$.

730. If the chords $AB$ and $CD$ of a conic be conjugate, and $ACB$ be a right angle, and a chord $DP$ meet $AB$ in $Q$; prove that the angle $PCQ$ is bisected by $CA$ or $CB$.

731. If $ABC$ be a triangle circumscribing a parabola and $abc$ the points at infinity on its sides, the tangents from

---

* LAMBERT *Insigniores Orbitæ Cometarum Proprietates* sect. I. lemma 18 §§ 51–53 (1761). The envelope of $LH$ is shewn to be the parabola touching the four given lines (Art. 28, Cor. 3). [The problem had been solved in another way in the *Arithmetica Universalis* prob. 52 (ed. 1707)—al. prob. 56]. Here we have obviously the anharmonic property of four tangents to a parabola; and by stating the result in the projective form that *the ratio of the ratios* $\frac{HI}{HL}$ *and* $\frac{KI}{KL}$ *is constant* we at once shew the property to be true for all conics. [See also the *Principia* lib. I. sect. v. lemma 27 Cor., where WREN and WALLIS are referred to for earlier solutions.]

† Briefly thus: the cross ratio of the common points of any two conics in the one is equal to that of their common tangents in the other.

any point $O$ to the parabola belong to the involution $O\{Aa\, Bb\, Cc\}$. Hence shew that the directrix of every parabola inscribed in a triangle passes through its orthocentre. [Art. 110.

732. The joins of four points on a conic **meet** any transversal in three pairs of points in an involution, to which the intersections of the transversal with the conic also belong. Hence **deduce (by removing the** transversal **to** infinity) that every conic **through a triad of points and** their orthocentre is a rectangular **hyperbola.**

733. If $AEB$ and $CDF$ be two triads of collinear points, the intersections of $(AF, CE)$, $(BF, ED)$, $(BC, DA)$ are in one straight line.*

734. In a hexagon inscribed **in a conic, if two pairs of** alternate sides are parallel **the third pair are parallel.**

735. In every hexagon **inscribed in a conic the two triangles** determined by the **two sets of alternate sides are in homology.** State the reciprocal theorem.

736. The Pascal lines of the sixty hexagons determined by a Pascal hexastigm pass by threes through twenty points; and the Brianchon points of the sixty hexagons determined by a Brianchon hexagram lie by threes on twenty straight lines.†

737. If two conics touch one another at $A$ and $B$, and if $LM$ be a chord of the outer which touches the inner conic; find the loci of the intersections of $AL$, $BM$ and $AM$, $BL$.

738. The chords joining four points on a conic to any fifth $P$ and to any sixth $Q$ intersect in four points lying on a conic through $P$ and $Q$.

---

\* Pappus *Collectio* lib. VII. prop. 139 (vol. II. p. 887, ed. Hultsch); Simson *De Porismatibus* p. 414; Chasles *Porismes* p. 77. Note that $AFBCED$ is a hexagon inscribed in a line-pair, so that Pascal's theorem is a generalisation of this lemma of Pappus.

† See Townsend's *Modern Geometry* II. 172. The terms hexastigm and **hexa**gram are here very appropriately **used** to denote the figures determined by six *points* and *lines* respectively, taken in any order. In the text however I have retained the term hexagram as a designation of Pascal's figure out of regard for historical considerations.

739. If a conic $S$ be inscribed in a triangle self-polar with respect to a conic $S'$, shew that triangles self-polar with respect to $S$ can be inscribed in $S'$.

740. Given the sum of the squares of the axes of a conic inscribed in a given triangle, the locus of its centre is a circle concentric with the polar circle of the triangle.*

741. Given five points on a conic, find (by cross ratio or involution) its second intersection with any straight line through one of the five points, and its two intersections with any other straight line; and determine its points at infinity and its asymptotes, real or imaginary.

742. Prove by cross ratio that five tangents determine a conic; and determine other tangents and their points of contact; and shew how to construct the tangents from any given point, real or imaginary.

743. Prove by involution that if three sides of a quadrilateral inscribed in a conic turn about three points in a straight line, the fourth side turns about a point in the same straight line; and hence shew how to inscribe in a conic a triangle whose three sides pass severally through three collinear points.

744. Prove Carnot's theorem, that if $aa'$, $bb'$, $cc'$, be the three pairs of points in which a conic meets the sides $BC$, $CA$, $AB$ of a triangle, then

$$Ab.Ab'.Bc.Bc'.Ca.Ca' = Ac.Ac'.Ba.Ba'.Cb.Cb'.†$$

Prove also that the same relation subsists when $A$, $B$, $C$ denote the *sides* of a triangle; $a$, $a'$, &c. the *tangents* from its vertices to a conic; and $Ab$ denotes the *sine* of the angle between any two lines $A$ and $b$.

745. The distances $pqr$ of any point on (or tangent to) a given conic from three fixed lines (or points) are connected

---

\* See the *Quarterly Journal of Mathematics* x. 130.
† This is an obvious corollary from Art. 16. It is given in Carnot's *Géométrie de position* § 236 (Paris, 1803) as a case of a more general theorem.

by a relation of the form
$$Pp^2 + Qq^2 + Rr^2 + P'qr + Q'rp + R'pq = 0,$$
where $P$, $Q$, &c. are constant coefficients.

746. If four tangents **to a conic parallel** to four chords $abcd$ through either focus **meet any** fifth tangent in points $ABCD$, then
$$\{ABCD\} = \{abcd\} \frac{pp'}{qq'},$$
where $p$, $p'$ and $q$, $q'$ are the perpendiculars upon the fifth tangent from two pairs of opposite intersections of the four tangents. If the latter be fixed $pp'$ varies as $qq'$. Hence deduce that the product of the focal perpendiculars **upon** any tangent to a conic is constant.\*

747. **If three** summits **of a quadrilateral circumscribing** a conic **slide** severally on three rays of a pencil, **the fourth slides** on a **fourth ray.** Hence shew how to circumscribe **to a** conic a triangle whose **three** vertices lie on three given radiants.

748. If upon a given arc $AB$ of a circle whose centre is $O$ there be taken any arc $Am$, and likewise an arc $Bn$ equal **to** $2Am$, then $O\{m\} = B\{n\}$. Hence deduce a solution **of the problem,** to trisect a given angle $AOB$. [Ex. 528.

749. The product of the **distances of** any point on a hyperbola from **a given** pair of parallels to the asymptotes varies as its distance from the **chord intercepted by the** parallels: and the product of the distances of **any point on a** parabola from two fixed diameters varies as its **distance from the chord** joining their extremities.

---

\* The distances **of any two** tangents from either focoid (Art. 123, Cor. 4) being in a ratio of equality, the products of the focal perpendiculars upon any two tangents are in a ratio of equality. The cross ratio of the range in which any tangent meets the sides of the quadrilateral $S\phi H\phi'$ is equal to $\dfrac{-4SY.HZ}{SH^2}$, where $SY$ and $HZ$ are the focal perpendiculars upon the tangent (*Oxf. Camb. Dubl. Messenger of Mathematics* IV. 94). Chasles calls the points of concourse of common tangents to two conics "points ombilicaux" (*Sections Coniques* chap. XIV.), with reference to the use of the term *Umbilicus* for focus noticed above on p. 5.

750. From Ex. 744 deduce a construction for a conic passing through four given points and touching a given straight line: and shew that the lines joining the vertices of a triangle circumscribing a conic to the opposite points of contact cointersect: and when four points on a conic and the tangent at one of them are given, shew how to draw the osculating circle at that point.

751. Through the centre of a conic and any conjugate triad with respect to it a hyperbola can be described having its asymptotes parallel to any given pair of conjugate diameters.

752. The system of radiants from any point parallel to the tangents to a parabola is homographic with the range in which these tangents meet any fixed tangent.*

753. If from a series of collinear points pairs of perpendiculars be drawn to two fixed straight lines, the joins of the feet of the several pairs of perpendiculars envelope a parabola touching the two fixed lines.

754. If any chord of a conic drawn from a fixed point $O$ upon it meets the sides of a given inscribed triangle in points $ABC$ and the conic again in $P$, shew that $\{ABCP\}$ is constant; and deduce a construction for the tangent at a given point to a conic of which four other points are likewise given.

755. If $ABC$ be the intersections and $abc$ the points of contact of three fixed tangents to a conic, the product of the distances of any tangent from $A$ and $a$ varies as the product of its distances from $B$ and $C$: the product of its distances from $b$ and $c$ varies as the square of its distance from $A$: the pairs of radiants from any point $O$ to $BC$ and $Aa$ determine an involution to which the tangents from $O$ to the conic belong: and these tangents with $Ob$ and $Oc$ determine an involution having $OA$ for one of its double rays.

756. Deduce from Brianchon's hexagon that when a quadrilateral circumscribes a conic the joins of its opposite points of

---

* For solutions of **Exx.** 741– 759, 765–800 see Chasles' *Sections Coniques* pp. 8–67, 72–100, 137—145, 160, 201, 209, 214—299, 321 &c.

contact pass through the intersection of its two diagonals; and that in a triangle circumscribing a conic, the three lines joining its vertices to the opposite points of contact meet in a point.

757. The three pairs of lines from the vertices of a triangle to the intersections of its opposite sides with a conic are tangents to one conic: and the lines from any two points to the vertices of a triangle meet its opposite sides in six points lying on one conic. Reciprocate these two theorems; and from the second of them deduce the property of the nine-point circle.

758. The ratio of the products of the distances of any point on a conic from the odd and even sides respectively of a given inscribed $2n$-gon is constant: and the products of the intercepts on any chord made by the odd and even sides are in the same ratio from whichever extremity of the chord the intercepts are measured.

759. The ratio of the products of the distances of any tangent to a conic from the odd and even summits respectively of a given circumscribed $2n$-gon is constant; and the ratio of the products of its distances from the summits and from the points of contact of any given circumscribed $n$-gon is constant.

760. If two angles of given magnitudes $PAD$ and $PBD$ turn about $A$ and $B$ as poles given in position, then if the intersection $P$ of one pair of their arms be made to describe a conic, the intersection $D$ of the other pair will in general describe a curve of the fourth order, having double points at $A$ and $B$ and at the limiting position of $D$ when the angles $BAP$ and $ABP$ vanish together: but the locus of $D$ will be of the third order if the angles $BAD$ and $ABD$ vanish together. If $P$ describes a conic passing through $A$, then $D$ describes a cubic having a double point at $A$ and passing through $B$.* This cubic

---

* This is Newton's *Curvarum Descriptio Organica* (note p. 264). The case at the end of Ex. 760 follows from the principle that a cubic proper cannot have two double points (Salmon's *Higher Plane Curves* § 42). This special case is given by Chasles (*Aperçu historique*, p. 337) as a *generalisation* of Newton's construction in the *Principia*.

degenerates into the line $AB$ and a conic through $A$ and $B$ in the case in which the original conic passes through both $A$ and $B$.

761. The nine-point circle of a triangle touches its inscribed and escribed circles at points lying on the ellipse which touches the sides of the triangle at their middle points.*

762. Reciprocate Maclaurin's description of a conic given in Art. 113, Cor. 2.

763. The sides of a quadrilateral inscribed in a conic meet the tangents at its opposite angles in four pairs of points lying on one conic.†

764. If a quadrilateral be circumscribed to a conic, the four pairs of lines joining its vertices to the opposite points of contact touch one conic.

765. If the sides of an $n$-gon turn severally about fixed points, whilst $n-1$ of its summits slide each on a fixed line; the $n^{\text{th}}$ summit describes a conic passing through the fixed points on the two adjacent sides.

766. Shew also that any two sides not adjacent intersect on a fixed conic through the points about which they turn.

767. If the arms $A$ and $B$ of an angle pass each through a fixed point, whilst its summit slides on a fixed line; shew that the join of the points in which $A$ meets one fixed line and $B$ another envelopes a conic touching the join of the fixed points.

768. If each summit of an $n$-gon slides on a fixed line, whilst $n-1$ of its sides pass severally through (or subtend given angles at) fixed points; the $n^{\text{th}}$ side envelopes a conic touching the lines on which its extremities slide; and every diagonal of the $n$-gon envelopes a conic.

769. Any two pairs of conjugate lines from a point $O$ to a conic determine an involution whose double rays are the tangents from $O$ to the conic. State the reciprocal theorem.

---

\* See **Salmon's** *Conic Sections*, § 345, **Exx.**
† Möbius *Barycentrische Calcul* § 281.

770. One point and one only of every conjugate triad with respect to a conic lies within the conic; and two sides of every self-polar triangle meet the conic.

771. The lines drawn from any point on a conic to two conjugate points $A$ and $B$ meet the conic at the extremities of a chord which passes through the pole of $AB$. State the reciprocal theorem.

772. If a quadrilateral be circumscribed to a conic, the extremities of any chord through the intersection of two of its diagonals lie on a conic passing through the extremities of both.

773. Any three pairs of points which divide the three diagonals of a quadrilateral harmonically lie on one conic.

774. If the extremities of two diagonals of a quadrilateral be conjugate points with respect to a conic, the extremities of the third will be likewise conjugate.

775. If two of the three pairs of joins of four points be conjugate lines with respect to a conic, the third pair will be conjugate with respect to it.

776. The pairs of chords drawn from a fixed point on a conic so as to make equal angles with a given line intercept a variable chord which passes through a fixed point.

777. The pairs of tangents to a conic from points on a straight line determine an involution on any transversal through its pole, or on any tangent to the conic.

778. The pairs of tangents to a parabola from points in the same straight line are parallel to conjugate rays of a pencil in involution.

779. Two tangents being drawn to a conic from any point on a fixed straight line, if $x$ and $x'$ be their distances from its pole, and $y$ and $y'$ their distances from a fixed point, shew that

$$\frac{x}{y} \pm \frac{x'}{y'} = \text{a constant}.$$

780. Any two ranges in involution on the same axis have one segment in common.

781. The locus of the middle point of a chord drawn from a fixed point to a conic is a conic through the point, and through the points of contact of the tangents from it to the original conic, and through the two points at infinity on that conic.

782. Find the envelope of a line which meets two fixed lines in a pair of conjugate points with respect to a given conic.

783. The envelope of the parallel from any point on a fixed straight line to the polar of the point with respect to a conic is a parabola touching the fixed line.

784. The locus of the intersection of a pair of conjugate lines with respect to a given conic, drawn each through a fixed point, is a conic, which passes through the two fixed points, and through the points of contact of the tangents from them to the original conic.

785. If two angles be circumscribed to a conic their two summits and their four points of contact lie on one conic.

786. Any transversal being drawn to a conic from a fixed point $O$, the perpendicular to it from its pole envelopes a parabola, which touches the polar of $O$ and the tangents to the conic at the feet of the normals to it from $O$.

787. Circumscribe to a given conic a polygon having each of its summits upon a given straight line.

788. The poles of a given straight line $L$ with respect to the system of conics through four given points is a conic, which with the line $L$ divides the six joins of the four points harmonically, and passes through their three intersections, and through the two points on $L$ which are conjugate with respect to every conic of the system: it also touches the sixteen conics which pass through the said conjugate points and touch by fours the sides of the four triangles determined by the given points.

789. If four conics pass through the same four points, the polars of any point with respect to them form a pencil whose cross ratio is constant, being equal to that of the tangents to the four conics at any one of their points of concourse: and reciprocally, if four conics touch the same four lines, the poles of any straight line with respect to them form a range whose cross ratio is constant, being equal to that of the points in which the four conics touch any one of their common tangents.

790. If two conics osculate at $O$, their tangents at the further extremities of any chord through $O$ intersect on the tangent at $O$, and conversely: and every two equal and coaxal parabolas osculate at infinity.

791. Two conics which have two pairs of conjugate diameters of the one parallel to two pairs of conjugate diameters of the other must be similar and similarly situated.

792. Deduce from Art. 114 (i) that parallel conics* have a common chord (real or imaginary) at infinity; and if also concentric they have double contact at infinity. Shew how to draw a conic which shall be parallel to a given conic, and shall also pass through three given points or touch three given lines.

793. Three fixed conics having four points in common, shew that if a variable pair of transversals be drawn from fixed points $O$ and $\omega$ to meet the three conics in triads of points $mAB$ and $mab$ respectively, the ratio of the ratios $\dfrac{OA.OB}{mA.mB}$ and $\dfrac{\omega a.\omega b}{ma.mb}$ is constant.† Hence deduce that a conic may be regarded as the locus of a point the square of the tangent from which to a fixed circle varies as the product of its distances from two fixed lines, which are common chords of the conic and the circle.

---

\* Similar and similarly situated conics may be called *parallel* since their curves are everywhere parallel at corresponding points: they have also been called "homothetic" (Chasles *Sections Coniques* § 373), which should rather mean "placed together." For another use of the term parallel see Gergonne's *Annales* xii. 1.

† Exx. 793 &c. have been extended to quadrics by Mr. Martin Gardiner in the *Quarterly Journal of Mathematics* x. 132—147.

794. If from **any point on** one of three conics which have four points in common a **tangent be drawn to** each of the remaining two, **the ratio** of the ratios **of these tangents** to the parallel diameters of their conics **is constant: and if** $OPQ$ be **the middle points of the** intercepts made by the three conics **on any** transversal, **then** $OP$ and $OQ$ are in **the ratio of the** parallel focal chords of the second and third conics.

795. Four fixed conics having four points in common **being met by a** variable transversal, viz. two of them in the pairs of points $aa'$ and $bb'$, and the third in **two** points of which $m$ **is one, and the fourth in two points of which** $n$ **is** one; shew that the ratio of the ratios $\dfrac{ma \cdot ma'}{mb \cdot mb'}$ and $\dfrac{na \cdot na'}{nb \cdot nb'}$ is constant.

796. If $ABCD$ be four conics such that the eight points of concourse of $AB$ and $CD$ lie on one conic, the eight points of concourse of $AC$ and $BD$ (or $AD$ and $BC$) lie on one conic.

797. When a point $O$ has the same polar with respect to three conics $ABC$, three pairs of the common chords of $AB$, $BC$, $CA$ respectively pass through $O$ and form a pencil in involution: and when two conics $A$ and $B$ have each double contact with a third conic $C$, a pair of the common chords of $A$ and $B$ are harmonic conjugates with respect to their chords of double contact with $C$.

798. The common tangents to three conics taken in pairs **form three** quadrilaterals: shew that the **three parabolas** inscribed in them have a common circumscribed triangle.

799. If through the intersections of two given conics $A$ and $B$ a third conic $C$ be drawn, and if from any point $O$ on $C$ there be drawn tangents $Oa$, $Oa'$ to $A$ and $Ob$, $Ob'$ to $B$; the lines $ab$, $ab'$, $a'b$, $a'b'$ and the four common tangents of $A$ and $B$ touch a fourth conic.

800. The locus of the point the pairs of tangents from which to two given conics form a harmonic pencil is a third conic, on which lie the eight points in which the given conics touch their common tangents. State the reciprocal theorem.

# CHAPTER XI.

## CONICAL PROJECTION.

**128.** Two figures $A$ and $B$ in any two planes are said to be in *Perspective* when a point $O$ can be found in space such that every radiant from it to a point on $A$ passes through a point **on** $B$, and conversely. Either figure is then said to be the *Central* or *Conical* **Projection** of the other on the plane of the former, the **point $O$ being called the** *Vertex* or the *Centre* of projection. When $O$ is at infinity the projection becomes parallel or orthogonal. [Art. 86.

Let $P$ and $Q$ be any two points in the plane of $A$, **and** $P'$ and $Q'$ their projections from the vertex $O$ upon the plane of $B$. Then evidently the lines $PQ$ and $P'Q'$ intersect upon **a fixed** straight line, viz. the common section of the planes of $A$ and $B$. **Now by** projecting the whole figure orthogonally upon any one plane, or by supposing the planes of $A$ and $B$ to become coincident, we see that if to every point $P$ of one figure corresponds **a single point $P'$** of another figure in the same plane, and conversely, and if $PP'$ passes through a fixed point; then every line $PQ$ in the **one figure** meets the corresponding line $P'Q'$ in the other upon **a fixed** straight **line.** For example, if the joins of the vertices of two triangles meet **in a point, the joins of** their opposite sides lie in one straight line. [Ex. 696.

Two figures thus **related in one plane are** said to be in *Perspective* or in *Homology*. **We shall in** general use the former term for this kind of correspondence, and the term *Projection* for the case of figures **in** perspective in space. The terms *Reversion* and *Homographic Transformation* will be explained in their place.

x 2

129. It is evident that Art. 88 applies to central as well as to parallel projection. Parallel lines however do not project from any vertex into parallels, except in the case in which they are parallel to the common section of the primitive plane with the plane of projection.

If through the vertex $V$ of projection (fig. p. 314) there be drawn the plane $Vab$ parallel to the plane of projection $AO'B$, and so as to meet the plane of the figure to be projected in the line $ab$; it is evident that all points on $ab$ will be projected to infinity, or in other words they will remain unprojected. For this reason $ab$ is called the *Unprojected Line*, and it is also said to be projected *to infinity*.

Since every point at infinity in the plane $AO'B$ corresponds projectively to some point on $ab$, we come again to the conclusion that all points at infinity in one plane lie in a straight line (Art. 17 Cor. 2). *The straight line at infinity is to be regarded as parallel to every other straight line in its plane*, since it intersects every such line at infinity: it is in fact coincident with the *circle of infinite radius*\* described about any point whatever in its plane. The line at infinity and the two *focoids* (Art. 123 Cor. 4) or *circular points* at infinity—so called because every circle in their plane passes through them—will be seen to be of peculiar importance in the projection and transformation of curves.

## THE FOCOIDS.†

### PROPOSITION I.

130. *Every circle in a given plane passes through the focoids, and conversely; and every two concentric circles in the same plane touch one another at the focoids.*

---

\* For a tangential equation to this circle, which is sometimes inadequately said to represent the focoids only, see Whitworth's *Trilinear Coordinates &c* Art. 382 (Cambridge 1866).

† This term is open to the objection that it combines a Latin word with a Greek ending: but we may perhaps be allowed to treat both as naturalised English expressions. In speaking of *the* focoids &c. we tacitly refer to a specified plane. Every plane not at infinity has its *two* focoids and its *one* line at infinity.

(i) Any number of right angles turning about their summits in one plane generate similar pencils in involution (Art. 110), whose imaginary double rays form two sets of parallels; that is to say, each set pass through one of two fixed imaginary points $\phi$, $\phi'$ on the line at infinity.

These are accordingly the *foci* of the involution which the arms of all right angles in one plane determine upon the line at infinity in that plane; and conversely *every right angle $AOB$ is divided harmonically by the lines $O\phi$ and $O\phi'$*.

If therefore $O$ be a variable point at which a fixed line $AB$ subtends a right angle, it follows from the harmonicism of $O\{A\phi B\phi'\}$ that the locus of $O$ is a conic through the points $AB\phi\phi'$ (Art. 113). That is to say, every *circle $AOB$* passes through the focoids, and conversely every conic through the focoids is a circle.

(ii) The centre $C$ being the pole of the line at infinity (which is the join of the focoids), it follows that the lines from $C$ to the focoids touch the circle at those points; and hence that all circles in one plane which have any point $C$ for their common centre touch one another at the focoids of that plane.

(iii) Or by §i and Art. 122 Cor. 5, all concentric circles in one plane touch one another at the focoids. This also follows from the consideration that any two diameters $CX$ and $CY$ of a circle which are at right angles are *conjugate lines* with respect to the circle, and the lines $C\phi$ and $C\phi'$* with respect to which they are harmonic conjugates must therefore touch the circle—and all circles having $C$ for centre must touch one another—at $\phi$ and $\phi'$.

*Corollary.*

Every rectangular hyperbola has for a pair of conjugate points with respect to it the focoids of its plane, since its points at infinity lie on two straight lines at right angles.

---

* These lines may be regarded as the *asymptotes* of the circle (Art. 114).

## PROPOSITION II.

**131.** *Every conic may be regarded as inscribed in the quadrilateral which has for opposite summits the real and imaginary foci of the curve and the focoids, and for diagonals the two axes of the conic and the line at infinity.*

(i) This is proved by the method of Art. 123 Cor. 4, where $S$ and $\Pi$ may be either the real or the imaginary foci.

[Scholium A.

(ii) Otherwise thus. Every two lines through $S$ which are conjugate with respect to the conic being at right angles (Art. 7), the lines $S\phi$ and $S\phi'$ which divide them harmonically are tangents to the conic (Art. 116 Cor. 2). That is to say, the lines joining the real or imaginary foci to the focoids touch the conic, as was to be proved.

## PROPOSITION III.

**132.** *Any two straight lines drawn at a given angle in a given plane and the lines joining their point of concourse to the focoids form a pencil of constant cross ratio.*

For if $ab$ be a fixed straight line, and $\omega$ any point at which it subtends an angle of given magnitude $\alpha$, then by a property of the circle $\omega ab$, the pencil subtended by $ab$ and the focoids at $\omega$ is of constant cross ratio; and the rays $\omega a$, $\omega b$ may be parallel to any two lines $OA$ and $OB$ inclined at an angle $\alpha$ in the same plane.

### *Corollary.*

Any plane figure may be moved about in any way in its own plane without changing its relation to the focoids, since every angle in the figure has an invariable relation to the focoids.

### SCHOLIUM A.

DESARGUES regarded the opposite extremities of an infinite line as coincident or consecutive points, and the asymptotes of a hyperbola as its tangents at infinity (Poudra's *Œuvres de Desargues* I. 103, 197, 210, 245). Hence we deduce (Scholium B, p. 153) that the hyperbola is a single curve, which spreads completely across its plane *without breach of continuity*. It follows logically that no transversal can be drawn in the same plane so as not to meet the hyperbola. Nevertheless it is obvious that some lines—its conjugate axis for example—do not (however far produced) meet the curve in geometrical points. Thus we are driven to the conception of ideal or *imaginary* points and chords of intersection, and are led to say that every straight line meets any hyperbola (or other conic) in two points real or *imaginary*, which coalesce in the case of tangency. Although this is here given merely by way of inference, the words of Desargues himself (used in another connexion) are very appropriate to this subject: "*L'entendement ne peut comprendre comment sont les proprietez que le raisonnement luy en fait conclure*" (*Œuvres* I. 195).

Boscovich has a very remarkable appendix to his treatise on conics, entitled *De Transformatione Locorum Geometricorum, ubi de Continuitatis lege ac de quibusdam Infiniti mysteriis* (Universæ Matheseos Elementa, tom. III. pp. 228—356), in which he brings out clearly and with an abundance of geometrical illustration the notions of positive and *negative* in direction: of geometrical *continuity*: of the transition from positive to negative through zero or *infinity*: of the *imaginary* chords of the hyperbola, whose squares are negative: and of the *quasi-elliptic* nature of the hyperbola, certain of the properties of which follow from properties of the ellipse by change of sign (§§ 678, 715, 758, 770, 808, 812, &c.). See also Scholium C, p. 101.

The discussion of these matters having been revived in the present century (Chasles *Rapport sur les progrès de la Géométrie* chap I. § 19, p. 60), Poncelet at length worked out his theory of *cordes idéales* (1820); and he shewed that all circles in one plane pass through the same two imaginary points $\phi$ and $\phi'$ on the line at infinity, and that a focus $S$ common to any two conics in one plane is a "centre of homology" or intersection of common tangents to the two conics. Hence it follows, by supposing one of the two conics to become a circle, that $S\phi$ and $S\phi'$ are tangents to every conic of which $S$ is a focus. See Gergonne's *Annales* XI. 73, XII. 234; Poncelet *Traité des Propriétés Projectives des Figures* §§ 89—98, 258, 367, 453 (Paris, 1822). Plücker extended this conception to plane curves of all orders, regarding as a "focus" of any curve the point of concourse of any two tangents drawn to it from the focoids, one from each (Crelle's *Journal* X. 84—91; Salmon's *Higher Plane Curves* § 138).

According to Plücker's definition, the tangents from the focoids $\phi$ and $\phi'$ to an ellipse (or other conic) determine by their opposite intersections *two pairs* of "foci." If $S$ be any one of the four, every pair of conjugate lines from $S$ to the conic form a harmonic pencil with $S\phi$ and $S\phi'$ (Art. 116, Cor. 2), and are therefore *at right angles*. This, which is of course a corollary from Desargues' theory of polars was proved for the real foci by De la Hire (*Sectiones Conicæ* Lib. VIII. prop. 23, p. 189. Paris, 1685). The two points on the transverse axis at distance $\pm\sqrt{(CA^2 - CB^2)}$ from the centre $C$ have been shewn to possess the property in question (Art. 7, Cor.);

312          CONICAL PROJECTION.

and by symmetry, the two points on the conjugate axis at distance $\pm \sqrt{(CB^2 - CA^2)}$ from $C$ must be the remaining two (or imaginary) foci. In order to prove that a point $S$ is a focus of a given conic, it suffices to prove that TWO *pairs of conjugate lines at right angles can be drawn to the conic from* $S$. [Art. 110.

## PROJECTION.

### PROPOSITION IV.

**133.** *All rows of points and pencils of rays are homographic with their projections.*

(i) For if $ABCD$ be any row of four points in the primitive plane, and $A'B'C'D'$ their projections from a vertex $V$ upon any other plane, it is evident that $\{A'B'C'D'\} = \{ABCD\}$. And if $O$ be any fifth point in the primitive plane and $O'$ its projection, then

$$O'\{A'B'C'D'\} = \{A'B'C'D'\} = \{ABCD\} = O\{ABCD\},$$

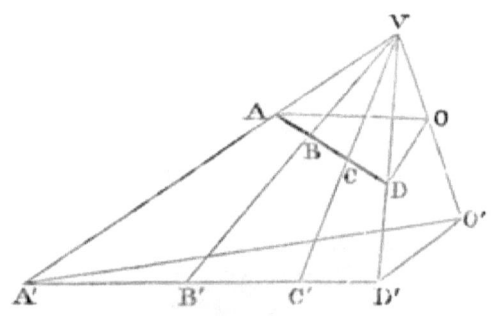

Thus every tetrad of radiants $OA$, $OB$, $OC$, $OD$ or of collinear points $ABCD$ is equicross with its projection; a result which may be briefly expressed by saying that *figures in perspective are homographic.*

(ii). More generally,* let the joins of any number (say six) of points $ABCDEF$ be connected by a homogeneous and symmetrical relation

$$l.AB.CD.EF + m.AC.BE.DF + n.AD.BE.CF = 0,$$

in which the terms differ from one another only in their

---

\* See Salmon's *Conic Sections*, Art. 351.

coefficients and in the *order* in which the letters $ABCDEF$ occur in them. And first let all the points lie in one straight line, and let $VP$ be the perpendicular upon it from the **vertex** of projection.

Then since

$$AB = \frac{VA \cdot VB}{VP} \cdot \sin AVB; \quad CD = \frac{VC \cdot VD}{VP} \cdot \sin CVD; \quad \&c.$$

the above **relation reduces, by the omission of a** common factor, to a relation between the sines of the angles which the joins of the six points subtend at $V$. It therefore still subsists when the points in question are replaced by their projections upon any plane.

And further, if any number of points $ABCDEF$ &c. lie on different straight lines, the perpendiculars upon which from $V$ are $VP$, $VP'$, $VP''$, &c., then any symmetrical and homogeneous relation between the joins of the points will still be projective, provided that it implicitly involves in every term the *same* factor $\dfrac{VA \cdot VB \cdot VC \cdot VD \cdot \&c.}{VP \cdot VP' \cdot VP'' \cdot \&c.}$. Thus Carnot's theorem (**Ex.** 744) is projective, so that when proved for the circle it is may be extended to all conics by projection.

### *Corollary.*

The properties of harmonic section, of poles and polars, and of involution are projective; so that it suffices to prove them for the simplest figure into which any figure to which they belong can be projected.

### PROPOSITION V.

**134.** *Any straight line in the primitive plane can be projected to infinity, and any two angles in that plane can at the same **time** be projected into angles of given magnitudes.*

(i) Draw any straight line $ab$ in the primitive plane, **and take** any plane $Vab$ through $ab$ for the "vertex-plane," in which the vertex $V$ of projection **is to** lie. Then it is evident that the line $ab$ projects to infinity upon any assumed plane of **projection** $ABO'$ parallel to the vertex-plane.

(ii) Two conditions now suffice to fix the position of $V$ in the vertex-plane.

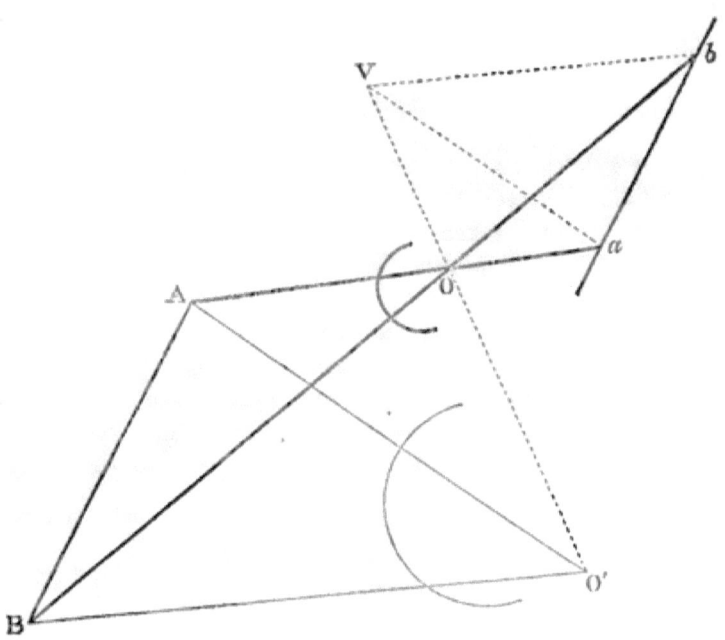

To project a given angle $AOB$ in the primitive plane into an angle of given magnitude $\alpha$, let the arms of $AOB$ meet the unprojected line in $a$ and $b$; and upon $ab$ describe in the vertex-plane a circular segment $aVb$ containing an angle equal to $\alpha$. Then the vertex $V$ may be taken at any point on this segment.

For the vertex-plane and the plane of projection (being parallel) are met by the plane $VOa$ in parallel lines $Va$ and $AO'$, and by the plane $VOb$ in parallel lines $Vb$ and $BO'$. Therefore, $O'$ being the projection of $O$,
$$\angle AO'B = aVb = \alpha,$$
or the projection $AO'B$ of the angle $AOB$ is of the assigned magnitude $\alpha$.

To project a second given angle in the primitive plane into an angle of given magnitude $\beta$, let its arms meet the unprojected line in $a'$ and $b'$: then the vertex $V$ must lie also on a **segment** described upon $a'b'$ in the vertex-plane so as to contain **an angle** equal to $\beta$: and the intersection of this with **the segment** on $ab$ completely determines the position of $V$.

### *Corollary* 1.

Project **any four collinear points** $ABCD$ **into** points $abcd$. Then **in the special case in which one of the latter** $d$ **is at infinity,**

$$\{ABCD\} = \{abcd\} = \frac{ab.c\infty}{a\infty.cb} = \frac{ab}{cb}.$$

If therefore we determine the point $D$ **on a given** straight line $ABC$ so that $\{ABCD\}$ may **be equal to a given** ratio, **and** if any straight line through $D$ be **taken as the** unprojected line, the projections of $AB$ and $CB$ will **be in the given ratio.** In like manner **a second point** $D'$ **on the** unprojected line is determined by the condition that the **segments of a second line** $A'B'C'$ shall project in another given ratio.

### *Corollary* 2.

Any pencil of rays in involution may be projected into a **rectangular** pencil **in** involution by projecting the angles between **any** *two pairs* **of** its conjugate rays into right **angles.** [Art. 110.

### *Corollary* 3.

*Any two points* $F$ **and** $F'$ *may be projected into the focoids of a given plane.* For **if** $AB$ and $CD$ be any two segments in the involution of which $F$ and $F'$ are the foci, we have only to project the **line** $FF'$ to infinity **and any** two angles $AOB$ **and** $CPD$ **in** the primitive plane **into** right angles (Art. 130 § i). This construction is *imaginary* when $F$ and $F'$ are real points.

### PROPOSITION VI.

**135.** *Any quadrilateral may be projected into any other quadrilateral of given form and magnitude.*

(i) To project **a given** quadrilateral $ABCD$ into a square, project one of its angles $BAD$ and the angle $AOD$ between its two diagonals into right angles, and its third diagonal $PQ$ to infinity. Thus the projection of $ABCD$ becomes a square, whose magnitude is determined by the distance of its plane from the vertex-plane.

(ii) To project a given quadrilateral $ABCD$ into another of given form, it suffices to project one of its angles $BAD$

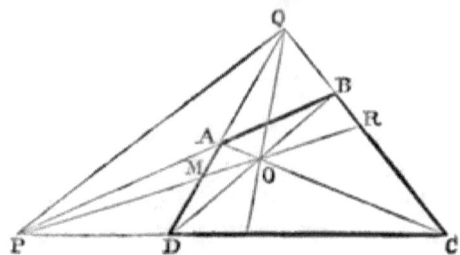

and the angle $AOD$ between its two diagonals into angles of certain given magnitudes, and the segments $AO$, $OC$ and $BO$, $OD$ into segments which are in certain given ratios.

[Art. 134 Cor. 1.

For in the projection—the same letters being used—if $AO$ be taken arbitrarily, the point $C$ is determined by the ratio $\dfrac{AO}{OC}$; and the position of the line $BOD$ is known; and from the angle $BAD$ and the ratio $\dfrac{BO}{OD}$ the points $B$ and $D$ are determined. The form of the projection being thus determined, its magnitude may be increased or diminished at pleasure by moving the plane of projection towards or away from the vertex-plane.

*Corollary.*

Any four points or lines in one plane may be projected into any other four points or lines in one plane.

PROPOSITION VII.

136. *A given conic may be projected into a conic having the projections of two given points for foci, or the one for centre and the other for a focus.*

(i) To project a given conic $Q$ and a given point $S$ within it into $Q'$ and $S'$ respectively so that $S'$ may be a focus of $Q'$: draw from $S$ any two pairs of lines conjugate with respect to $Q$, and project the angles contained by them into right angles. Thus $S'$ becomes a focus of $Q'$, being a point such that *every pair of conjugate lines drawn* from it to $Q'$ are at right angles.

[Art. 134 Cor. 2.

We may at the same time project a given point $C$ in the plane of $Q$ into the centre $C'$ of $Q'$, viz. by taking the polar of $C$ with respect to $Q$ for the unprojected line.

(ii) Otherwise thus. Let $CS$ and the tangent at any assumed point $P$ to the conic $Q$ meet the polar of $S$ in $X$ and $R$

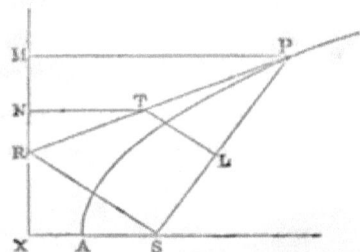

respectively. Then if the polar of $C$ be projected to infinity and each of the angles $RXS$ and $RSP$ into a right angle, the points $C$ and $S$ will be projected into a centre and focus of $Q'$, as before.

(iii) By properly choosing the point $C$ in the foregoing constructions, we may project $Q$ so that any two points $S$ and $H$ within it project into $S'$ and $H'$ the real foci of $Q'$.

For if $SH$ meets $Q$ in $A$ and $B$, and if the double points of the involution determined by the couples $AB$ and $SH$ be the point $C$ on $SH$ and the point $G$ on its complement; then in the projection, the double point $C'$ bisects *every segment* $S'H'$, $A'B'$, &c. of its involution, since in conjunction with the second double point (in this case at infinity) it divides every such segment harmonically. [Art. 112.

Hence $S'$ and $H'$ are equidistant from the centre $C'$ of $Q'$, and since $S'$ is a focus $H'$ is likewise a focus, as required.

(iv) *The system of conics inscribed in a given quadrilateral* $SFHF'$ *may be projected into confocal conics* by projecting $F$ and

$F'$ into the focoids of the plane of projection (Art. 134 Cor. 3). This construction is *imaginary* when the quadrilateral $FSHF'$ is real: but the foregoing constructions are always real, the points $S$ and $H$ being taken within $Q$.

### PROPOSITION VIII.

**137.** *A given conic may be projected into a circle having the projection of a given point for centre: a system of conics through four given points may be projected into coaxal circles: or a system of conics touching one another at two given points into concentric circles.*

(i) By taking the point $C$ at $S$ in Art. 136 we project the given conic into a conic having the same point (not at infinity) for both centre and focus; that is to say, we project it into a circle having the projection of a given point for centre.

(ii) Otherwise thus. Take the polar of any point $C$ for the unprojected line: through $C$ draw any chord $ACA'$, and

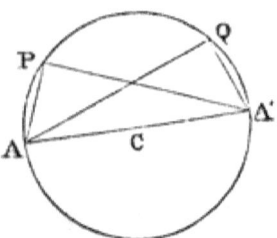

project the angles which it subtends at two assumed points $P$ and $Q$ on the given conic into right angles. Then in the projection, the same letters being used,

$$CA = CA' = CP = CQ,$$

or the projection is a circle about $C$ as centre.

Thus the *two angles* determine the species of the projection, and the unprojected line may be taken arbitrarily.*

(iii) Hence, by projecting any conic into a circle and one of its chords $FF'$ into the line at infinity, we may project

---

* This may also be deduced from a consideration of the circular sections of a cone described arbitrarily on any given conic as base (Salmon's *Conic Sections* Art. 365).

any two points $F$ and $F''$ on a conic into the focoids, as was otherwise shewn in Art. 134 Cor. 3.

It follows that all conics through **two given points may** be projected into circles, and all conics through four given points into *coaxal circles*, and all conics touching one another at **two** given points into *concentric circles*. [Art. 130 § ii.

### PROPOSITION IX.

138. *The arms of any angle of constant magnitude in a given plane may be projected into rays of a pencil of constant cross ratio, whose other two rays pass each through a fixed point.*

For the arms of a constant angle **and the** lines joining their intersection to the focoids form a pencil of constant cross ratio (Prop. III.), which **projects upon any plane into a pencil of constant cross ratio, two of whose rays pass through the projections of the focoids. Note that this pencil is** *harmonic* when the constant **angle is a** *right angle*. [Art. 130 § i.

139. In the following examples of the projection of angle-properties* the theorem to the right follows in each case from that opposite to it on the left, as appears conversely by projecting the points $FF''$ into the focoids.

| | |
|---|---|
| The tangent to a circle is at right angles to the radius to its point of contact. | Any chord $FF''$ of a conic is cut harmonically by any tangent and the line joining its point of contact to the pole $C$ of $FF''$. |
| Confocal conics intersect at right angles. | If two conics be inscribed in a quadrilateral of which $FF''$ are a pair of opposite summits, the tangents at any one of their common points cut $FF''$ harmonically. |
| The locus of the point of concourse of two tangents to a conic which intersect at right angles is a concentric circle; or in the case of the parabola the locus is the directrix. | The locus of the point of concourse of two tangents to a conic which divide a given line $FF''$ harmonically is a conic touching the former at $FF''$; or if $FF''$ touches the original conic, the locus is the join of the points of contact of the second tangents to it from $F$ and $F'$. |
| The locus of the intersection of tangents to a parabola which meet at a | The locus of the intersection of tangents to a conic which divide a given |

---

* See Salmon's *Conic Sections* §§ 356-8; Rouché et de Comberousse *Géométrie* § 1175.

given angle is a hyperbola having the same focus and directrix.

The envelope of a chord of a conic which subtends a constant angle at one of the foci is another conic having the same focus and directrix.

If $P$ be any point on a given conic, $S$ any fixed point, and $SPT$ an angle of constant magnitude, the envelope of $TP$ is a conic having $S$ for a focus.

If the point $P$ be taken on a given straight line (instead of a conic), the envelope of $TP$ becomes a parabola having $S$ for focus.

finite line $FF'$ touching the conic in a constant cross ratio is a conic touching the former at the points of contact of the second tangents to it from $F$ and $F'$.

If tangents $SF$ and $SF'$ be drawn to a conic from given points $F$ and $F'$, the envelope of a variable chord $AB$ such that $S\{AFBF'\}$ is constant is a conic touching the former at its points of contact with $SF$ and $SF'$.

If $SFF'$ be fixed points, and $P$ a variable point on a conic through and $F'$, the envelope of a line $PT$ such that $P\{SFTF'\}$ is constant is a conic touching $SF$ and $SF'$.

If $P$ be taken on a given straight line (instead of a conic), the envelope of $TP$ becomes a conic inscribed in the triangle $FSF'$.

## PERSPECTIVE.

140. The relation of *Perspective* in one plane may be treated either as a limiting case of the projective relation (Art. 128), or independently as follows.*

From a fixed centre of perspective $S$ in the plane of a given figure draw radiants to all points $p$ of the figure, and let these radiants meet a fixed axis of perspective in the same plane in points $R$ (fig. p. 10). Then if on every radiant $SR$ there be taken a point $q$ such that

$$\{SpRq\} = \text{a constant,}$$

the locus of $q$ is said to be in *Perspective* with the locus of $p$.

Taking any two positions of $SR$, we have

$$\{Sp'R'q'\} = \{SpRq\};$$

and therefore $pp'$ and $qq'$ always intersect on the axis of perspective $RR'$ (Art. 104). Hence also we see that to every straight line $pp'$ in the one figure corresponds a straight line $qq'$ in the other; and to every range $\{p\}$ in the one a homographic range $\{q\}$ in the other. Figures in perspective *in plano* are therefore homographic, and they possess the same properties as figures projectively related in space.

---

* See Chasles *Sections Coniques* p. 169.

It may be shewn that if two figures in perspective in relief be turned about the line of intersection of their planes, their centre of perspective describes a circle in a plane perpendicular to that line.* [Ex. 850.

#### SCHOLIUM B.

The method of Projection—which is implicitly contained in the ancient theorem of Art. 103—was freely used by DESARGUES. It was used also by NEWTON, under the name *Generatio curvarum per Umbras*, in his Enumeration of Lines of the Third Order, where he remarks (p. 25, ed. Talbot): "And in the same manner as the circle, projecting its shadow, generates all the conic sections, so the five divergent parabolas, by their shadows, generate all other curves of the second genus. And thus some of the more simple curves of other genera might be found, which would form all curves of the same genus by the projection of their shadows on a plane."

Desargues also proved the fundamental property (Ex. 696) of triangles in perspective, whether in relief or *in plano* (*Œuvres* I. 413, 430). The term "homologie," for perspective in one plane, was introduced by Poncelet, and is now generally used by French writers. But since the term is in itself inexpressive, an inconvenient distinction has to be made between *homologue* and *homologique* (Rouché et De Comberousse *Géométrie* §§ 1094, 1167).

## REVERSION.

**141.** Take fixed points $S$, $O$ and a fixed straight line $MN$: and through the fixed points draw any two straight lines intersecting at some point $R$ on $MN$, and also a pair of parallels meeting $RO$ and $RS$ in $P$ and $p$ respectively (p. 10). Then $P$, $p$ may be called *Reverse Points*: $O$ and $S$ the *Origins of* reversion: and $MN$ the *Base Line*. When the locus of $P$ is a conic having $S$ and $MN$ for focus and directrix, we have seen that the locus of the reverse point $p$ is the eccentric circle of $O$; and we have derived properties of the conic from properties of this circle.† We now proceed to treat the subject of reversion more generally. The original figure from which a reverse figure is derived may be called its *Obverse*.

---

\* Chasles *Géométrie Supérieure* §§ 368–9; Cremona *Géométrie Projective* § 90.
† See Arts. 4–6, 16 and Exx. 6–10.

## PROPOSITION X.

142. *To any straight line drawn in a given direction corresponds a reverse line passing through a fixed point on the base line.*

Let $O$ and $\omega$ be the origins of reversion: $P$ and $p$ any two reverse points: $\omega m$ and $PM$ a pair of parallels, meeting the base line in $M$ and $m$.

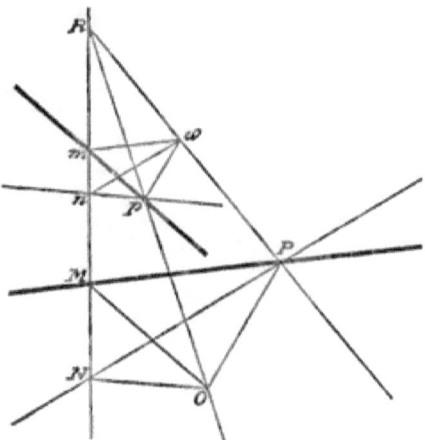

Then, if $R$ be the point on the base line at which $P\omega$ and $Op$ intersect,

$$OP : \omega p = PR : \omega R = PM : \omega m,$$

and therefore $OM$ and $pm$ are parallel.

Hence, if $\omega m$ be a fixed line and $P$ a variable point on any assumed line parallel to $\omega m$, the locus of $p$ is the straight line drawn through the fixed point $m$ on the base line parallel to $OM$.

### Corollary 1.

The point at infinity on any system of parallels $PM$ corresponds to a reverse point $m$ on the base line. All points at infinity in the same plane are therefore to be regarded as lying in one straight line, of which the base line is the reverse.

Furthermore *the direction of the line at infinity is indeterminate.* For, as $pm$ turns about the same point $m$ on the base line, the reverse line $PM$ remains parallel to $\omega m$: and ultimately, when $pm$ coalesces with the base line, $PM$ becomes the *line at infinity*, which may accordingly be regarded as parallel to any assumed line $\omega m$.

## Corollary 2.

If the arms of any angle $MPN$ and of the reverse angle $mpn$ meet the base line in $M$, $N$ and $m$, $n$ respectively, then

$$\angle m\omega n = MPN; \text{ and } \angle MON = mpn.$$

**Notice** that to every angle $PSQ$ subtended *at either origin* $S$ (fig. Art. 4) corresponds an *equal* reverse angle $pOq$ subtended at the other. For example, the angles $PSO$, $pOS$ are **equal**, in the figure of Art. 6.

### PROPOSITION XI.

**143.** *Any straight line being taken as base line, any two given angles may be reversed into angles of given magnitudes.*

For the angle $MPN$ (Prop. x. Cor. 2) reverses into an angle of given magnitude $\alpha$, if the origin $O$ be taken on the circular segment $MON$ described on $MN$ so as to contain an angle equal to $\alpha$. By a like construction a second angle may be reversed into an angle of given magnitude $\beta$. And if $O$ be taken at the intersection of the two segments, the two angles will reverse simultaneously into angles equal $\alpha$ and $\beta$ respectively.

The applications of this general theorem are precisely analogous to those of the corresponding theorem in Conical Projection. [Prop. v.

### Corollary.

From any origin $O$ a given conic may, by properly choosing the base line, be reversed into a conic through two given points at infinity, whose magnitude is then determined by the position of the reverse origin $\omega$. Or if the base line be given, the origin $O$ may be determined by reversing the angles between two assumed pairs of lines $PA$, $PB$ and $PC$, $PD$—which may be drawn conjugate with respect to the given conic—into angles $\alpha$ and $\beta$ respectively. By properly determining the origins and the base line together we may reverse *any conic and point U and P into any conic and point U' and P'*. For example, if α and β be right angles and the polar of $P$ with respect to $U$ be taken as base line, $U'$ becomes a circle whose centre is $P'$. [Prop. VIII.

**144.** The following are some applications of the property of reverse figures that *all angles subtended at the origin in the one figure correspond to equal angles subtended at the reverse origin in the other.* [Art. 142 Cor. 2.

*a. A variable chord of a conic which subtends a right angle at a given point envelopes a conic having that point for a focus.*

For if the given point $O$ and its polar be taken as origin and base line, the reverse conic has its centre at the reverse origin ω (Art. 142 Cor. 1); and a variable chord of the latter conic which subtends a right angle at ω envelopes a concentric circle (Ex. 289), of which the obverse is a conic having $O$ and the base-line for a focus and directrix.

*b. A variable chord of a conic which subtends a right angle at a given point on the curve passes through a fixed point on the normal thereat.*\*

For if a conic through $O$ be reversed into a circle through ω, every chord of the former which subtends a right angle at $O$ has for its reverse a diameter of the circle, and therefore passes through the fixed point which is the obverse of the centre of the circle. Note that the tangents and also the normals at $O$ and ω are reverse lines.

Hence, to reverse a conic from any point $O$ upon it as origin into a circle, we must have as base line *the polar BC of the point of concourse of all chords which subtend right angles at O.*

*c.* Let $DOE$ be a fixed angle inscribed in a conic, $P$ any point on the curve, $B$ and $C$ the points in which $PD$ and $PE$ meet the polar of the point of concourse of all chords which

---

\* This theorem of Frégier—p. 276, note, and *Correspondance sur l' Ecole Royale Polytechnique* tome III p. 394, 1816—is a limiting case of §a. See Scholium D, p. 285.

subtend right angles at $O$: *then will the angle $BOC$ be equal or supplementary to the angle $DOE$.*\*

For if with $O$ as origin and $BC$ as base line **the conic be reversed into a circle**, then (with the same notation) the points $B$ and $C$ are removed **to infinity**, and the theorem follows at once from the equality **of the angles $DOE$ and $DPE$** in the same segment of the **circle**.

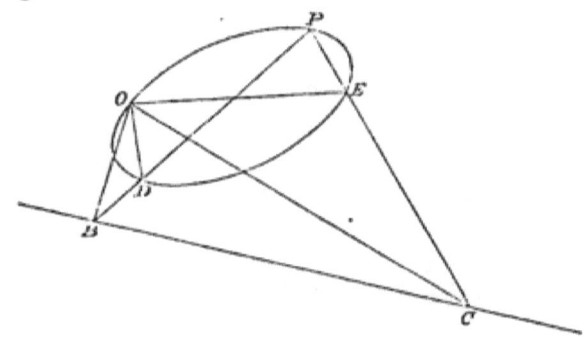

*d.* If a straight line $PDB$ turning about **a fixed point $P$** meet the arms of **a** constant angle $BOD$, which turns about a fixed point $O$, in $B$ and $D$; then if the point $B$ moves along a straight line $BC$, *the point $D$ describes a conic through $O$ and $P$.* For when $BC$ is the line at infinity the locus of $D$ is evidently a circle through $O$ and $P$; and therefore by reversion, the locus of $D$ in the general case is a conic through $O$ and $P$.

This is a limiting form of Newton's *Descriptio Organica* (Art. 113 Cor. 1), since the line through $P$ may be regarded as a vanishing angle $BPD$.

*e.* **Every** range $\{ABCD\}$ and its reverse $\{abcd\}$ subtend similar pencils $O\{ABCD\}$ and $\omega\{abcd\}$ at the origins, and are therefore homographic. All the properties **of cross** ratio may therefore be extended **from** the circle **to** the general conic by reversion.

**145.** *The Orthocentre.*

*a.* Let the sides of **a** triangle $ABC$, the reverse of $A'B'C'$,

---

\* See *Mathematical Questions from the* EDUCATIONAL TIMES, vol. I. pp. 33, 40 (Question 1409).

meet the base line in points $D$, $E$, $F$, and let $aD$, $bE$, $cF$ be segments of the base line which subtend right angles at $\omega$. Then $Aa$, $Bb$, $Cc$ are reverse to the perpendiculars of the triangle $A'B'C'$ (Art. 142 Cor 2) and cointersect at the reverse $P$ of its orthocentre.

It is hence evident that the sides of any triangle $ABC$ and the radiants from any point to its vertices determine an involulution $\{aD, bE, cF\}$ on any transversal. [Art. 110.

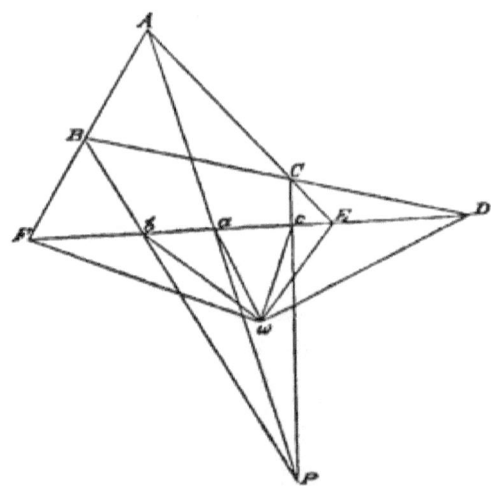

*b.* If the triangle $ABC$ envelopes a fixed conic touching the base line, the obverse of which is a *parabola*, the point $P$ traces a straight line, the reverse of the directrix of the parabola. [Art 29 Cor. 1.

Or if, starting with the parabola and taking its directrix as base line, we reverse it into a circle about $\omega$ as centre, the point $P$ is removed to infinity. Hence, if the sides of a triangle $ABC$ touch a circle, and meet any fourth tangent to it in $abc$, and if the diameters parallel to the polars of $abc$ meet the fourth tangent in $DEF$, the lines $AD$, $BE$, $CF$ are parallel. In other words:

*If the sides of a triangle $ABC$ touch a circle, and if the parallel tangents meet any seventh tangent in $DEF$, the lines $AD$, $BE$, $CF$ are parallel.*

More generally:

*If from three collinear points XYZ pairs of tangents be drawn to a conic, and if ABC be the triangle formed by one tangent from each pair and DEF the points in which the remaining three tangents meet any seventh tangent, the lines AD, BE, CF meet at a point in a straight line with XYZ.*

### 146. *The Normal.*

The reverse of the normal at any point $P$ to a conic is the line through the reverse point $p$ which with the tangent at $p$ intercepts on the base line a length which subtends a right angle at the reverse origin $\omega$.  [Art. 142 Cor. 2.

### 147. *Conjugate Diameters.*

If a conic be reversed into the eccentric circle of $\omega$, it may be seen that a pair of its conjugate diameters inclined at angles $\alpha$ and $\pi - \alpha$, reverse into lines through the pole of the base line with respect to the circle and which contain angles $\pi - \alpha$ and $\alpha$.

### 148. *The Asymptotes.*

If a conic meets the base line in $M$ and $N$, the asymptotes of its reverse correspond to the tangents at $M$ and $N$, and are therefore parallel to $MO$ and $NO$, where $O$ is the origin (Art. 142). We may therefore determine the eccentricity of the reverse conic by making the angle $MON$ of any assumed magnitude, real or imaginary.

#### SCHOLIUM C.

REVERSE lines $OP$ and $p\omega$ through the origins (which may be supposed to lie on the same side of the base line) being reverse in direction, figures are consequently, in a manner, *turned over* in this transformation, so that an original figure and its derivative may be regarded as obverse and reverse respectively. Thus in Art. 4, if the circle be divided by axes through $O$ parallel and at right angles to the base line, its first and third quadrants must be turned over or interchanged, and likewise its second and fourth, in order that they may become similarly situated with the sectors of the conic to which they severally correspond.

If reverse points $P$ and $p$ be referred to rectangular axes of coordinates, the base line being the common axis of $x$ and the axes of $y$ being drawn through $O$ and $\omega$ respectively, then if $X$, $Y$ be the coordinates of $P$ and $x$, $y$ those of $p$, and if $OD$ and $\omega d$ be the ordinates of $O$ and $\omega$, it may be shewn that $Yy = \omega d \cdot OD$, and $X : -x = Y : \omega d$. Hence an equation of any degree between $x$ and $y$ implies an equation of the same degree between $X$ and $Y$.

Reversion is a special case of the following transformation. Take fixed origins $O$ and $\omega$, and a fixed director line (or plane) corresponding to each: from any point $P$ draw $POD$ to meet the $O$-director and $P\omega d$ to meet the $\omega$-director: then the point of concourse of $\omega D$ and $Od$ corresponds to $P$. The construction in the text results from supposing one of the directors to be at infinity. The analysis for the general case is fully given in a section by Prof. Cayley contributed to my article on the Homographic Transformation of Angles in the *Quarterly Journal of Mathematics* XIV. 25—39.

## HOMOGRAPHIC TRANSFORMATION.

### PROPOSITION XII.

149. *Any two plane homographic figures of the same species are capable of being placed in perspective.*

We have seen that any two plane figures in perspective are so related that to every range in the one corresponds a homographic range in the other (Prop. IV). Conversely, any two plane figures thus related are capable of being placed in perspective.

(i) For if $ABCD$ be four fixed points and $P$ a variable point in a plane figure, and $A'B'C'D'P'$ the corresponding points in a homographic figure, it is evident from the relation,

$$P\{ABCD\} = P'\{A'B'C'D'\},$$

that by projecting the points $ABCD$ into $A'B'C'D'$ (Art. 135) we at the same time project every point $P$ into its correspondent $P'$.

(ii) The same result may also be arrived at as follows.

Let $A$ be a given plane figure, regarded as moveable in any way in its plane, and $B$ a fixed homographic figure in the same plane. Then to the focoids $\phi$ and $\phi'$, regarded as belonging to $A$, correspond fixed points $F$ and $F'$ related to $B$.

[Art. 132.

Let $F\phi$ and $F'\phi'$ meet in the point $O$ related to $B$, and let $O'$ be the corresponding point in $A$. Also let $P$ and $Q$ be any two points in $B$, and $P'$ and $Q'$ the corresponding points in $A$.

Move the figure $A$ a certain distance in a certain direction until $O'$ coincides with $O$, and then turn it about $O$ until the points $POP'$ are brought into one straight line.* Then, since $A$ and $B$ are homographic,

$$O\{PQFF'\} = O\{P'Q'FF'\};$$

and therefore, since three rays in the one pencil coalesce severally with three in the other, the fourth ray $OQ$ coalesces with the fourth $OQ'$, or every two corresponding points $Q$, $Q'$ are in a straight line with $O$, the required centre of perspective of $A$ and $B$. It then follows from Art. 140 that $A$ and $B$ may be placed perspective in space.

## Corollary.

Since figures homographic with the same figure are homographic with one another, and since any conic and an assumed point in its plane may be projected into a circle and its centre (Art. 137), and conversely; it follows that *any conic and point in one plane may be projected into any other conic and point in one plane.* [Prop. XI. Cor.

### SCHOLIUM D.

Transformation is a convenient (if not strictly accurate) expression for the derivation of one figure from another in accordance with an assigned law of correspondence. The general idea of homographic transformation may be found in a passing remark of DESARGUES (*Œuvres* I. 214), who, having enunciated the fundamental property of the polar planes of a sphere, concludes by stating curtly that it may be extended to surfaces which are related to the sphere as the ellipse is to the circle: " Semblable propriété se trouve à l'egard d'autres massifs qui ont du rapport à la boule, comme les ouales autrement ellipses en ont au cercle, mais il y a trop à dire pour n'en rien laisser."

In the tract on *Plani-coniques* appended to his *Nouvelle méthode en Géométrie &c.* (Paris 1673), De la Hire derived the general conic

---

\* See Salmon's *Higher Plane Curves* Art. 330.

from the circle by a geometrical transformation *in plano*. A similar transformation appears to have been arrived at by Le Poivre about thirty years later (Chasles *Sections Coniques* p. 174). NEWTON shewed how to transform curves of all orders (*Principia* Lib. I. sect. v. lemma 22) by substitutions of the form $X = \frac{ab}{x}$ and $Y = \frac{ay}{x}$ (cf. Schol. C), and added two examples of the application of his method (*loc. cit.* props. 25, 26). The "collinear" figures of Möbius "sont aussi des figures homographiques les plus générales" (Chasles *Sections Coniques* p. 165). Möbius proved *inter alia* that any four points in a plane may be projected into any other four points in a plane (*Barycentrische Calcul* p. 327, 1827). For a general exposition of the principle of Homography see Chasles' *Mémoire* on duality and homography, at the end of his *Aperçu historique*. See also his *Géométrie Supérieure* pp. 362—412, and Townsend's *Modern Geometry* chaps. 19—22.

## EXAMPLES.

[*It is left to the reader in some cases to modify the enunciations of the pairs of theorems in the double columns so as to bring them into exact correspondence*].

801. The diagonals of a parallelogram and the lines bisecting its two pairs of opposite sides form a harmonic pencil.

Each diagonal of a complete quadrilateral is divided harmonically by the remaining two diagonals.

802. If two triangles be similar and similarly placed, the joins of their corresponding vertices meet in a point.

If the intersections of the three pairs of sides of two triangles lie in one straight line, the joins of the opposite vertices cointersect.

803. Any two pairs of parallels through points $P$ and $Q$ meet any transversal in an involution having its centre on $PQ$.

The three pairs of joins of any four points in a plane determine an involution on any transversal.

804. The centres of the diagonals of a complete quadrilateral are in one straight line. [p. 256.

An infinity of pairs of straight lines can be found which divide the three diagonals of a quadrilateral harmonically.

805. Parallel chords of a circle are bisected by a straight line through its centre.

Concurrent chords of a conic are divided harmonically by their common point and its polar.

806. If two of the three pairs of opposite sides of a hexagon inscribed in a circle are parallel, the third pair are parallel.*

The three pairs of opposite sides of any hexagon inscribed in a conic have their intersections in one straight line.

807. A system of coaxal circles meet any transversal in pairs of points in an involution. [Art. 109.

All the conics through four given points meet any transversal in pairs of points in an involution.

---

* See Gergonne's *Annales* IV. 79.

# EXAMPLES. 331

808. Four circles can be drawn touching three given straight lines.

Four conics can be drawn through two given points and touching three given lines.

809. Four conics can be drawn through **three** given points so as to have a given point for a focus.

Four conics can be **drawn through** three given **points so as to touch two** given lines.

810. The diameters of a **circle subtend** a pencil in involution at **any point on the** circumference.

A system of concurrent chords of a conic subtend a pencil in involution at **any point on the curve.** [p. 276.

811. **Given two pairs of lines conju**gate with respect **to** a circle, the locus of its **centre is** the rectangular hyperbola **circumscribing** the quadrilateral of which **the conjugate** lines are opposite sides.*

Given a chord $FF'$ of a conic and two pairs of lines conjugate with respect to it, the locus of the pole of $FF'$ is a conic with respect to which $F$ and $F'$ are a pair of conjugate points.

812. Given three pairs of lines conjugate with respect to a circle, the positions of its centre constitute an orthocentric tetrastigm.

Through two given points four conics **can** be drawn so as to have three given pairs of lines conjugate **with respect to** them; **and their common chord is divided harmonically by every conic through its four poles with respect to them.**

813. Every circle through the centre of a rectangular hyperbola circumscribes an infinity of triangles self-polar with respect to the hyperbola.

If two triangles be self-polar with respect to a conic, their six angular points lie on a conic.

814. **If a** triangle $PQR$ right angled at $P$ be inscribed in a rectangular hyperbola, **the perpendicular from** $P$ **to** $QR$ is the tangent **at** $P$.

If $F$ and $F'$ be conjugate points with respect to a conic, $PQ$ and $PR$ any two chords **which** divide $FF'$ harmonically; then $QR$ **and** the tangent at $P$ divide $FF'$ harmonically.

815. The directions of two sides of a triangle inscribed in a circle being given, the envelope of the third side is a concentric circle.

**If two sides of** a triangle inscribed in a conic **pass each** through a given point, the envelope **of the** third side is a conic touching the **former at two points on the** join of the **given points.**

816. **The** envelope of the polar of any point on a circle with respect to a concentric circle is a concentric circle.

The envelope of the polar of any point on a conic with respect to a second having the same focus and directrix is a third having the same focus and directrix.

---

* From the centre $O$ of the circle draw a perpendicular $OP$ to one of the lines, and let it meet the conjugate line in $Q$; and draw $OP'$ perpendicular to one of the second pair of lines, and let it meet the fourth line in $Q'$. Then since $OP \cdot OQ = $ (radius)$^2$ $= OP' \cdot OQ'$, the locus of $O$ is a conic through the four vertices of the quadrilateral; and it is easily seen that the orthocentre of any three of them is a point on the locus.

817. Parallel chords of a circle are cut in a constant ratio by a concentric ellipse touching the circle at the extremities of the perpendicular diameter.

Concurrent chords of a conic are divided in a constant cross ratio by every conic having double contact with the former upon the polar of the point of concurrence.

818. Every parallelogram inscribed in a circle is rectangular.

The intersections of the two diagonals and of the opposite sides of any quadrilateral are a conjugate triad with respect to every conic circumscribing the quadrilateral.

819. The diagonals of every parallelogram circumscribed to a circle meet at right angles at its centre.

The diagonals of a complete quadrilateral are a conjugate triad with respect to every conic inscribed in it.

820. The centres of all the rectangular hyperbolas circumscribed to a given triangle lie on its nine-point circle.

Given four points on a conic, the locus of the pole of a given line is a conic, &c. [Ex. 788.

821. The circumscribed circle of every triangle which circumscribes a parabola passes through its focus.

If two triangles circumscribe a conic, their six summits lie on a conic.

822. The envelope of the polar of a given point with respect to a system of confocal conics is a parabola touching their axes and having the given point* for a point on its directrix. [Ex. 379.

The envelope of the polar of a given point with respect to the system of conics inscribed in a quadrilateral is a conic touching its three diagonals; and the chord of contact of the second tangents to this conic from the extremities of any diagonal of the quadrilateral is the line joining the given point to the point of concourse of the remaining two diagonals.

823. If from a fixed point $O$ tangents $OP$ and $OQ$ be drawn to any one of a system of confocal conics, the circle through $OPQ$ passes through a second fixed point. [Exx. 340, 380.

If from a fixed point $O$ tangents $OP$ and $OQ$ be drawn to any one of a system of confocal conics, the conic through their foci and $OPQ$ passes through a fourth fixed point.†

824. Given three concentric circles, any tangent to one of them is divided into segments of constant lengths by the remaining two.

If three conics touch one another at the same two points, any tangent to one of them is divided in a constant cross ratio by the remaining two.

825. Four fixed tangents to a parabola divide any fifth tangent into segments whose ratios are constant. [Ex. 726.

Four fixed tangents to a conic divide any fifth tangent in a constant cross ratio.

---

* The tangents at this point to the two confocals through it touch the parabola.

† Project the common foci of the first system into the focoids of the plane of the second. See the *Quarterly Journal of Mathematics* x. 287.

826. Any two pairs of points which divide the diagonals of a rectangle harmonically lie on a circle.

If the extremities of each of two diagonals of a quadrilateral are conjugate points with respect to a conic, **the extremities of the third are conjugate with respect to it.**[*]

827. If $POp$, $QOq$, $ROr$, $SOs$ be any four concurrent chords of a conic, **the conics through** $OPQRS$ **and** $Opqrs$ have a common **tangent at** $O$.

828. If $FF'$ be **a** common chord of **two** given conics, its **pole with** respect to any **conic** which touches **both** of them **and** passes through $F$ and $F''$ has for its locus a conic touching the tangents at $F$ and $F'$ to the given conics.

829. If a conic touches the **sides** $SF$ and $SF'$ of a **given** triangle and also two other **given lines, the second tangents** to it from $F$ and $F''$ meet on **a fixed straight line.**

830. Given, in addition to a chord of a conic, two tangents, or one tangent and one point, find in each case the locus of the pole of the given chord.

831. **If two** conics have double contact, the cross ratio of **four of the** points in which any four tangents to the one **meet the other** is equal **to that** of the remaining four points, **and also to that of the points of contact** taken in the same sense of rotation.[†]

832. Extend by projection **Newton's theorem,** that the diameter of **a** quadrilateral **is the centre-locus of all** conics inscribed therein.[‡] [p. 282.

833. The circle through any triad of points conjugate with respect to a conic is orthogonal to its orthocycle (pp. 274, 280). Is this theorem projective?

---

[*] This theorem and its reciprocal are due to Hesse (Crelle's *Journal* XX. 301, 1840).

[†] See Salmon's *Conic Sections* Arts. 276, 354.

[‡] Its analogue in space was given in Gergonne's *Annales* XVII. 200.

834. From a fixed point $O$ tangents $OP$ and $OQ$ are drawn to any conic inscribed in a given quadrilateral, and through $P$ and $Q$ are drawn the straight lines which with $PO$ and $QO$ respectively divide the three diagonals of the quadrilateral harmonically. Shew that the lines so drawn touch the fixed conic which is the envelope of $PQ$. [Ex. 822.

835. If $AB$ be one of the diagonals of this quadrilateral, the conic through $ABOPQ$ passes through a fourth fixed point $O'$, such that $AO$, $AO'$ and $BO$, $BO'$ divide the remaining two diagonals harmonically. Shew also that the three positions of $O'$ corresponding to the three diagonals of the quadrilateral lie in one straight line.

836. If from a fixed point $O$ tangents $OP$ and $OQ$ be drawn to any one of a system of confocal conics, and if the normals at $P$ and $Q$ meet in $N$; the locus of the orthocentre of the triangle $NPQ$ is a straight line, and the locus of the orthocentre of $OPQ$ is a rectangular hyperbola having one asymptote parallel to the central distance of $O$.* What do these theorems become by projection?

837. Given the orthocentre of a triangle inscribed (or circumscribed) to a given conic, the product of the segments of its perpendiculars is constant. Hence shew that if one solid angle contained by three planes mutually at right angles can be inscribed (or circumscribed) to the surface of a given cone of the second degree, an infinity of such angles can be inscribed (or circumscribed) to it.

838. The locus of the point in space from which triads of lines mutually at right angles can be drawn to triads of points on a given conic is a sphere.

839. If the conic be supposed to vary, yet so as always touch the sides of a given quadrilateral, the sphere will pass

---

* See the *Quarterly Journal of Mathematics* x. 289.

through a fixed circle in a plane at right angles to the plane of the quadrilateral.*

840. $ABC$ is a triangle inscribed in a conic: $Aa, Bb, Cc$ are chords drawn through a point $O$: and $Ab, Bc, Ca$ meet the polar of $O$ in $PQR$ respectively. Shew that the lines from any point on the conic to the points $PQR$ respectively meet the sides of $ABC$ in three collinear points.

841. Rays being drawn from a fixed point on a conic, shew that the intercepts upon them between the conic and a fixed tangent may all be projected upon a given line through the fixed point, from another point on the conic, into segments of the same length.

842. If a point $P$ on a conic be connected with two fixed points $F$ and $F''$ in its plane, all the chords which are divided harmonically by $FP$ and $F''P$ are concurrent; and the locus of their point of concourse, as $P$ varies, is a conic touching the first at two points on $FF''$.

843. If a conic passes through two given points and touches a given conic at a given point, its chord of intersection with the given conic passes through a fixed point.

844. $ABCD$ being four points on a conic, $E$ and $F$ are the poles and $O$ is the point of concourse of $AB$ and $CD$. Through $E$ is drawn a straight line meeting $CD$ in $M$ and the conic in $Q$ and $R$; and upon this line a point $P$ is taken a fourth harmonic to $QMR$. Shew that the locus of $P$ is the conic through $ABCDEF$, the tangents to which at $E$ and $F$ pass through $O$; and that the tangents from $O$ to the first conic pass through the four points in which the common tangents to the two conics touch the second.

---

* On Exx. 837–9 see Picquet's *Étude géométrique des Systèmes Ponctuels et Tangentiels de Sections Coniques* §§ 72, 73, 85, 86. Picquet now uses the term *orthoptic circle* (p. 42) to denote the orthocycle, and the term *orthoptic summits* (p. 41) of the "pencil" of conics inscribed in a quadrilateral to denote the two fixed points on their orthocycles. Gaskin's discovery of these points was anticipated by Plücker (*Analytisch-geometrische Entwicklungen* II. 198, 1831).

845. If a variable conic has double contact with each of two fixed conics, find the loci of the points of concourse of its common tangents with the fixed conics.

846. If $aa'$, $bb'$, $cc'$ be parallel chords of a conic and $p$ any seventh point on the curve, prove that the three points $(ap, b'c')$, $(bp, c'a')$, $(cp, a'b')$ lie on a straight line parallel to the chords.

847. If from each of four points on a circle perpendiculars be drawn to the joins of the remaining three, the feet of these perpendiculars lie by threes on four concurrent lines. Generalise this theorem by reversion, or otherwise.

848. Any two conics may be regarded as homographic figures in which any three points on the one correspond to three points taken arbitrarily on the other.*

849. Shew how to place reverse figures in perspective, and adapt the constructions of Arts. 141—148 to the case of figures in perspective in one plane.

850. The construction in Art. 134 (ii) for fixing the position of $V$ in the vertex-plane is independent of the angle between that plane and the primitive plane: as this angle varies the vertex remains fixed in its plane: it therefore describes a circle in a plane perpendicular to $ab$, or to the intersection of the plane of projection with the primitive plane.

---

* Chasles *Sections Coniques* p. 167.

NOTE.

The undermentioned Examples (cf. p. 141) are taken from the EDUCATIONAL TIMES *Reprint*:

Ex. 441 (I. 52); 442 (XI. 31); 443 (V. 101); 444 (VII. 49); 445 (XXX. 90); 450 (XII. 27); 454 (VIII. 72); 455 (III. 35); 466 (XXX. 90); 516 (I. 55); 517 (V. 56); 518 (I. 53); 520 (V. 36); 521 (VII. 96); 522 (XI. 59); 523 (XVIII. 32); 524 (XVIII. 54); 526 (XXI. 35); 527 (XXIV. 87); 560 (XVII. 35); 562 (XVIII. 52); 563 (XVIII. 98); 578 (VI. 71); 607 (XV. 88); 611 (XXV. 23); 612 (XXV. 21); 613 (IV. 70); 620 (XXIV. 43); 638 (IV. 97); 640 (XIII 60); 644 (V. 87); 645 (XI. 104); 680 (IV. 69); 681 (X. 32); 682 (XVII. 60); 684 (XXII. 22); 685 (XXII. 51); 687 (XXIV. 101); 688 (XXVI. 69); 689 (XXVIII. 95); 690 (XXXI. 18); 714 (XI. 60); 727 (XIII. 26); 737 (XXV. 61); 738 (II. 6); 835 (XIII. 61); 840 (III. 39); 841 (IX. 79); 842 (XI. 100); 843 (XII. 54); 844 (XIV. 80); 845 (XIX. 101); 846 (XXIII. 94); 847 (XXX. 64. XXVIII. 57).

# CHAPTER XII.

## RECIPROCATION AND INVERSION.

150. WE now return to the method of Reciprocal Polars, of which some account was given in Chapter x. [p. 268.

Take any fixed conic as *Director*, and let any straight line or point be said to *correspond* or be *reciprocal* to its pole or polar **with respect to the director**. It is evident that to the intersection of any two lines corresponds the join of the reciprocal points, **and that to a system of concurrent lines** correspond a system of collinear points. [Art. 17 Cor. 1.

To any curve, regarded **as** the envelope of its tangents, corresponds the locus of their poles with respect to the director; and **to the** join of any two *consecutive* points on either curve corresponds the join of two consecutive tangents to the other. Hence, to **every** tangent and its point of contact in either figure **correspond a point and the** tangent thereat **in the other**.

It is hence evident **that if** two curves touch one another in one or more points, their reciprocals **touch one** another in the same number of points.

Notice that *the* **tangents to** *the* **director** *at its* **intersections** *with any curve are also tangents to the* **reciprocal curve**.

### PROPOSITION I.

151. *The degree of any curve not having singular points is equal to the class of its reciprocal with respect to a* **conic**, *and conversely.*

(i) For if $U$ be any curve and $u$ its reciprocal, $A$ any straight line and $a$ its reciprocal; then to every point in which $A$ meets $U$ corresponds a tangent to $u$, and every such

z

tangent passes through $a$. The number of points in which an arbitrary line $A$ meets $U$ is therefore equal to the number of tangents that can be drawn from any one point to $u$, as was to be proved.

It follows that the reciprocal of a *conic* is a conic, as was otherwise shewn in Art. 117; and that the reciprocals of all conics touching the same four lines, or *having the same foci*, are conics passing through the same four points. [Prop. VI.

## PROPOSITION II.

**152.** *To every point and its polar with respect to a conic correspond a straight line and its pole with respect to the reciprocal conic: to every conjugate* **triad** *of points in the* **one** *figure a conjugate triad of lines in the other: and to every pencil or range in the one a homographic range or pencil in the other.*

(i) For if $ab$ and $ac$ be the tangents from any point $a$ to a conic, then to their points of contact $b$ and $c$ correspond a pair of tangents $AB$ and $AC$ to the reciprocal conic; and to the points of contact $B$ and $C$ correspond reciprocally the tangents at $b$ and $c$ to the original conic. [Art. 150.

It follows that the join of $B$, $C$ corresponds to $a$, and the join of $b$, $c$ to the point of concourse $A$ of the tangents at $B$ and $C$.

Hence to any point $a$ and its polar $bc$ in the one figure correspond a line $BC$ and its pole $A$ in the other; and therefore to every conjugate triad of points in either corresponds a conjugate triad of lines in the other.

(ii) It has been shewn in Art. 116 that every row of points and their polars with respect to any director are homographic.

*Corollary.*

Since the points of concourse of a conic and its reciprocal correspond to their common tangents, *the* **cross** *ratio of the four common points of any* **two** *conics in either is equal to that of their common tangents* **in** *the other.* [Ex. 727.

## PROPOSITION III.

**153.** *Given in a plane two straight lines and their poles with respect to a conic, the ratio of the distances of any point in the plane from the given lines varies as the ratio of the distances of its polar from their poles.*[*]

(i) Let the polars of any two points $O$ and $P$ with respect to a conic whose centre is $C$ meet $CO$ in $L$ and $M$ respectively, and let $PN$ be an ordinate to the diameter $CO$.

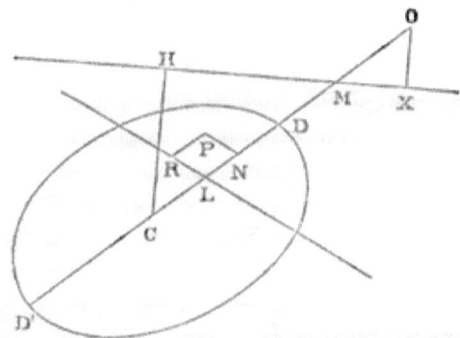

Then since $\qquad CL.CO = CM.CN,$[†]

or $\qquad CM + OM : CM = CL + NL : CL ;$

therefore $\qquad OM : CM = NL : CL = PR : CL,$

if $PR$ be drawn parallel to $CL$ to meet the polar of $O$. That is to say, *the distances of $O$ and $C$ from the polar of $P$ are as the distances of $P$ and $C$ from the polar of $O$.* [Ex. 325.]

(ii) If $OX$ and $CH$ be perpendiculars to the polar of $P$, it follows that

$$PR : CL = OX : CH;$$

and the same proportion will hold if $PR$ and $CL$ be now supposed *perpendicular* to the polar of $O$.

In like manner, taking any second position $o$ of $O$ whilst $P$ remains as before, we have

$$Pr : Cl = ox : CH.$$

---

[*] Chasles *Aperçu historique* p. 590, 1875.

[†] If $CM$ meets the conic in $D$ and $D'$, and if $\infty$ be the polar of $DD'$, the point $M$ and its polar $P\infty$ divide $DD'$ harmonically. Hence $CM.CN = CD^2$, whatever be the position of $P$. This follows also by orthogonal projection from Ex. 281.

Hence, whatever be the position of $P$, the ratio $\dfrac{PR}{Pr}$ is equal to $\dfrac{CL}{Cl} \cdot \dfrac{OX}{ox}$, where $CL$ and $Cl$ are constant for given positions of $O$ and $o$; that is to say, *the ratio of the perpendiculars from a variable point $P$ to the polars of two fixed points $O$ and $o$ varies as the ratio of the perpendiculars from $O$ and $o$ to the polar of $P$.* We may of course, as a special case, suppose either point and its polar to become a point on the curve and the *tangent* thereat.

### Corollary.

It is hence evident that any homogeneous relation between the distances of a variable point $P$ from any number of fixed straight lines implies a homogeneous relation of the same degree between the distances of the polar of $P$ with respect to a conic from the poles of the fixed lines. Thus from the *Locus ad quatuor lineas*, $PQ.PR = k.PS.PT$ (Scholium C, p. 266), we deduce a relation of the same form between the distances of the tangent at $P$ from two pairs of fixed points, opposite vertices of a quadrilateral circumscribed to the conic. [Ex. 707.

## POINT RECIPROCATION.

**154.** If $O$ be the centre of a circle, $OA$ the perpendicular from it to any straight line $L$ and $a$ the point on $OA$ or its prolongation such that $OA.Oa = (\text{radius})^2$; then $a$ is the pole or reciprocal of $L$ with respect to the circle. The same point $a$ may also be determined by regarding $O$ merely as a fixed point, without reference to the circle, and taking $OA.Oa$ equal to a constant quantity $c^2$. This last construction is called *reciprocation with respect to a point*, the point being called the *origin*. When $c^2$ is *negative* every line or point and its reciprocal lie on opposite sides of the origin. The construction is then equivalent to reciprocating with respect to an *imaginary* circle. [Prop. v. (ii).

Notice that to *the foot of the perpendicular* $A$ from the origin to any straight line $L$ corresponds the line through the reciprocal point $a$ parallel to $L$ or at right angles to $Oa$;

and that to each line in *a system of parallels* corresponds a point on the common perpendicular to them from the origin.

## PROPOSITION IV.

**155.** *Any two straight lines contain an angle equal to that subtended by their point-reciprocals at the origin; and the distance of any straight line from the origin varies inversely as the distance of its reciprocal therefrom.*

The first part of the proposition follows immediately from the perpendicularity of every straight line to the line joining its reciprocal to the origin.

The second part is merely another way of stating that the product $OA.Oa$, in the construction of Art. 154, is equal to a constant $c^2$.

### Corollary 1.

From the relation $Oa = \dfrac{c^2}{OA}$ we deduce, that to any straight line $L$ through $O$ corresponds the point *at infinity* in the direction at right angles to $L$, and conversely. All points at infinity in the same plane therefore lie on the reciprocal of the origin, and are consequently to be regarded as collinear. To the polar of the origin with respect to the original conic corresponds *the centre of the reciprocal conic*, the polar of the line at infinity with respect to it.  [Art. 17 Cor. 2.

### Corollary 2.

To the tangents $OP$ and $OP'$ from the origin to any conic correspond the points at infinity in the directions at right angles to $OP$ and $OP'$; the *eccentricity* of the reciprocal conic is therefore determined by the angle $POP'$, the supplement of the angle between its asymptotes. Hence the reciprocal will be a hyperbola, an ellipse or a parabola according as $O$ is taken without, within or upon the original conic. Thus we see again that *every parabola touches the line at infinity*, the reciprocal of the origin. The reciprocal of a conic with respect to any point at which it subtends a right angle is a *rectangular hyperbola*. In any case the *axis* of the reciprocal conic is parallel to the bisector of the angle $POP'$.

### Corollary 3.

From either point of concourse of their orthocycles as origin any two conics $U$ and $V$ reciprocate into rectangular hyperbolas $u$ and $v$; and conversely every conic through the four intersections of $u$ and $v$, being itself a rectangular hyperbola (Art. 69), is reciprocal to a conic subtending a right angle at the origin and touching the four common tangents to $U$ and $V$. Hence *the orthocycles of all the conics which touch four given lines have two points in common.* [Art. 123 Cor. 1.

### PROPOSITION V.

**156.** *The reciprocal of a circle with respect to any point is a conic having that point for a focus, and conversely.*

(i). If the director be a circle about the origin $O$ as centre, any other circle $C$ meets it at the focoids $\phi$ and $\phi'$, and therefore has for its reciprocal a conic touching $O\phi$ and $O\phi'$, the tangents to the director at the focoids. [Art. 130.

This conic therefore has $O$ for a focus; and *its $O$-directrix* (the polar of $O$) corresponds to the pole of the line at infinity with respect to $C$, that is to say, it *corresponds to the centre of $C$.*

(ii). Otherwise thus. Reciprocate a conic from either focus $H$ as origin, and let $V$ be the point corresponding to any tangent to the conic, and $Z$ the projection of $H$ upon that tangent.

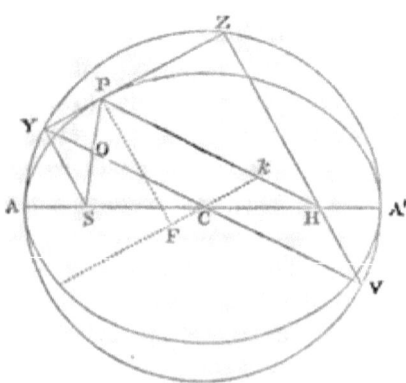

Then if $CB$ be the conjugate semiaxis, and $-CB^2$ be taken as the constant of reciprocation,

$$HV.HZ = -CB^2,$$

or $V$ lies on the major auxiliary circle, which is accordingly a reciprocal of the conic with respect to either focus. Hence, whatever be the constant of reciprocation, the reciprocal of a conic with respect to either focus is a circle, and conversely.*

In the limiting case of the *parabola*, produce $SY$ and $SA$ (fig. Art. 26) to $Z$ and $B$, so that $SY.SZ = SA.SB =$ a constant. Then $Z$ lies on the circle upon $SB$ as diameter, which is accordingly a reciprocal of the parabola with respect to $S$, and conversely.

(iii). In the diagram, the *eccentricity* of the conic is equal to $\dfrac{CH}{CA}$, and its *latus rectum* to $\dfrac{2CB^2}{CA}$.

Moreover, if $X$ be the foot of the $H$-directrix,

$$HC.HX = -CB^2, \qquad [\text{Art. 35 Cor. 3}$$

and therefore (1) the $H$-directrix is reciprocal to the centre of the *circle*, and (2) the polar of $H$ with respect to the circle—which in this case coincides with the $H$-directrix (Art. 35 Cor. 1) —is reciprocal to the centre of the *conic*.

If the constant of reciprocation be changed the relative *magnitude* of the conic and the circle will alter, but the following relations will still be found to subsist. The origin is now denoted by $O$, and the centre of the *circle*—which will in general be distinct from that of the conic—by $C$.

$$\text{eccentricity} = \frac{\text{distance of } C \text{ from } O}{\text{radius of circle}}.$$

$$\tfrac{1}{2} \text{ latus rectum} = \frac{\text{constant of reciprocation}}{\text{radius of circle}}.$$

centre of circle $\equiv$ reciprocal of $O$-**directrix.**

centre of conic $\equiv$ reciprocal of polar of $O$ with respect to circle.

---

* Another proof has been given by Laquière, *Nouvelles Annales* xx. 42, 1861; and another by Salmon, *Conic Sections* Art. 308.

*Corollary* 1.

One inscribed and three escribed circles can be drawn to a given triangle, and their radii are connected by the relation,

$$\frac{1}{r} = \frac{1}{r_1} + \frac{1}{r_2} + \frac{1}{r_3}.$$

Hence, by reciprocation with respect to any point $O$, four conics can be described having a given point for a focus and passing through three given points, and the latera recta of three of them are together equal to the latus rectum of the fourth.

*Corollary* 2.

The point $V$ in the diagram corresponds to the tangent $YZ$; the point $C$ to the $H$-directrix; the line $VC$ to the point of concourse $D$ of $YZ$ with the $H$-directrix; the point of concourse $Y$ of $VC$ and $YZ$ to the line $DV$. But $Y$ lies on the circle; therefore $DV$ touches the conic. The second point $V'$ in which $DV$ meets the circle corresponds to the second tangent from $Y$ to the conic. This tangent is evidently parallel to the opposite tangent $DV$; therefore $YHV'$ is at right angles to $DV$ and $H$ is the orthocentre of the triangle $DYV$. [Ex. 330.

### PROPOSITION VI.

157. *All the circles of a coaxal system reciprocate from either of their limiting points into confocal conics.*

(i). Whatever be the position of the origin $O$ in the plane of the circles, their reciprocals have $O$ for a common focus.

If the origin be taken at either limiting point* of the system of circles, it has the same polar with respect to them all, and therefore the line at infinity has the same pole with respect to all their reciprocals. That is to say, the latter have a common *centre* as well as one focus in common. They are therefore confocal, as was to be proved.

---

* These points are the limits of the system "par rapport à l'*infiniment petit*," and the radical axis and the line at infinity are its limits "par rapport à l'*infiniment grand*." When a circle becomes infinite it degenerates in **general** into a straight line at a finite distance *together with the line at infinity*. See Poncelet's *Traité des Propr. Projectives* p. 49 (1822); Townsend's *Modern Geometry* 1. 199.

(ii). The system of conics through four given points reciprocate with respect to any one of the system as director into conics touching the tangents to the director at the given points. The construction § i. corresponds to the case in which the focoids are two of the four given points, and the director reduces to a point-circle.

## Corollary.*

The further limiting point $O'$ of the circles corresponds to the *minor axis* of the confocals, and the line bisecting $OO'$ at right angles to their *further focus*. The orthogonal circles through $O$ and $O'$ reciprocate at the same time into *parabolas touching the minor axis of the confocals, and whose directrices pass through the further focus of the confocals*; and every common tangent to one of the confocals and one of the parabolas subtends a right angle at $O$.

### SCHOLIUM A.

RECIPROCATION presupposes the idea of an envelope, which originated, according to Montucla (*Hist. des Mathématiques* tome II. 120, 1758), with Florimond DE BEAUNE (1601—1651), a zealous advocate of the new Cartesian geometry. In a letter to De Beaune dated 20 fev. 1639, Descartes writes: "Pour vos lignes courbes, la propriété dont vous m'envoyez la démonstration me paroît si belle que je la préfère à la quadrature de la parabole trouvée par Archimède; car il examinoit une ligne donnée, au lieu que *vous déterminez l'espace contenu dans une qui n'est pas encore donnée*. Je ne crois pas qu'il soit possible de trouver généralement la converse de ma règle pour les tangentes, &c." (*Œuvres de Descartes*, ed. Cousin, tome VIII. p. 105, Paris 1824). Huyghens was the discoverer of evolutes (Scholium, p. 221): Tschirnhausen of caustics (Chasles *Aperçu historique* p. 110, 1875).

MACLAURIN (*Geometria Organica*, sect. III. pp. 94 &c., London, 1720) propounded the theory of pedal and negative pedal curves. Notice, as a converse of Art. 38, that a conic may be regarded as the envelope of the arm $YZ$ of a right angle inscribed in its auxiliary circle, whose other arm $VZ$ passes through a focus $H$. Hence by projection, if two sides of a triangle inscribed in a given conic pass through fixed points $C$ and $H$ respectively, the envelope of the third side is a conic having double contact with the former on the line $CH$.

---

* This corollary was suggested by Mr. R. R. Webb, Fellow of St. John's College.

We have seen that NEWTON proved the general tangent-property of conics, which was eventually presented in a **new form**, as a property of "double" or "anharmonic" **ratio**, by **Steiner** and Chasles (pp. 262, 295. Cf. Ex. 726, note). Chasles' **second** proof of Ex. 724 (*Quetelet's Correspondance &c.* **v.** 289), viz. by reciprocation with respect to a parabola, is interesting **as an** application of reciprocation to *metric* properties. See also Bobillier in *Gergonne's Annales* XVIII. 185; Poncelet, *Propriétés Projectives* II. 431 (1866); Booth, *New Geometrical Methods* vol. I. chap. 29.

The *Principle of Duality* was first fully brought out by Poncelet's method of reciprocal polars (Scholium E, p. 290). For some controversies on the discovery of the principle see his *Propriétés Projectives* II. 351—396. It has **since** been illustrated by the **coordinate methods of Möbius, Plücker,** Booth, &c. See also Chasles' *Aperçu historique* pp. 572—694, 1875; Townsend's *Modern Geometry* chap. 23. **Figures** which correspond according to the **law** of duality **have been** called by Chasles (**p.** 587) *Correlative* figures. They **also** be called *Dual* figures. Any two dual figures are such that to every point in **either** corresponds a straight line in the other, and to every range **in** either a homographic pencil in the other, as is the case, for example, with reciprocal figures.

## MINOR DIRECTRICES.

### PROPOSITION VII.

**158.** *With either focus and directrix of an ellipse as origin and base line the major auxiliary circle reverses into a similar ellipse having for its minor auxiliary circle the reverse of the original ellipse.*

(i). **For in** Art. 4 it is evident that the major auxiliary circle **of the** ellipse reverses into an ellipse having the circle about $O$ for its minor auxiliary circle; and **by comparing the** segments of the latus rectum of the obverse with the segments of **the major axis of the reverse we see that** $\frac{CB^2}{CA} \div CB$ in the one is equal to $\frac{CB}{CA}$ in the **other.** The two ellipses are therefore *similar.*

(ii). **Let the** annexed diagram represent the reverse figure, and let $CA_0$ be the major semiaxis of the obverse ellipse, and $F$ the point **on the minor axis of** the reverse corresponding to $C$. Then $F$ and its polar (the base line) are said **to** be a *Minor Focus* and *Directrix* of the reverse.

It is evident by parallels that
$$OF : OB : OD = CS : CA_0 : CX,$$
or
$$OF = \frac{CS}{CA_0} OB = e . OB,$$
where $e$ is the common eccentricity of the two ellipses.

(iii). In the **case of the** Hyperbola, if $\pi - 2\chi$ and $\chi$ be the

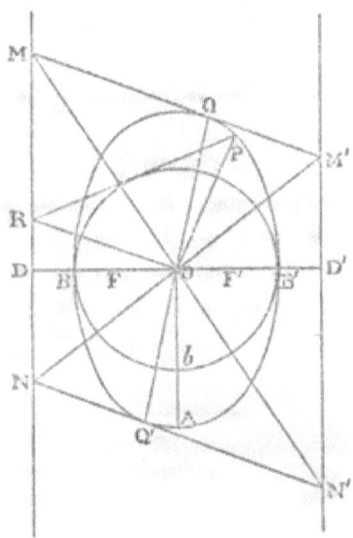

angles of the obverse **and reverse** hyperbolas respectively, and $F$ as before the point corresponding to the **centre** $C$ of the obverse, it may be shewn in like manner that $F$ now lies on the *major* axis* of the reverse at a distance equal to $\dfrac{OA}{\sin \chi}$ from its centre $O$.

In the Rectangular Hyperbola the "major" and minor foci and **directrices** are coincident. In the Circle the **minor** foci coincide with the centre and the minor directrices are at infinity.

(iv). The properties of the minor directrices may also be arrived at by *Reciprocation*. It is easily seen that the reciprocal of an ellipse with respect to its major auxiliary circle is a similar ellipse having that circle for its minor auxiliary circle, &c.

---

* Notice, in justification of the term *minor* axis in the general hyperbola, that the **square of this** axis, being negative, is always less than that of the transverse axis.

Notice, in illustration of this remarkable identity of results arrived at by reciprocation and reversion, that *if an ellipse and a hyperbola have the same axes, each is its own reciprocal with respect to the other.* [Ex. 326.

### Corollary 1.

From $S$ in the obverse draw $SY$ to meet the major auxiliary circle and draw $S\infty$ at right angles to $SY$; then $Y\infty$ touches the conic (Art. 38). Hence, in the reverse figure, if $OP$ be any radius of the conic and $OR$ be drawn at right angles to it to meet the minor directrix, the envelope of $PR$ is the minor auxiliary circle.

### Corollary 2.

Every chord drawn through $S$ to the major auxiliary circle of the obverse has its pole on the base line and makes equal angles with the tangents to the circle at its extremities. Hence any two parallel tangents $QM$, $QN$ drawn to the reverse conic and terminated by the minor directrix subtend equal angles at $O$.

#### SCHOLIUM D.

The theory of Minor Directrices[*] is due to BOOTH (*New Geometrical Methods* vol. I. 269), who investigated their properties by the method of reciprocation. In some cases he makes use of a double reciprocation, first reciprocating a figure $A$ into $B$ and then $B$ into $A'$. But since figures reciprocal to the same figure are homographic with one another, it should be possible to derive the properties of $A'$ directly from those of $A$. Take for example the property that *if a fixed straight line and the tangents from any point $P$ upon it to an ellipse about $O$ as centre meet one of its minor directrices in $Q$ and $T$, $T'$ respectively, then* $\tan\frac{1}{2}TOQ \cdot \tan\frac{1}{2}T'OQ$ *is constant.* Regard the conic as the reverse of a circle, as in Prop. VII., and let $qtt'$ be the points *at infinity* corresponding to $QTT'$. Then the angles $TOQ$, $T'OQ$ are equal to $tSq$, $t'Sq$ or $tpq$, $t'pq$ respectively; and it remains only to prove for the circle that $\tan\frac{1}{2}tpq \tan\frac{1}{2}t'pq$ is constant. Compare the longer method of double reciprocation by which Booth establishes the proposition (loc. cit. p. 276). Similar remarks apply to his double reciprocation of umbilical quadrics (p. 208).

Taking $O$ and $\omega$ as origins, let it be required to reverse three given points $PQR$ in space into given points $pqr$ respectively

---

[*] They have also been called *secondary* directrices (*Proc. Royal Irish Academy* III. 503).

(Scholium, p. 328), each pair of points $Pp$, &c. being supposed to lie in a plane through the origins. Each pair $Pp$ determine a point $(OP, \omega p)$ on the $O$-director and a point $(Op, \omega P)$ on the $\omega$-director; and thus the three pairs $Pp$, $Qq$, $Rr$ completely determine the two director planes. This construction is equivalent to reversing *five* given points $O\omega PQR$ into five given points $\omega Opqr$. Cf. Chasles, *Aperçu &c.* p. 754. If one director-plane be now removed to infinity, every two corresponding lines through the origins become parallel, as *in plano* (Art. 142).

From a given point $O$ on an ellipsoid draw triads of chords $OA$, $OB$, $OC$ at right angles, and let the *fixed* point of concourse of the planes $ABC$ (cf. p. 324, note) and its polar plane be called the Frégier point and plane of $O$. Then, if $O$ be taken as origin and its Frégier plane as director, the ellipsoid reverses into a sphere. If $O$ be not on the surface, we may reverse the ellipsoid into a spheroid about $\omega$ as *centre*, and thus shew that the envelope of the planes $ABC$ (Booth I. 97) is a quadric of revolution having $O$ for a focus. [Art. 144.

159. *Examples of Reciprocation.*

We shall now give some illustrations of the method of applying the principles established above. The following theorems will be seen to be reciprocal:

If two vertices of a triangle slide on fixed straight lines whilst the sides pass each through a fixed point, the locus of the third vertex is a conic passing through the fixed points on the adjacent sides. [p. 264.

If two sides of a triangle pass through fixed points whilst the vertices slide each on a fixed straight line, the third side envelopes a conic touching the lines on which its extremities slide.

If two sides of a triangle inscribed in a conic pass each through a fixed point, the envelope of the third side is a conic having double contact with the former on the join of the fixed points.*

If two vertices of a triangle circumscribed to a conic slide each on a fixed line, the third describes a conic having double contact with the former upon the tangents to it from the intersection of the fixed lines.

The diameter of a quadrilateral is the centre-locus of all conics inscribed therein. [Ex. 832.

Given four points on a conic, the polar of a fixed point passes through a fixed point conjugate to the former with respect to every conic through the four given points. [p. 278.

The envelope of the polar of a given point with respect to a system of confocal conics is a parabola touching their axes, &c. [Ex. 822.

Given four points on a conic, the locus of the pole of a fixed straight line is a conic, &c. [Art. 125.

---

* This is easily proved by projecting the conic and one of the fixed points into a circle and its centre (Art. 38).

350   RECIPROCATION.

× The six centres of similitude of three arbitrary circles lie by threes on four straight lines.

† The locus of the centre of a circle which touches two given circles is a conic having their centres for foci.

The centre of any one of the eight circles which touch three given circles may be determined as a point of concourse of two conics, each of which has for foci the centres of two of the given circles.*

A variable chord drawn through a fixed point $O$ to a conic subtends a pencil in involution at any point on the curve. [Art. 120.

The product of the focal perpendiculars upon any tangent to a conic is constant.

If three conics have two common tangents (or a common focus), their six chords of intersection pass by threes through the same four points.

The polar of the centre of a circle touching two given circles with respect to either of them envelopes another circle.

The polar of the centre of a circle touching three given circles with respect to any one of the three may be determined as a common tangent of two other circles. The centre itself may then be determined from its polar.

Parallel tangents to a conic (or tangents from points on a given straight line) determine an involution on any fixed tangent.

The square of the distance of any point on a conic from a fixed origin varies as the product of its distances from two fixed right lines.

For if the polars of any two points $P$ and $F$ be taken with respect to a *circle whose centre is* $O$, and if $Pf$ be a perpendicular to the polar of $F$ and $Fp$ a perpendicular to the polar of $P$, it is easily seen (Art. 153 § I.) that $OF \cdot Pf = OP \cdot Fp.$† If therefore $F$ and $F'$ be the foci of a conic and $P$ any point on the curve, it follows that $OF \cdot Pf \cdot OF' \cdot Pf' = OP^2 \cdot Fp \cdot F'p'$, or $OP^2$ varies as the product of the perpendiculars from $P$ to the reciprocals of $F$ and $F'$.

160. *Angles. Confocal Conics.*

We shall next give some examples of the reciprocation of angles (Art. 155), and of confocal conics.

At any point on a circle the tangent is at right angles to the radius.

The polar of any point with respect to a circle is at right angles to the diameter through the point.

Any point on a conic and the intersection of the tangent thereat with the directrix subtend a right angle at the focus.

The pole of any straight line with respect to a conic and the point of concourse of the line with the directrix subtend a right angle at the focus.

---

\* See Salmon's *Conic Sections* Art. 317.

† For an independent proof see Macdowell's *Exercises in Euclid &c.* Art. 256.

| | |
|---|---|
| ✓ Orthogonal tangents to a conic intersect on a concentric circle. | A chord of a conic which subtends a right angle at any fixed point envelopes a conic having that point and its polar for a focus and directrix. |
| ✓ Confocal conics intersect everywhere at right angles, and the tangents to two confocals from any point, taken alternately, include equal angles. [Art. 50. | Every common tangent of two circles subtends a right angle, and the opposite intercepts on any transversal subtend equal angles, at either limiting point. |
| If $XX'$ and $YY'$ be two pairs of collinear points on two circles, the tangents at $XX'$ intersect those at $YY'$ in four points lying on a third coaxal circle.* | If tangents be drawn from any point to two confocal conics, the four joins of the alternate points of contact touch a third confocal. |
| Given three pairs of lines conjugate with respect to a circle, every conic through the four positions of its centre is a rectangular hyperbola. [Ex. 812. | Given a focus $O$ of a conic and three pairs of points conjugate with respect to it, there are four positions of the $O$-directrix, and the orthocycle of every conic touching the four passes through $O$. |
| Tangents being drawn in a given direction to a system of confocal conics, their points of contact lie on a rectangular hyperbola. | The tangents to the circles of a coaxal system at their points of concourse with a given transversal through either limiting point $O$ envelope a conic, whose orthocycle passes through $O$. |

**161.** *The Parabola.*

*a.* The reciprocal of a parabola with respect to any point $O$ is a conic through $O$, and if $O$ be the focus of the parabola the reciprocal is a circle through $O$, and conversely. Hence the following theorems are reciprocal:

| | |
|---|---|
| ✓ If $AOB$ be a right angle inscribed in a circle the hypotenuse $AB$ passes through the centre. | The locus of the vertex of a right angle circumscribed to a parabola is the directrix. |
| ✗ The locus of the vertex of a right angle circumscribed to a parabola is the directrix. | A chord of any given conic which subtends a right angle at a fixed point $O$ on the curve passes through a fixed point $F$ on the normal at $O$. |

Thus by a double reciprocation we deduce Frégier's theorem (Art. 144 §*b*) from a property of the circle. Notice that the *Frégier-point* $F$ corresponds to the *directrix* of the parabola, the normal at $O$ to the point at infinity on the parabola, the further extremity of the normal at $O$ to the tangent at the vertex of the parabola, and that the tangent at $O$ is parallel to the directrix.

---

\* See Townsend's *Modern Geometry* 1. 264.

*b.* A triad of points $ABC$ reciprocate with respect to their orthocentre into a triad of lines parallel to $BC$, $CA$, $AB$, and having the same orthocentre. Also a parabola reciprocates from any point $O$ on its directrix into a rectangular hyperbola through $O$. Hence the following are reciprocal theorems:

| | |
|---|---|
| The orthocentre of any triangle circumscribed to a parabola lies on the directrix. | The orthocentre of any triangle inscribed in a rectangular hyperbola lies on the curve. |

The former is a special case of Brianchon's theorem (p. 290), the latter of Pascal's (p. 175).

*c.* If the reciprocals of three points $ABC$ with respect to $O$ be the lines $abc$, then to any point $A'$ on $BC$ corresponds the line through $bc$ making an angle equal to $AOA'$ with $a$. Hence—

| | |
|---|---|
| The perpendiculars of any triangle circumscribed to a parabola meet on the directrix. | If from a fixed point $O$ on a conic there be drawn any three chords $OA$, $OB$, $OC$ and the three lines at right angles to them, and if the latter meet $BC$, $CA$, $AB$ in $A'$, $B'$, $C'$ respectively, then $A'$, $B'$, $C'$ lie on a straight line passing through the Frégier point of $O$. |

## 162. *The Minor Directrices.*

*a.* Let the tangent at $Q$ to a conic meet the minor directrices in $M$ and $M'$, and let $Fp$ and $F'p'$ be perpendiculars to this tangent and $QT'$ and $QT''$ perpendiculars to the minor directrices. Then the angle $QOM$ is equal to $QOM'$ (Art. 158 Cor. 2), and therefore

$$OM : OM' = QM : QM' = QT' : QT'' = Fp : F'p',$$

$\dfrac{QT'}{Fp}$ being to $\dfrac{QT''}{F'p'}$ in a *constant* ratio (Prop. III.) which is evidently a ratio of equality when $Q$ is taken on either axis.

*b.* It may be seen that at any point $Q$ on the conic,

$$QT \cdot QT'' : OQ^2 = OB^2 : e^2 \cdot OA^2,$$

since if $Q$ be a point such that $\dfrac{QT \cdot QT''}{OQ^2}$ is constant the locus of $Q$ is a conic (Art. 159), and the above proportion requires that this conic should meet the axes in the same points as former. It then follows that

$$Fp \cdot F'p' : OQ^2 = OY^2 : CA^2,$$

if $OY$ be the central perpendicular to $MM'$.

*c.* If $C$ be the pole of $MM'$ with respect to the **minor** auxiliary circle, so that $OM$ is at right angles to $FC$ and $OM'$ to $F'C$, and $FC + F'C = 2OB$ (Art. 158 § iv), then, with the help of similar triangles, it may be shewn that, if $X$ be the intersection of $MM'$ with **the major axis**,

$$FC : OM = OF : OX = F'C : OM';$$

and therefore $DM + D'M' = 2OX = e(OM + OM')$, $D$ and $D'$ being the points in which the minor directrices meet the minor axis.

*d.* The following are examples **of reversion,** or the same results may be obtained by reciprocation.

| | |
|---|---|
| Any chord of a circle is at right angles to the diameter through its pole. | Any chord of a conic and the line joining the minor focus $F$ to the pole of the chord meet the $F$-directrix in points which subtend a right angle at the centre. |
| If $AB$ be fixed points on a circle and $C$ any other point upon it, the angle $ACB$ has one of two constant and supplementary values. | The arms of any angle $ACB$ inscribed in given segment of a conic intercept on the minor directrices lengths which subtend constant and supplementary angles at the centre. |

In like manner it may be shewn that the two pairs of **opposite sides of a** quadrilateral inscribed in a conic make intercepts **on** either minor directrix which subtend supplementary **angles at the centre.**

*e.* The major auxiliary **circles of a system of conics** having a focus $S$ and its directrix in common may be reversed into concentric conics having the same minor directrices (Prop. VII.). Hence, the circles being coaxal and having $S$ for a limiting point:*

| | |
|---|---|
| The opposite intercepts **made by any** two circles on any transversal **subtend** equal angles at either limiting **point.** [Art. 160. | The opposite intercepts made by any two concentric conics having the same minor directrices upon any transversal subtend equal angles at the centre. |

*f.* Each focus and directrix of a Rectangular Hyperbola being at the same time a minor focus and directrix (Art. 158 § iii),

---

* See Macdowell's *Exercises in Euclid &c.* Art. 251.

we obtain in this case the following property, corresponding to that of the ellipse in Art. 158 Cor. 1:

If $PD'DP'$ be a chord of a rectangular hyperbola touching its auxiliary circle and meeting the directrices in $D$ and $D'$, *the diameters through $P$ and $D$ (or $P'$ and $D'$) are at right angles.* Hence it follows also that $PD$ and $P'D'$ subtend equal angles at the centre.

## INVERSION.

163. We shall conclude with a slight sketch of the method of Inversion.

If $O$ be a fixed point and $P$ a variable point in a given plane, and if a point $p$ be taken on $OP$ or its complement such that $OP.Op$ is equal to a constant $c^2$, then $p$ is said to be an *inverse* of $P$ with respect to the pole $O$, and the locus of $p$ is said to be the inverse of the locus of $P$. It is evident that *a straight line through the pole is its own inverse.*

To curves intersecting in points $PQR$ &c. correspond curves intersecting at the inverse points $pqr$ &c.; and therefore to curves having contact of any order at $P$ correspond inverse curves having contact of the same order at the inverse point $p$.

164. *The inverses of any two curves intersect at the same angles as the original curves.*

For if $PQ$ be any two points on a curve and $pq$ the inverse points, then $OP.Op = OQ.Oq$, and therefore the angles $OPQ$ and $Oqp$ are equal. Hence, supposing $P$ and $Q$ to coalesce, the *tangents* at the inverse points $P$ and $p$ are equally inclined (on opposite sides) to the radius vector $OPp$. It then follows that the tangents to any two curves at a common point $P$ are inclined at the same angles as the tangents to the inverse curves at $p$.

165. *The inverse of a straight line not passing through the pole is a circle through the pole, and conversely; and the inverse of a circle not passing through the pole is a circle.*

(i). To a given line draw a perpendicular $OX$ from the pole of inversion, take any point $D$ on the given line, and let the point $x$ be inverse to $X$ and $d$ to $D$. Then since $Ox.OX = Od.OD$, and since $OXD$ is a right angle, therefore $Odx$ is a right angle, or the locus of $d$ is a circle having its diameter through $O$ at right angles to the given line.

Conversely *the inverse of a circle through $O$ is a straight line at right angles to its diameter through $O$*.

It is further evident that the *tangent* to any curve at $P$ inverts into a circle through $O$ touching the inverse curve at the inverse point $p$; unless $P$ coincides with $O$, in which case the tangent at that point is also an *asymptote* to the inverse curve.

(ii). The inverse of a circle not passing through $O$ is a circle. For if $OPQ$ be drawn to meet the given circle in $P$ and $Q$, and if the point $p$ be inverse to $P$, and $q$ to $Q$, then since $OP.OQ$ is constant and $OP.Op$ is likewise constant; therefore $Op$ varies as $OQ$, or the locus of $p$ is similar to the locus of $Q$, which is a circle.

(iii). Notice that by making the constant of inversion equal to $OP.OQ$ we may invert the given circle into *itself*. If $QQ'$ be a common tangent to two circles and $M$ its middle point, then with $M$ as pole and $MQ^2$ as the constant of inversion each of the circles inverts into itself. Again, if $O$ be the centre and $c$ the radius of a circle orthogonal to a set of coaxal circles, then with $O$ as pole and $c^2$ as the constant of inversion the whole system inverts into itself.

166. *The nine-point circle of any triangle touches the inscribed and escribed circles.*

Let $ABC$ be a triangle, $I$ the inscribed circle touching $BC$ in $Q$, and $E$ the escribed circle opposite to $A$ touching $BC$ in $Q'$. Bisect $BC$, $CA$ in $M$ and $M'$ respectively,\* draw $AP$ perpendicular $BC$, and let the nine-point circle $N$ meet $AP$ in $D$, which will be the further extremity of its diameter through $M$. Then, with $M$ as pole and $MQ^2$ (equal to $MQ'^2$) as the constant of inversion, $I$ inverts into itself, $E$ into itself, and $N$ into a *straight line* at right angles to $MD$.

---

\* See the lithographed figure No. 5. On the above proof see p. 191, note.

This straight line meets $BC$ in a point $R$ such that $MP.MR = MQ^2$, that is to say, in the point of concourse of $BC$ with the line joining the centres of $I$ and $E$;* and it makes an angle with $BC$ equal to $MDP$, or $MM'P$, or $B \sim C$, and may therefore be shewn to coincide with the second tangent from $R$ to $I$ and $E$.

And since the inverse of $N$ thus touches the inverse of $I$ and the inverse of $E$, therefore $N$ touches $I$ and $E$, which are their own inverses; that is to say, it touches the inscribed circle and each of the escribed circles of the triangle $ABC$, as was to be proved.

To determine the points of contact, let $RS$ be the second tangent from $R$ to $I$ (or $E$), produce $MS$ to meet the circle again in $S'$, then $S'$ (the inverse of $S$) is the required point in which $N$ touches $I$ (or $E$).

### 167. *The Cardioid.*

The inverse of a parabola with respect to its focus is a cardioid having its cusp at the origin. Hence the following are inverse theorems:

| | |
|---|---|
| The sum of the reciprocals of the segments of any focal chord of a parabola is constant. | The length of any cuspidal chord of a cardioid is constant. |
| Every focal chord of a parabola is divided harmonically by the focus and the directrix. | The locus of the middle points of the cuspidal chords of a cardioid is a circle through the cusp. |
| The tangents to a parabola at the extremities of a focal chord which makes an angle $a$ with the axis are inclined at angles $\frac{a}{2}$ and $\frac{\pi}{2} - \frac{a}{2}$ to the chord. | The tangents to a cardioid at the extremities of a cuspidal chord inclined at an angle $a$ to the axis make angles equal to $\frac{a}{2}$ and $\frac{\pi}{2} - \frac{a}{2}$ with the chord. [Art. 164. |

Hence *the tangents to a cardioid at the extremities of any cuspidal chord intersect at right angles;* and it may now be shewn that the locus of their intersection is the circle concentric with the bisector of all cuspidal chords and of thrice its radius.

| | |
|---|---|
| The tangents to a parabola at the extremities of any focal chord intersect at right angles on the directrix. | The two circles through the cusp which touch a cardioid each at one extremity of any cuspidal chord meet at right angles on a fixed circle through the cusp. |

---

* See Macdowell's *Exercises in Euclid &c.* Art. 86.

From a fixed point $O$ draw $OY$ to any point $Y$ on a given straight line; then the envelope of the line through $Y$ at right angles to $OY$ is a parabola.

The intersections of any three tangents to a parabola lie on a circle through the focus.

If $ABCA'B'C'$ be any six points on a parabola, the intersections of $(AB, A'B')$, $(BC, B'C')$, $(CA, C'A')$ lie in a straight line.

If $ABCA'B'C'$ be any six tangents to a parabola, the joins of $(AB, A'B')$, $(BC, B'C')$, $(CA, C'A')$ meet in a point.

From a fixed point $O$ on a given circle draw any chord $Oy$; then the envelope of the circle on $Oy$ as diameter is a cardioid having its cusp at $O$.

Any three circles touching a cardioid and passing through its cusp meet in three other points lying on a straight line.

If $abca'b'c'$ be any six points on a cardioid whose cusp is at $O$, the intersections of the three pairs of circles $(Oab, Oa'b')$, $(Obc, Ob'c')$, $(Oca, Oc'a')$ lie on one circle.

If $abca'b'c'$ be any six circles touching a cardioid and passing through its cusp, the three circles through the intersections of $(ab, a'b')$, $(bc, b'c')$, $(ca, c'a')$ are coaxal.

## 168. *Circles of Curvature.*

The osculating circle at any point $P$ on a curve inverts from any point $O$ into the circle (Art. 165) which osculates the inverse curve at the inverse point $p$. [Art. 163.

But if the former circle passes through $O$ it inverts into a straight line, and $p$ becomes a point of zero curvature, or of *inflexion.* Hence the following are inverse theorems:

Three points can be found on an ellipse whose osculating circles meet at a given point on the curve, and these three points lie on a circle through $O$.

The inverse of an ellipse with respect to any point upon it is a curve having three points of inflexion, which lie in a straight line.

It may be added that a parabola inverts from its vertex as pole into a *cissoid;* a central conic from either focus into a *limaçon,* and from its centre into an *oval of Cassini* or, if an equilateral hyperbola, into a *lemniscate of Bernoulli;* a conic from any point upon it into a *circular cubic* having a node at the pole; and a conic from any other point in its plane into a *trinodal quartic* having its nodes at the focoids and the pole.

### SCHOLIUM C.

For the principle of Inversion Chasles (*Rapport* pp. 140—2) refers to Ptolemy, and to Quetelet (1827); and for a general statement of the method to Bellavitis (1836). In 1843—4 it was propounded afresh by Ingram and Stubbs (*Transactions of the Dublin*

*Philosophical Society* vol. I. 58, 145, 159; *Philosophical Magazine* XXIII. 338, XXV. 208). It has been applied by Dr. Hirst to attractions (*Phil. Mag.* 1858), and embodied by Peaucellier in his linkages. Cf. also *Camb. and Dublin Math. Journal* VIII. 47; *Oxf. Camb. Dubl. Messenger of Mathematics* III. 228; Booth's *New Geometrical Methods* vol. I. chap. 30; Salmon's *Higher Plane Curves*, Arts. 348 &c.; *Proc. London Math. Soc.* V. 105, VII. 91. And for a complete exposition of the method as applied to the straight line and circle see Townsend's *Modern Geometry*, chaps. 9, 24. This method, unlike projection and reciprocation, enables us to deduce properties of the higher curves from those of a lower order, and is thus peculiarly effective as an instrument of discovery and research.

## *MISCELLANEOUS EXAMPLES.*

{851. If a triangle is self-polar to a parabola (p. 281), the three lines joining the middle points of its sides touch the parabola, and conversely. [Ex. 715.

852. Two rectangular hyperbolas being such that the axes of the one are parallel to the asymptotes of the other, and the centre of each lies on the other; shew that any circle through the centre of either meets the other again in a conjugate triad with respect to the former.

853. If two angles of given magnitudes turn about their summits $A$ and $B$ as poles, then (1) if one pair of their arms remain constantly parallel, the other pair intersect at a constant angle and thus describe a *circle C* through the poles; and (2) if one pair of their arms intersect on a fixed straight line $D$ as director, the other pair by their intersection describe in general a *conic* through the poles. The points at infinity on the conic correspond to the intersections of $C$ and $D$; the axes of the conic are parallel to the positions which the parallel arms in case (1) assume when the arms describing the circle intersect at the extremities of the diameter at right angles to $D$; and if the director be any line in a system of parallels the axes of the conics described are parallel.

854. If the tangent at $O$ to a rectangular hyperbola be met at right angles in $P$ by a chord $QR$, the diameters

bisecting $OQ$ and $OR$ bisect the angles between the diameters to $O$ and $P$.*

855. If $AA'$, $BB'$, $CC'$ be the three pairs of summits of a quadrilateral circumscribed to a parabola whose focus is $S$, then
$$SA \cdot SA' = SB \cdot SB' = SC \cdot SC'.$$

856. Reciprocate the theorem that the feet of the focal perpendiculars upon the tangents to a parabola are collinear.

857. Given two conics, find a conic with respect to which they are polar reciprocals.

858. The tangent to a circle at any point makes with any chord through the point an angle equal to the angle in the alternate segment. What does this proposition become by reciprocation with respect to any origin?

859. The problem, to inscribe in a given conic a $2n$-gon whose $n$ pairs of opposite sides shall pass in any assigned order through $n$ given points, is always indeterminate or impossible.†

860. If two circles be drawn meeting a conic in $OABC$ and $OA'B'C'$ respectively, every two of the vertices of the triangles $ABC$ and $A'B'C'$ subtend at $O$ an angle equal to that between the opposite sides. Conversely, if the vertices of two triangles inscribed in a conic be thus related to a point $O$, then $O$ lies on the conic.

---

* The nine-point circle (Art. 64 Cor. 4) of $OQR$ passes through $O$.

† This question and its solution were suggested by Prof. TOWNSEND. Starting from an arbitrary point $P$ on the conic as one vertex, draw $n$ successive sides of the polygon through the $n$ points taken in the assigned order, and the other $n$ sides through the same points taken in the reverse order; and let the points on the conic thus arrived at be $Q$ and $Q'$ respectively. As $P$ varies, the three systems of points $P$, $Q$, $Q'$ are homographic (Art. 120 Cor. 2), and therefore also the two systems $P + Q$ and $Q' + P$. If $Q$ and $Q'$ once coincide, one pair of homologous points in the two homographic systems $P + Q$ and $Q' + P$ are interchangeable, and therefore every pair are interchangeable (Townsend's *Modern Geometry* II. § 360), or $Q$ and $Q'$ always coincide. If then $Q$ and $Q'$ coincide for any *one* position of $P$ the solution is indeterminate, and if not it is impossible.

861. The circumscribed circles of any two triangles $PQR$ and $P'Q'R'$ circumscribing a parabola meet in a point $O$ (other than the focus) such that the angle subtended at $O$ by any two of the vertices of the triangles is equal to that between the opposite sides.* Hence shew that the seven points $OPQRP'Q'R'$ lie on a conic.

862. If three conics have one point in common, their nine common chords which do not pass through it touch one conic.† Conversely, if three triangles circumscribe a conic, the three conics which circumscribe them by pairs have one point in common. [Ex. 861.

863. Any three parabolas, taken in pairs, have three triads of common tangents, whose nine intersections lie on a conic. What does this become by projection and reciprocation?

864. If $ABC$ be the triangle formed by three tangents to a parabola whose focus is $S$, the inclination of $BC$ to the axis is equal to the angle subtended by $SA$ at the circumference of the circle. Hence shew that the square of the radius of the circle is equal to $\dfrac{Sa \cdot Sb \cdot Sc}{p}$, if $p$ be the parameter and $abc$ the points of contact of the tangents.

865. Two triangles $ABC$ and $A'B'C'$ inscribed in a circle being such that every two of their vertices subtend at the circumference an angle equal to that between the opposite sides, shew that, if $O$ be the centre and $OS$ a given radius of the circle,

$$\angle SOA + SOB + SOC = SOA' + SOB' + SOC',$$

and conversely. Hence deduce that, if $ABC$ be a variable triangle circumscribed to a given parabola whose focus is $S$, and inscribed in a fixed circle whose centre is $O$, the sum of the angles $SOA$, $SOB$, $SOC$, measured in the same sense

---

\* If $O$ be a point on the arc $SP$ (p. 56) and $r$ the point of contact of $PQ$, the angle $ROS$ is equal to $RPQ + SPr$, and therefore to $SrQ$. In like manner $R'OS$ is equal to $Sr'Q'$; and therefore $ROR'$ is equal to the angle $(PQ, P'Q')$.

† Picquet, *Etude géométrique des Systèmes Ponctuels &c.* pp. 27, 51 (Paris 1872).

of rotation is constant; and the sum of the angles which $BC$, $CA$, $AB$ make with the axis is constant. Shew also that the product $SA.SB.SC$, or $Sa.Sb.Sc$ (Ex. 864), is constant, and that the product of the focal perpendiculars upon the three tangents is constant.

866. If a triangle $ABC$ inscribed in a given ellipse envelopes a fixed parabola, the sum of the eccentric angles of its vertices is constant;* and the circles $ABC$ pass through a fixed point on the ellipse and have a common radical axis.

867. A variable conic through four given points $ABCD$ meets a fixed conic through $D$ at the vertices of a variable triangle, which envelopes a fixed conic inscribed in $ABC$. State the reciprocal theorem.

868. The tangents to an ellipse from any point on a minor directrix intercept on the major axis a length which varies as the central distance of the point.

869. Any chord $PQ$ of an ellipse and the tangents at $P$ and $Q$ meet the minor directrices in pairs of points $RR'$, $MM'$ and $NN'$ respectively such that, if $O$ be the centre, the angle $ROR'$ is equal to $\frac{1}{2}(MOM' + NON')$; and the central distances of the points $MM'NN'$ and their perpendicular distances from $PQ$ are to one another severally in the same ratio.

870. Prove by reciprocation with respect to a point, that the sum of the reciprocals of the perpendiculars from any point $O$ within a circle to the tangents from any point on the polar of $O$ is constant. Also prove that the reciprocals of equal circles with respect to the same point have equal parameters, and the reciprocals of coaxal circles with respect to any point on the radical axis have equal minor axes.†

---

* This theorem, which is due to Mr. R. Pendlebury, Fellow of St. John's College, follows from Ex. 865 by orthogonal projection.

† If $OAB$ be drawn from the origin to meet a given circle in $A$ and $B$, then $\frac{k^2}{OA}$ and $\frac{c^2}{OB}$ are equal to the perpendiculars from a focus of the reciprocal conic to a pair of parallel tangents. The length of the minor axis is therefore determined by the product $OA.OB$.

871. Deduce the property of the focus and directrix of a conic from the *Locus ad quatuor lineas*.\*

872. If $P$ and $Q$ be opposite intersections of the common tangents to two conics, any two lines $OA$ and $OB$ which are conjugate with respect to both conics are harmonic conjugates with respect to the lines $OP$ and $OQ$. Deduce this from Newton's property of the diameter of a quadrilateral (Art. 124), viz. by projecting $OA$ or $OB$ to infinity. Also state the reciprocal theorem.† [I. 19.

873. An ellipse being drawn through the centre $O$ of a circle, shew that the lines from $O$ to a pair of opposite intersections of their common tangents are equally inclined to the tangent to the ellipse at $O$. [I. 33.

874. Find the locus of the centre of a conic which cuts four given finite straight lines harmonically, or which passes through two given points and cuts two given finite straight lines harmonically. [I. 62.

875. If a given polygon be moved about in its plane so that two of its sides touch each a fixed circle, every side of the polygon touches a fixed circle. [I. 68.

876. Deduce from Ex. 738, that if a curve has one tetrad of foci (p. 311) lying on a circle it has three other such tetrads, and the four circles cut one another orthogonally. [II. 10.

877. If four circles be mutually orthogonal, their centres form an orthocentric tetrastigm, and one at least of the circles is imaginary. [II. 10.

---

\* First deduce the *Locus ad tres lineas* $\alpha\beta = k\gamma^2$, and let the two tangents be drawn from the focus $S$. These are represented by the Cartesian equation $x^2 + y^2 = 0$, and the perpendiculars to them from $(x, y)$ are proportional to $x \pm \sqrt{(-1)}\, y$. Therefore $x^2 + y^2$ varies as $\gamma^2$, or the distance of $(x, y)$ from $S$ varies as its distance from the polar of $S$.

† Exx. 872 &c. are taken from the EDUCATIONAL TIMES *Reprint*, the volume and page of which are specified in each case. See also pp. 141, 336.

878. **Find the locus** of the pole of a common chord **of two** fixed conics with respect **to a conic** having double contact with each of them. [II. 46.]

879. If two triangles circumscribed **(or inscribed) to a conic** be in perspective, **every radiant** through **their centre of** perspective **meets** their **sides in three pairs of points in** involution. Reciprocate **this theorem, and point out its relation to** Steiner's **property of the directrix of a** parabola inscribed in **a triangle.**
[II. 50.]

880. If two conics meet any transversal in pairs of points $AB$ and $A'B'$ respectively, the foci **of** the involution determined by $AA'$ and $BB'$ (or $AB'$ and $A'B$) lie **on a** third **conic passing** through the intersections of **the former two.** [II. 91.]

881. **Given that one focus of a conic to which a given** triangle is self-conjugate **lies on a given straight line, find the** locus of its second focus, and **deduce Ex.** 715. [III. 33.]

882. Given two points $P$ and $Q$ on a conic, find a third point $O$ upon it such that $OP$ and $OQ$ may divide a given finite straight line **in a given** cross ratio. [III. 47.]

883. **Let** *abc* **be the middle** points of the sides of a triangle, $O$ the centre of its circumscribing **circle** and $O'$ its orthocentre. Then if $Oa$, $Ob$, $Oc$ be produced **to** $ABC$ respectively so that $OA = 2Oa$, $OB = 2Ob$, $OC = 2Oc$, the sides of the triangle $ABC$ and of the **original** triangle **touch** one conic, which has their common nine-point **circle** for its major auxiliary circle and the points $O$ and $O'$ for foci. [III. 53.]

884. Through four given points draw a conic such that the chord which it intercepts on a given line shall be of given length, or shall subtend a given angle at a given point. [III. 84.]

885. If $p$, $p'$ be variable points collinear with a fixed point $A$, and so situated that the segment $pp'$ always subtends a right angle at another fixed point $M$, prove the following properties of *corresponding* loci of $p$ and $p'$. Right lines equidistant from the middle point **of** $AM$ correspond to similar conics passing

through $A$ and cutting $AM$ perpendicularly at $M$. These conics are similar ellipses, or parabolas, or similar hyperbolas according as the common distance of the primitive lines from the middle point of $AM$ is greater than, equal to, or less than $\frac{1}{2}AM$. The circles which pass through $A$ and $M$, taken in pairs, are corresponding loci, as also are the circles which pass through $M$ and have their centres on $AM$. [III. 91.

886. The critical conic of a quadrilateral being defined as the circumscribed conic which projects into a circle when the quadrilateral is projected into a square, shew that, if $AA'$, $BB'$, $CC'$ be the three pairs of summits of a quadrilateral, a conic can be found having double contact at points lying on $AA'$ with the critical conic of $BB'CC'$, double contact at points on $BB'$ with the critical conic of $CC'AA'$, and double contact at points on $CC'$ with the critical conic of $AA'BB'$. [III. 92.

887. If a straight line meet the sides $BC$, $CA$, $AB$ of a triangle in $PQR$ respectively, and $O$ be any point in the same plane, the tangents at $O$ to the conics $OAPBQ$ and $OAPCR$ are harmonic conjugates with respect to $OA$ and $OP$. [IV. 44.

888. Shew how to prove the principal properties of the lemniscate by inversion. [IV. 47.

889. Prove that the "characteristics" of a system of conics satisfying four conditions are unaltered when, in place of passing through a given point, each conic is required to divide a given finite segment harmonically. [IV. 56.

890. Given four straight lines in a plane, we may project one of them to infinity and the remaining three into the sides of an equilateral triangle. Is it possible to project two given triangles at once into equilateral triangles, or a conic and a triangle into a circle and an equilateral triangle? [IV. 88.

891. The envelope of the circles on a system of parallel chords of a conic as diameters is a conic having its foci at the extremities of the diameter conjugate to the chords. Find where any circle touches the envelope. [V. 40.

892. A variable conic being drawn touching three given conics, if the normals at the three points of contact cointersect investigate the locus of their point of concourse. [v. 50.

893. Four conics being drawn through the same four points so that their tangents thereat form four harmonic pencils, shew that if one conjugate pair of the conics be a circle and an equilateral hyperbola the other pair must have equal eccentricities. [v. 103.

894. Through a given point $O$ on a hyperbola two chords are drawn each at right angles to an asymptote, and from a variable point $P$ on the curve perpendiculars $PM$ and $PN$ are drawn to the two chords through $O$. Shew that $MN$ passes through a fixed point $F$; find the locus of $F$ for different positions of $O$ on the hyperbola; and determine the hyperbola for which the locus reduces to a point. [VI. 45.

895. Three conics being described so that each of them passes through the same point $O$ and through the extremities of two of the diagonals of the same complete quadrilateral, prove that the remaining three points of concourse of the conics lie upon their tangents at $O$. [VI. 54.

896. If $ABCD$ be four points on a conic, the intersections of $AB$ and $CD$ with any two tangents lie on a conic touching $AC$ and $BD$. [VI. 56.

897. The axes of every conic circumscribed to a quadrangle, which is itself inscribed in a circle, are parallel to two fixed right lines, viz. the asymptotes of the equilateral hyperbola (Ex. 518) which is the centre-locus of all conics circumscribed to the quadrangle.* [VI. 88.

898. If $P$ be any point on a circle, $A$ and $B$ fixed points on a diameter and equidistant from the centre, the envelope of a transversal which is cut harmonically by the circles

---

\* If $ABC$ be the common self-polar triangle of all the circumscribed conics; $Ox$ and $Ox'$ the axes of any one of them, so that $Ox \infty' $ is a self-polar triangle with respect to it; the points $ABCO\infty\infty'$ lie on a conic, which in this case is the rectangular hyperbola $ABCO$.

described with $A$ and $B$ as centres and $AP$ and $BP$ respectively as radii is the conic which has $A$ and $B$ for foci and touches the circle. [VII. 34.

899. Draw the minimum chord of a given angle which can be cut in a given ratio by a given line. [VII. 41.

900. If $Q$ and $R$ be the foci of any ellipse inscribed in a triangle $ABC$, deduce from Ex. 322 that,

$$AQ.AR.BC + BQ.BR.CA + CQ.CR.AB = BC.CA.AB.$$
[VII. 43.

901. If $Q$ be the intersection of the polars of any point $P$ with respect to two given parallel conics (Ex. 792, note), the locus of the middle point of $PQ$ is their radical axis. Hence shew that, if $DEF$ be the feet of the perpendiculars of any triangle $ABC$, and $AL, BM, CN$ be parallels to $EF, FD, DE$, meeting $BC, CA, AB$ in the points $LMN$ respectively, then the axis of perspective of the triangles $ABC$ and $DEF$ bisects each of the segments $AL, BM, CN$. [VII. 78.

902. Construct geometrically the four chords of contact with a given conic of the four inscribed conics which pass through three given points. [VII. 92.

903. Triads of parallels being drawn through the vertices $ABC$ of a given triangle to meet the opposite sides in $abc$, shew that the envelope of the axis of perspective of the triangles $ABC$ and $abc$ is the maximum ellipse that can be inscribed in $ABC$. [VII. 94.

904. If from any point on a conic parallels be drawn to the diameters bisecting the sides of any inscribed triangle, the lines so drawn meet the corresponding sides of the triangle in three collinear points. Extend this theorem by projection, and also reciprocate it. [VIII. 44.

905. Prove by inversion, that the circles having for diameters three chords $AB, AC, AD$ of a circle intersect again by pairs in three collinear points. [VIII. 48.

906. If through a pair of opposite intersections $AA'$ of four fixed tangents to a given conic there be drawn a pair of lines conjugate to the conic, the locus of their point of concourse is a conic passing through $AA'$ and through the points of contact of the four tangents. [VIII. 62.

907. If $A$ be any point within or without a conic, $B$ any point on its polar, $CD$ a fixed straight line, $BC$ and $BD$ tangents cutting $CD$ in $C$ and $D$ respectively; shew that the intersections of $AD$, $BC$ and $AC$, $BD$ lie on a fixed straight line, which meets $CD$ on the polar of $A$. [VIII. 63.

908. If $DP$ and $DQ$ be a pair of tangents to a conic and $ABC$ a self-polar triangle, any conic through $ABCD$ cuts $PQ$ harmonically. Hence shew that the perpendicular $DM$ to either axis bisects the angle $PMQ$. [VIII. 110.

909. Deux droites qui divisent harmoniquement les trois diagonales d'un quadrilatère rencontrent en quatres points harmoniques toute conique inscrite dans le quadrilatère.
[IX. 62, XII. 50.

910. La condition qu'une conique divise harmoniquement les trois diagonales d'un quadrilatère circonscrit à une autre conique, coincide avec la condition que la première conique soit circonscrite à un triangle conjugué à l'autre. [IX. 74.

911. The degree of the locus of the foci of a system of conics subject to four conditions is three times as great as that of the locus of their centres. [X. 63.

912. If $A$ and $B$ are fixed points with regard to a conic of which $ACD$ is a variable chord, shew that the polar of $A$ meets $BC$ and $BD$ in points $E$ and $F$ such that $AB$, $DE$, $CF$ cointersect. [X. 81.

913. Given three points $ABC$ and a straight line through each, shew how to cut the three lines by a fourth in points $PQR$ such that the lengths $AP$, $BQ$, $CR$ may be equal. [XI. 19.

914. If three equidistant lines parallel to an asymptote of a hyperbola meet the curve in $ABC$, prove by involution (or

otherwise) that any parallel to the other asymptote is divided harmonically by the sides of the triangle $ABC$ and the curve.

[XI. 20.

915. The six points which, in conjunction with any common transversal, divide harmonically the six sides of a tetrastigm, lie on a conic passing also through the three intersections of the opposite sides of the tetrastigm; and the three straight lines which join the six points in opposite pairs cointersect at the pole of the transversal with respect to the conic. [XI. 21.

916. Reciprocate the theorem, that if the orthocentre of a triangle inscribed in a parabola lies on the directrix, the circle to which the triangle is self-polar passes through the focus.

[XI. 32.

917. Extend by projection and also reciprocate the following theorem. Given two parallel conics (Ex. 792, note) $A$ and $B$, two circles can be drawn having double contact with $A$ and $B$ respectively and meeting their common chord in the same two points. [XI. 43.

918. If $O$ and $O'$ be the limiting points of a system of coaxal circles, and if with $O$ and $O'$ respectively as one focus two conics be described osculating any circle of the system at one and the same point, their corresponding directrices will coincide. [XI. 74.

919. Given three points $ABC$ and a conic, the envelope of a chord $PQ$ such that $A\{BPQC\}$ is harmonic is a conic touching $AB$ and $AC$ at points lying on the polar of $A$. [XI. 83.

920. Find the envelope of a transversal on which two given conics intercept segments having a common middle point $M$, and find the locus of $M$. [XI. 84.

921. Any tangent to a conic is divided in involution by three other tangents and the radiants to their intersections from either focus $S$. Prove that the double points of this involution, as the tangents vary, subtend a pencil in involution at $S$. [XI. 105.

922. $PQ$ being a chord of a conic equally inclined to the axis with the tangent at $P$, a circle is drawn through $PQ$ cutting the conic again in $RS$. Shew that the point on the circle harmonically conjugate to $P$ with respect to $R$ and $S$ lies upon the chord of the conic supplemental to $QP$. What does this become by inversion? [XII. 90.

923. Any two parallel tangents to a conic meet the tangents from a given point $O$ in points $T$ and $T'$ respectively such that $OT.OT'$ is constant. [XIII. 44.

924. Prove by inversion, that if three circles meet two and two in $AA'$, $BB'$, $CC'$, and $O$ be any point in their plane, the circles $OAA'$, $OBB'$, $OCC'$ are coaxal. [XIV. 102.

925. Given in a conic two tetrastigms $PQRS$ and $pqrs$ whose corresponding chords pass by fours through the same three points, shew that a conic may be drawn touching $Pp$, $Qq$, $Pr$, $Ss$ at $pqrs$ respectively. [XIV. 104.

926. Find the constant ratios which five fixed radiants in space determine on a variable transversal plane;* and deduce the anharmonic property of four radiants in one plane. [XV. 26.

927. Prove, generalise and reciprocate the theorem, that the bisectors of the angles between the two pairs of opposite sides of a trapezium inscribed in a circle are at right angles.
[XV. 36.

928. The envelope of a transversal cut harmonically by two given similarly situated parabolas is a third parabola (Ex. 800). [XV. 86.

929. The tangents to a conic from a variable point on a fixed straight line $L$ meet the tangent at a given point $A$ in $R$ and $R'$. Shew that the relation between $AR$ and $AR'$ is of the form $a.AR.AR' + b.AR + c.AR' + d = 0$ (Ex. 777); and determine the positions of $L$ in order that (1) the sum of the intercepts $AR$ and $AR'$, or (2) the sum of their reciprocals may be constant. [XVI. 59.

---

* See the *Messenger of Mathematics* vol. v. 94 (1876).

370      MISCELLANEOUS EXAMPLES.

930. Two sides of a triangle circumscribed to a given circle being fixed, the three lines joining its angles severally to the points of contact of the escribed circles with the opposite sides meet in a point, whose locus is a hyperbola having the fixed sides for asymptotes. [XVI. 62.

931. Given three fixed straight lines $lmn$ and three fixed points $LMN$ in a straight line, the lines from a current point on $l$ to $M$ and $N$ meet $m$ and $n$ in four points, the conic through which and $L$ envelopes a conic touching $m$ and $n$ at their intersections with $l$. [XVI. 98.

932. The first positive and negative pedals of an equilateral hyperbola are reciprocal polars with respect to it. [XVI. 106.

933. Given a point on one arm of a constant angle inscribed in a circle, find the envelope of the other arm.
[XVI. 110.

934. Circles being described on the two halves of a diameter of a given circle as diameters, shew that the perpendicular radius of the given circle is trisected by the centre and circumference of a fourth circle touching the three; and deduce a new theorem by reciprocation. [XVII. 23.

935. Deduce from Ex. 785, that if $BC$ be a chord of a circle and $A$ its pole, the conic through $ABC$ which touches the circle at a point $D$ has its curvature at $D$ twice as great as that of the circle. [XVII. 109.

936. An ellipse having double contact with a fixed ellipse $E$ has one focus $F$ fixed: shew that the other focus describes an ellipse confocal with $E$ and passing through $F$. [XVIII. 70.

937. The area of the triangle formed by the polars of the middle points of the sides of a triangle with respect to any inscribed conic is equal to the area of the given triangle.
[XVIII. 107.

938. From two fixed points on one of a series of confocal conics tangents are drawn to a variable conic of the series: if they meet the fixed conic again in $QR$ and $Q'R'$, shew that the locus of the point $(QR, Q'R')$ is a conic. [XIX. 51.

939. The ratio of the product of the diameters of two circles to the square of one of their common tangents inverts into an equal ratio when the circles are inverted from any point as pole. Hence deduce Feuerbach's property* of the nine-point circle. [XIX. 54.

940. In a quadrilateral whose diagonals intersect at right angles shew how to inscribe a conic having their intersection for a focus. [XIX. 69.

941. A point $C$ being taken on the diameter $AB$ of a semicircle, semicircles are described on $AC$ and $BC$ as diameters. Also a series of circles are described, the first touching the three semicircles, and every $n^{th}$ circle touching the $n-1^{th}$ and the semicircles on $AB$ and $AC$. Prove that, as $C$ varies, the loci of the centres of the several circles are ellipses having a common focus. [XIX. 88.

942. A line being drawn from the focus of a conic to meet the tangent at a constant angle, find where the locus of the point of concourse touches the conic. [XIX. 111.

943. The tangent at any point of a cardioid meets the curve again in two points $PQ$ the tangents at which divide the double tangent $AB$ harmonically; and the locus of the point of concourse of the tangents at $PQ$ is a conic passing through $AB$ and touching the cardioid at one real and two imaginary points. [XX. 34.

944. If a lamina moves in its own plane so that two given points of it describe each a fixed straight line, any other point of the lamina describes an ellipse. [XX. 89.

945. If $ABC$ be three points on a parabola, $A'B'C'$ the intersections of the tangents thereat, and $abc$ the centres of the circles $BCA'$, $CAB'$, $ABC'$; prove that the circle $abc$ passes through the focus. [XXI. 72.

---

* On this see Schröter in Neumann's *Mathematische Annalen* vol. VII. 517–530 (Leipzig 1874), where the property is cited from Feuerbach's *Eigenschaften einiger merkwürdigen Punkte des geradlinigen Dreiecks* (Nürnberg 1822). The nine-point circle itself has been improperly called Feuerbach's by Baltzer and others.

946. The tangent to the evolute of a parabola where it cuts the parabola is also a normal to the evolute. [XXI. 79.

947. Find at what points on a conic the angle between the tangent and the chord drawn to a fixed point on the curve is greatest or least. [XXII. 29.

948. If $OP$ and $OQ$ be tangents to a conic, $R$ the middle point of $PQ$, and $O'$ the point harmonically conjugate to $O$ with respect to the foci on the circle through $O$ and the foci, shew that $OP.OQ = OR.OO'$; and deduce that if $O$ and the foci be fixed the circles $OPQ$ are coaxal.* [XXIII. 17.

949. If $DD'$ be a fixed diameter of a conic and $AB$ any two conjugate points in an involution on the tangent at $D'$, then $DA$ and $DB$ meet the conic again upon a chord which passes through a fixed point. [XXIII. 55.

950. If $AB$ be the base of a segment of a parabola and $P$ any point on the curve, the locus of the orthocentre of $APB$ is a line parallel to $AB$. Hence shew how to describe a parabolic segment of given base and height by points. [XXIII. 61.

951. A plane figure moves so that two fixed straight lines in it always pass through two fixed points: find the envelope of any straight line in the figure. [XXIII. 67.

952. One focal chord of a conic meets the tangents at the extremities of another in $A$ and $B$. If straight lines $ACD$ and $BEF$ be drawn perpendicular to $AB$ and meeting the curve in $CDEF$, prove that $CE$ and $DF$ meet $AB$ at a point $P$ on the directrix; that $CF$ and $DE$, $AF$ and $DB$, $AE$ and $BC$ meet on the polar of $P$; that the intercepts $CD$ and $EF$ subtend equal angles at the focus $S$; that $SA : SC : SD = SB : SE : SF$; that $CF$ and $DE$ meet $AB$ in two points $G$ and $H$, having properties like those of $A$ and $B$; and that of the four intersections of the tangents from $A$ and $B$ two lie on the polar of $P$ and two on the directrix. [XXIV. 21.

---

* Compare Exx. 322, 337, 340, 380.

953. Given a conic inscribed in a parallelogram, if any tangent to the conic meets the sides opposite to an angle $A$ in $B$ and $C$, prove that the triangle $ABC$ is of constant area.
[XXIV. 51.

954. Six circles pass through twelve points on a conic in the order,

(a) $A_1 A_2 A_3 A_4$,    (b) $B_1 B_2 B_3 B_4$,    (c) $C_1 C_2 C_3 C_4$,
(d) $A_1 A_2 B_3 C_4$,    (e) $B_1 B_2 C_3 A_4$,    (f) $C_1 C_2 A_3 B_4$;

prove that the circles $abc$ meet the circles $def$ in six new points which lie on the circumference of another circle. [XXIV. 75.

955. Prove that there are eight chords of an ellipse normal to the curve at one extremity and to the central radius vector at the other. [XXIV. 83. XXV. 73.

956. If $ABCD$ be a quadrilateral inscribed in a conic, $F$ and $G$ the intersections of its opposite sides; prove that every conic through $ACFG$ has with the given conic a chord of intersection which passes through a fixed point, viz. the pole of $BD$. [XXIV. 93.

957. If $PP'$ be points on equal circles whose centres are $O$ and $O'$ respectively, and if the lines $OP'$, $O'P$ be parallel, find the envelope of the line bisecting both. [XXV. 53.

958. What is the condition that the conic $\alpha\beta = k\gamma^2$ may touch the conic $\alpha\beta = \gamma^2$ externally? [XXV. 88.

959. Given five points $ABCDE$ on a conic, shew that there is a sixth point on it the parallel through which to $AB$ passes through the fourth point of concourse of the circle $CDE$ with the conic. [XXVI. 17, 103.

960. If six lines taken in the orders $1231'2'3'$, $123'1'2'3$, $12'31'23'$, $12'3'1'23$ respectively form hexagons each inscribed in a conic, each pair of the conics have a common chord in the same straight line with a common chord of the opposite pair; and nine of the common chords are the sides and the joins of the opposite vertices of two triangles in perspective, one of which is inscribed in the other. [XXVI. 21.

961. If two points be such that the tangents to a parabola from the one are at right angles to the tangents from the other, the loci of the two points are in perspective. What does this become by projection and reciprocation? [XXVI. 94.

962. The joins of $n$ points on a conic intersect again in three times as many points as there are combinations of $n$ things taken four together, and of these intersections one third lie within and two thirds without the curve. [XXVI. 101.

963. If the three pairs of opposite summits of a quadrilateral be severally conjugate with respect to a conic, the joins of the poles of its diagonals cut the conic in a hexagon to which the sides of the quadrilateral are Pascal lines.
[XXVII. 105.

964. Shew that there are in general eight positions of a chord of an ellipse which meets the curve at given angles at its two extremities. [XXVIII. 63.

965. Three conics $S_1 S_2 S_3$ being such that the polar reciprocal of any one with respect to another is the third, a triangle $ABC$ is inscribed in $S_1$ and circumscribed to $S_2$. Prove that the triangle determined by the points of contact is self-polar to $S_1$ and circumscribes $S_3$; and that the tangents to $S_1$ at $ABC$ form a triangle self-polar to $S_2$ inscribed in $S_3$.
[XXVIII. 97.

966. If $S$ be the focus of a conic inscribed in a triangle $ABC$, and any tangent meet the focal chords perpendicular to $SA$, $SB$, $SC$ in $PQR$ respectively, prove that $AP$, $BQ$, $CR$ meet in a point. [XXVIII. 99.

967. A variable circle being drawn through two given points, through one of which pass two given lines; find the envelope of the chord joining the other points in which the circle meets them. [XXIX. 24.

968. If four conics $SABC$ have one focus and a tangent $D$ in common, and if a common tangent to each of the pairs $(SA)$, $(SB)$, $(SC)$ meet a directrix of $ABC$ respectively upon the tangent $D$; the common tangents of $BC$, $CA$, $AB$ meet at three points in a straight line. [XXIX. 43.

969. If three conics touch one another and have a common focus, the common tangent of any two cuts the directrix of the third in three points on one straight line. [XXIX. 69.

970. Prove the following pairs of reciprocal properties of a system of two conics:

*a.* When two conics are such that two of their four common points subtend harmonically the angle determined by the tangents at either of the remaining two, they subtend harmonically that determined by those at the other also.

*b.* When two conics are such that two of their four common tangents divide harmonically the segment determined by the points of contact of either of the remaining two, they divide harmonically that determined by those of the other also.

*c.* The associated conic, envelope of the system of lines divided harmonically by the two original conics, breaks up in the former case into the point-pair determined by the eight tangents to them at their four common points.

*d.* The associated conic, locus of the system of points subtended harmonically by the two original conics, breaks up in the latter case into the line-pair determined by their eight points of contact with their four common tangents. [XXIX. 88.

971. If the sides of a variable triangle pass severally through three fixed points in a straight line, whilst one vertex moves on a straight line and a second describes a given curve; prove that the locus of the third vertex is homographic with the given curve. [XXIX. 96.

972. The triangles whose vertices are two triads of points on a conic intersect in nine points, such that the join of any two not on the same side is a Pascal line of the six vertices.
[XXX. 25.

973. If a system of conics having a common focus envelop a given curve, and have their eccentricities proportional to the focal distances of the poles of their directrices with respect to a circle about the common focus as centre, the locus of the poles is a parallel of the reciprocal of the given curve with respect to the circle. [XXX. 93.

974. Prove and also reciprocate the following theorem. If a circle $A$ touches a circle $B$ internally at $P$, and if the tangent to $A$ at any point $Q$ meets $B$ in $R_1$ and $R_2$, then $\angle R_1PQ = \angle R_2PQ$.  [XXXI. 65.

975. Two lines being drawn through any point $P$ on an ellipse to meet the major axis in $A$, $B$ and the minor axis in $a$, $b$ respectively; shew that if $PA = PB = \frac{1}{2}$ minor axis, the intersection of $Ab$ and $Ba$ is the Frégier-point of $P$.  [XXXII. 48.

976. If a circle and a rectangular hyperbola intersect in four points, the line joining their centres is bisected by the centroid of the four points.  [XXXII. 48.

977. If $PQ$ is a chord normal at $P$ to an ellipse, and $N$ the intersection of the normal at $Q$ with the tangent at $P$, then $PN$ is to the projection of the semi-diameter $CP$ upon it as the square of $PQ$ to the square of the conjugate semi-diameter.
[XXXII. 58.

978. The focal radii to the points in which a fixed tangent to an ellipse meets a variable pair of conjugate diameters intersect on a fixed circle.  [XIII. 33. XXXII. 81.

979. If four parallel chords of an ellipse $\alpha\beta\gamma\delta$ be met by a straight line in $abcd$ respectively, shew that

$\alpha^2 . bc . cd . db + \gamma^2 . da . ab . bd = \beta^2 . cd . da . ac + \delta^2 . ab . bc . ca$.
[XXXIII. 27.

980. If $P$ be a current point on a given segment $AB$, the ellipses of given eccentricities described with $AP$ and $BP$ respectively as foci intersect upon a fixed ellipse whose foci are $A$ and $B$.  [XXXIII. 52.

981. If $Pp$ and $Qq$ be chords of a parabola parallel to the tangents at $q$ and $p$ respectively, and $Oo$ the poles of $PQ$ and $pq$; shew that $\triangle OPQ = 27 \triangle opq$, and that, if $pq$ be parallel to a fixed line, the locus of the intersection of $PQ$ with the tangents at $p$ and $q$ is a similarly situated parabola.  [XXXIII. 58.

982. The locus of the foot of the perpendicular from any point on a given diameter of a conic to its polar is a rectangular hyperbola.  [XXXIII. 76.

983. Any focal **chord** being drawn to a hyperbola, the circle on the portion of it intercepted by the tangents at the vertices as diameter touches the hyperbola. [XXXIII. 110.

984. The envelope of the **axes** of a conic which touches four fixed tangents to a **circle is a parabola**.*

985. **Shew how to solve the** problems of **the two** mean proportionals and of **the trisection** of an angle by the intersections of **a circle** and a parabola.†

986. The axes of a conic, any chord and **the** normals at its extremities touch one parabola. Deduce **a** construction for the centre of curvature at any point.‡

987. If the tangent and **normal at any** point of a conic meet the major and **minor axes in** $TG$ and $tg$ respectively, the radius of curvature **at the point subtends a right angle at** $(Gt, gT)$. To what does this reduce in the case of the parabola?

988. If from any point of a conic a line equal to the radius of curvature be drawn normally outwards, the circle upon it **as diameter is** orthogonal **to the** orthocycle. What does this become in the case of the parabola? and what in **the case** of the rectangular hyperbola? ‖

---

\* One triangle $ABC$ is self-polar to every conic inscribed in the given quadrilateral; and the axes of any one of them produced to infinity determine a second such triangle $O\infty\infty'$ with respect to it. The conic inscribed in $ABC$ and $O\infty\infty'$ is a fixed parabola, whose directrix is easily seen to be the diameter of the quadrilateral, since the orthocentre of $ABC$ is the centre of the circle.

† See Descartes *Geometria* lib. III p. 91 (ed. Schooten, 1659).

‡ Exx. 862, 986–992, &c. are to be found, with or without solutions, in Steiner's posthumous work *Vorlesungen über synthetische Geometrie*, Theil. II pp. 80, 206–212, 222–3, 242 (ed. 2, 1876). On his theorem Ex. 993 see *Nouvelles Annales* XIV. 103 (1855); Housel *Introd. à la Géométrie Supérieure* p. 231 (Paris 1865).

‖ In the rectangular hyperbola the diameter of curvature at any point is equal to the normal chord, as Mr. Wolstenholme thus proves. Take on the curve three points $ABC$ and their orthocentre $O$; then $OA$ produced to meet the circle through $ABC$ again is bisected by $BC$, and its halves, when $ABC$ coalesce, become the normal chord and the diameter of curvature at $A$.

989. If $a\alpha$, $b\beta$, $c\gamma$ be three couples in an involution, shew that
$$\{\alpha abc\}.\{\beta bca\}.\{\gamma cab\} = \{a\alpha\beta\gamma\}.\{b\beta\gamma\alpha\}.\{c\gamma\alpha\beta\} = -1,$$
where $\{\alpha abc\}$ denotes the cross ratio $\dfrac{ab}{a\overline{b}} : \dfrac{ac}{a\overline{c}}$. Also if $abcde$ be any five lines in a plane, shew that
$$\{e.abcd\}.\{c.abde\}.\{d.abec\} = 1.$$

990. If $OA$, $OB$, $OC$, $OD$ are concurrent normals to a conic, the parabola which touches the tangents at $ABCD$ touches the axes and the polar of $O$, has the diameter through $O$ for its directrix,* and is the polar reciprocal of the rectangular hyperbola through $ABCD$ (Art. 114 Cor. 1) with respect to the original conic. Determine the focus of the parabola.† [Ex. 379.

991. The tangents from any point of an ellipse to its auxiliary circles are equal to the real and imaginary semi-axes of the confocal hyperbola through the point.

992. If $abcd$ and $\alpha\beta\gamma\delta$ be two tetrads of points on a conic, the joins of $(ab, \alpha\beta)$ and $(cd, \gamma\delta)$; $(ac, \alpha\gamma)$ and $(bd, \beta\delta)$; $(ad, \alpha\delta)$ and $(bc, \beta\gamma)$ meet in a point. And if $a\alpha$, $b\beta$, $c\gamma$ be concurrent chords of a conic, then
$$\{\alpha ab\gamma\} + \{\beta bca\} + \{\gamma ca\beta\} = 1.$$

993. If $a\alpha$, $b\beta$, $c\gamma$ are the foci of three conics inscribed in the same quadrilateral, then
$$ac.a c : a\gamma.a\gamma = bc.\beta c : b\gamma.\beta\gamma.$$

994. If a conic $A$ circumscribes a conic $B$ harmonically, then $B$ is harmonically inscribed to $A$; the reciprocals of $A$ are

---

\* As a second proof that $O$ lies on the directrix, Mr. Pendlebury remarks that the normals are also normals at points $abcd$ to the reciprocal of the conic with respect to $O$, so that the parabola has for its reciprocal the rectangular hyperbola $Oabcd$, and therefore subtends a right angle at $O$ (Art. 155 Cor. 2). Thus also we see that the normals meet the original conic again in points $A'B'C'D'$ which lie on a rectangular hyperbola through $O$; since the reciprocals of $ABCD$ touch a parabola having for directrix the diameter through $O$ to the reciprocal of the original conic, and (by symmetry) the reciprocals of $A'B'C'D'$ touch a parabola having the same directrix.

† The parabola is the same for all confocal conics.

harmonically inscribed to the reciprocals of $B$; and the centre of perspective of any triangle inscribed in $A$ and its reciprocal with respect to $B$ lies on $A$.*

995. A conic is harmonically inscribed to every circle orthogonal to its orthocycle; and a rectangular hyperbola harmonically circumscribed to a circle passes through its centre.

996. **The asymptotes of** a conic are conjugate **lines with respect to any parabola** harmonically inscribed to it.

997. Given that a focus of one conic is a point $O$ on the orthocycle of another, if one of the conics be harmonically inscribed to the other, **it touches the polar** of $O$ with respect to the latter.

998. Describe **the conic with respect to which five given pairs of lines are conjugate; and the conics which pass through** 4, 3, 2, 1 or 0 given points and **are harmonically circumscribed** to 1, 2, 3, 4 or 5 given conics.

999. The orthocycles of the conics which touch two given lines $SA$ and $SB$ at given points $A$ and $B$, including the circle on $AB$ as diameter and the point-circle at $S$, are coaxal.†

1000. The **number** of conics touching five given conics **is** 3264.‡

---

\* On **Exx.** 994–8 see Picquet's *Etude géométrique* &c. **pp.** 58, 91, 108, 131–3; Prof. H. J. S. Smith *On some Geometrical Constructions* (Proceedings of the London Mathematical Society vol. II. 85–100). **One** conic is **said to be** harmonically inscribed or circumscribed to another when **it is** inscribed or circumscribed to a triangle self-polar with respect to the latter.

† Gaskin *Geometrical Construction &c.* p. 31. Hence (Ex. 577 note) the theorem, lately pointed out by Mr. R. W. Genese and Mr. Torry, that *the directrix of a conic is a common chord of the conic, its orthocycle and a point-circle at the focus*. Notice that every straight line through a focoid, as being an asymptote or self-conjugate diameter of a circle (pp. 142, 309), is *at right angles to itself*.

‡ See Salmon's *Conic Sections* (end); Halphen *Proc. London Math. Soc.* vol. IX. 149 and X. 87; and the original memoir by Chasles, *Détermination du nombre des sections coniques qui doivent toucher cinque courbes données d'ordre quelconque, ou satisfaire à diverses autres conditions* (Comptes Rendus LVIII. 225, 1864).

# INDEX.

Adams, vi, 15, 220.
Algebraic geometry, xlvi, **lvi, lxi, lxxiii**.
Almagest, l, lxii, lxxix.
Analogy, Kepler's principle of, lviii; Boscovich on, lxxiv.
Analysis, geometrical, xix, xxxi, xxxii.
Angle. See Trisection.
Angles, projection of, 319; reciprocation of, 341, 350; reversion of, lxxxvi, 323—8.
Anharmonic properties of conics, lxxxiii, 262—7 See Cross ratio.
Antipho, xxx, **xli**, **lx**.
Apollonius, xlii—l, **lxxxiv**, 72, 82, **154, 195**; on foci, 81, **111**; on **concurrent** normals, xlvii, 265.
Application of areas, xxv, **xliii**.
Archimedes, xxxv—xlii, lix, 59.
Archytas, xxxi.
Aristæus, xxxiii, xlvi.
Asymptotes, xliv, 143—154, 327; known to Menæchmus, xxxii; tangents at infinity, lxii, lxix, 142; of the equilateral hyperbola, 169; of the circle, 309, 379.
Auxiliary circles of conics having a common focus and directrix are coaxal, 353.
Axes, of the projection of a conic, lxiii. See Self-polar triangle.
Bellavitis, on inversion, 357.
Bernoulli, on the latus rectum of a section of the scalene cone, 211.
Besant, 138, 213.
Booth, 156, 166, 189, 346; on the right cone, 200; on minor directrices, 348.
Boscovich, vi, lxxi—lxxvii, 3, 90, 105, 311.
Bosse, lxi, lxiv.
Brianchon, lxxviii—lxxxi; his hexagram, 289—291, 295, 352; and Poncelet, 175, 191, **282**, lxviii.
Brougham, vi.
Cantor, xxxiv, xlii.
Cardioid, 356, 371.
Carnot, lxxviii, 256, 291, 298; his theorem projective, 313.
Cartesian, mechanical description of **the**, 178.
Casey, 293.

Caustics, 345.
Cayley, lxx, 175, 328.
Central conics, 75—112.
Centre, the pole of the line at infinity, lxiii; of the parabola, lx, 26, 44.
Centre-locus of a conic, given four tangents, lxviii, 282, 333; given four points, 283, 365; of an equilateral hyperbola, given three points, 171.
Centroid of a quadrilateral, 284.
Chasles, lii, lxxxii—lxxxv, 266, 320, 330, 339, 379; problems from his Sections Coniques, 300, 336.
Circle, a conic whose directrix is at infinity, 7, 22; focus of a conic regarded **as a** point-circle, 210; line-circle in**cludes** the line at infinity, lxxv, 344; or coincides with it, 308; every circle **passes** through the focoids, 309; reci**procal** of a circle with respect to a point, 342. **See** Quadrature.
Circular **points at** infinity, **308. See** Focoids.
Circumscribed polygons, **139, 140, 243**.
Clifford, 186, 257
Coaxal circles, transformed into **other** coaxal circles, lxxxvi; determine an involution on any transversal, 258; conics through four points project into, 318; confocals reciprocate into, 344, 351. See Orthocycle. Auxiliary circles.
Complement of a line, lxxv, 77.
Concentric circles, touch at the focoids, 309; conics having double contact project into, 319.
Concurrent chords of a conic subtend an involution at any point **on** the curve, 276. See Frégier.
Cone, sections of the, **192—206**; problems **on** the scalene, **211—3**, 334. See Sections.
Confocal conics, intersect at right angles, **84**, 351; locus of vertex of right angle which touches two, 89; problems on, 132—140, 163; conics touching four lines project into, 317; reciprocate into coaxal circles, 344, 351; transformed into other confocals, lxxxvi, 332.

# INDEX. 381

Conic, the general, 14—35; Newton's organic description of a, lxvi, lxxxiii, 136, 264, 325, 358; Maclaurin's description of a, 264; determined from five data, lxv, lxxxi, 136, 164, 175, 279—283, 288—290, 379; subject to four conditions, lxviii, 275—285; conic and point projected into a circle and its centre, 318; or into any other conic and point, 329.
Conjugate diameters, xlix, 95—102, 233, 327; of hyperbola, 149, 151, 169; construction of a conic from given pair of, 125, 152; determine a pencil in involution, 259, 265.
Conjugate hyperbola, lxxv, 101, 153.
Conjugate lines and points with respect to a conic, lxiii, 270, 278, 281; every two conjugate lines through a focus are at right angles, 270, 310, 312. See Hesse.
Conjugate triads with respect to a conic, 273; lie on circles orthogonal to the orthocycle, 274; or which pass through the centre, 171, 273. See Self-polar.
Conoids, xl, 213.
Continuity, history of, lviii, lx, lxxiii, lxxxi, 311.
Coordinates, used by Apollonius, xliii; in space, 1.
Correlative figures, 346.
Cotes, edits Newton's Principia, lxv; his theorem of harmonic means, lxxi, 276.
Cremona, lxxxiii, 265, 292—4, 321.
Cross ratio, 249—290; projectivity of, 251, 312, 328; history of, lii, lxiv, lxxxiii—lxxxv. See Anharmonic.
Cube, its duplication reduced to the problem of the two mean proportionals, xxviii, 189.
Cubics, Newton on, 301, 321.
Cunynghame, 177.
Curvature, 214—222, 279, 377; coordinates of the centre of, xlii, xlviii; Steiner's property of concurrent circles of, 228, 236, 357; circles of curvature invert into circles of, 357.
Curves, generated by compounded motions, xl; regarded as limits of polygons, xxx, lx; organic description of, lxx, lxxxvii, 178.
Dandelin, discovers the focal spheres, 204; his proofs of Pascal's and Brianchon's theorems, 287.
Davies, lxxxvi, 213, 257.
De Beaune, on envelopes, 345.
Degeneration, of conic into line or line-pair, 77, 144, 171, 278, 285. See Circle.
De la Hire, lxiv, lxxi, 112, 161, 311; the orthocycle discovered by, 90, 117; on transformation, 329.
Delambre, li.
Delian problem, xxviii, 189. See Cube.
Desargues, lx—lxiv, lxxx; on involution, 261, 277; polars, 291; transformation, 329. Descartes, lxi, 189, 266, 345, 377.
Determining ratio, lxxi, 1. See Directrix.

Diameter, of a conic, 23; of a quadrilateral, 138, 256. See Centre-locus.
Director circle, two uses of the term, 90, 165. See Orthocycle.
Directrix, history of the, liv, lxv, lxxi; the polar of the focus, lxxi, 15; a conic, its orthocycle and a point-circle at either focus intersect on the corresponding, 379; of parabola inscribed in a triangle, 57. See Steiner.
Double contact, conics having, 279. See Concentric circles.
Double reciprocation, 348.
Dual figures, 346.
Duality, discovery of the principle of, lxxviii, 290, 346.
Eccentric circle, 3, 9, 28, 321, lxxvi; works founded upon the, lxxii.
Eccentricity, use of the term, 211; of conics in the cone, 197.
Educational Times, problems from, 141, 336, 362—377.
Egyptian geometry, xvii, xxii, xxvi.
Eisenlohr, xxii, xxvi.
Eleven-point conic, 284, 365; degenerates into nine-point circle, 171, 285.
Ellipse, names of the, xliii, 195; area of the, xli, 234. See Central conics.
Elliptic compasses, lviii, 114, 178.
Envelopes, 345.
Equicross, the term, 250.
Equilateral hyperbola, 167—177, 342, 352; conjugate to the focoids, 278, 309.
Euclid, xix, xxxv; on conics, xlvi; Enc. I. 47 proved by dissection, xxiii. See Porisms.
Eudemus, xviii, xxiv, xxix.
Endoxus, xix, xxxii; his cubature of the cone, xxxviii.
Euler, 211, 242, 247.
Eutocius, xxxiii, xxxvi, xxxix, xlii, 45, 194.
Evolutes, xlviii, 221; homographic pairs of, lxxxvi, 358.
Exhaustions, xxxiv, xxxvii, xli, lix.
Fagnani's theorem, 140.
Faure, 186.
Feuerbach's property of the nine-point circle, 355, 371.
Figure on the axis, 82.
Fluxions, lx, lxxi.
Focal spheres, 196—205.
Foci, Apollonius on, xlv, 111; named by Kepler, liv, lvii; Desargues on, lxiii; regarded as point-circles, 210; Poncelet on, lxxxi, 311; of the projection of a conic, lxiii; Plücker's definition of, 311; foci of an involution, 259, 261, 309. See Conjugate lines. Confocal conics. Directrix.
Focoid, the term, 281, 308.
Focoids, Poncelet's discovery of the, lxxxi, 311; their relation to the foci of conics, 281, 299, 310; and other curves, 311; projection of any two points into, 315; all circles pass through and concentric circles touch at the, 309; constant relation of a figure moving in its own plane to the, 310.

## 382 INDEX.

Focus of parabola, liv, lxxi; not discovered by Apollonius, xlv, 81. See Kepler.
Frégier, theorem that a chord which subtends a right angle at a given point on a conic passes through a fixed point on the normal, 122, 176, 276, 324, 351; its analogue in space, 349.
Frégier-point, 349, 351, 376.
Frisch, lvi.
Gardiner, 305.
Gaskin, 165, 280, 293, **295, 335, 379**; theorem that the circle through any conjugate triad with respect to a conic is orthogonal to the orthocycle, 186, 274.
Geminus, xxiv, xliii, 194.
Genese, 163, 379.
Geometry, no royal road to, **xx**.
Graves' theorem, 133.
Gregory St. Vincent, lxxi, lxxix, 166, 189.
Halley, on the parabola, xlix, lxxxiv; editions by, xlii, lxxxiv.
Halphen, 379.
Hamilton, 206; his determination of the focus and directrix in the cone, **lxxii, 204**.
Hammond, 178.
Harmonic, the term, xxvi, xlv, liv; ranges and pencils, liv, lv, lxxix, 254, 313; property of a quadrilateral, 254—6; points on or tangents to a conic, 276.
Harmonically inscribed and circumscribed conics, 379.
Heilberg, xxxv.
Henrici, 252.
Hesse, theorem that if two pairs of summits of a quadrilateral are conjugate to a conic the third are conjugate, 333.
Hexagon, inscribed in a line-pair, liii, 297; Pascal's, 286—8; Brianchon's, 289—291.
Hexastigm, 297.
Hippocrates, xix, **xxvii—xxx**, 189.
Hirst, 358.
Homographic, the term, lxxxv, 250.
Homographic, figures may be placed in perspective, 312, 328; correspondence of points and lines in reciprocal figures, 269, 338. See Cross ratio.
Homology, lxiv, lxxxv, 292, 307, 321.
Homothetic conics, 305. See Parallel.
**Horne**, proof of the anharmonic properties of conics, 267.
Hultsch, li.
Huyghens, 221, **345**.
Hyperbola, why so called, 82, 195; a continuous curve, 10, 310; a quasi-ellipse, lxxv, 101, 153, 235; degenerate forms of, lvii, 285. See Central conics.
Ideal chords, 311.
Imaginary, transition from the real to the, lxxv; diameters of a hyperbola, 101, 153, 180; circular points at infinity, 308; foci, 310, 312.
Infinite chords of a conic, ratios of the, lxxvii, 149, 163.
Infinitesimals, method of, lx.

**Infinity**, the line at, 32, 308, 322; parallels **meet** at, lix, lxii; change of sign on passing through, lxxiv. See Opposite infinities.
Ingram, 357.
Inversion, 354—8, 364, 371.
Involution, lii, lxii, 257—281.
Joachimstal, 228.
Join of points or lines, 252.
Kempe, on linkages, lxxxvii.
Kepler, vi, lvi—lx; on the further focus and the centre of the parabola, lviii, lx; his doctrine of the infinite, lix; of the infinitesimal, lx; of continuity, lviii, lxxiii.
Lagrange, **vi**.
Lambert, on the parabola, lxxxv, 57, 296; theorem in elliptic motion, 237, 248; in parabolic motion, vi, 247.
Lamé, 278.
Laquière, 343.
Latus rectum, according to Apollonius, **82**; in the scalene cone, 211.
Leibnitz, lx, lxxi, 222.
Lemniscate, 357, 364.
Le Poivre, 330.
Leslie, lxxii, 125, **135**, 164.
Levett, 57.
Limiting forms of conics. See Degeneration.
Line at infinity, 32; parallel to every straight line in its plane, 308, 322; a factor of every line-circle, lxxv, 344; its relation to the conics, 144, 316, 341.
Linkages, lxxxvii.
Loci, the earliest writer on, xxvii.
Locus ad quatuor lineas, xlv; Newton's proof of the, lxvi, 266; proof by orthogonal projection, 235; theorems of Desargues and Pascal deduced from the, 277, 287; property of focus and directrix deduced from the, 362; reciprocal of the, lxxxiv, 293, 340, 346.
Logarithms, geometrical representation of, 166.
Lunes of Hippocrates, xxix.
Maccullagh, 246, 248.
Macdowell, 260, 292, 350—8.
Maclaurin, lxxi, lxxx, 128, 276; his construction of a conic, 264; theory of pedal curves, 345; on attractions, lxxxii.
Main, 219, 222.
Maxima and minima, Apollonius on, xlvii.
Mean proportionals, problem of the two, xxviii, xxxi, xxxix, xlviii, **45, 189**, 377.
Mechanical proofs of geometrical theorems, xxxvii, 283—4.
Menæchmus, xxix, xxxi, 45, 194.
Menelaus, theorem of the six segments, l.
Minor axis of hyperbola, 76, 347.
Minor directrices, 346—8, 352.
Möbius, lxxxiii, 257, 302, 330.
Monge, lxxiii, lxxviii, 256.

## INDEX.                                      383

Montucla, lvi.
Moore, 290.
Mulcahy, lxxxvi, 293.
Neil, his rectification of the semi-cubical parabola, 221.
Newton, lxiv—lxxi; his property of the tangents to conics, lxviii, lxxix, lxxxiv, 346; organic description of curves, lxix, lxxxiii, 136, 301; property of the diameter of a quadrilateral, lxviii, 282, 333; on the Locus ad quatuor lineas, lxvi, 266; rational transformation, lxvi, 330; the equilateral hyperbola, 172; the projection of cubics, 321.
Nine-point circle, 191; Feuerbach's property of the, 302, 357, 371; Casey on the, 293: of a right-angled triangle, 355. See Eleven-point conic.
Normals, concurrent, xlii, xlvii, xlix, 123, 224, 228, 265, 378.
Ombilic, 299.
Opposite infinities adjacent, lix, lxii, lxxv, 310.
Organic description, earliest use of the method of, xxxi; of curves, xxxiii, lxix, lxxxvii, 301; of surfaces, lxx; of the rectangular hyperbola, 177; of Cartesian ovals, 178. See Conic.
Orthocycle, the term, 280; characteristic property of the, 88—90, 351; Gaskin's theorem that the circle through any conjugate triad with respect to a conic is orthogonal to the, 186, 274; Plücker's theorem that the orthocycles of all conics touching the same four lines are coaxal, 280, 335, 342. See Directrix.
Orthogonal projection, 229—242; applied to curvature, 221, 235.
Orthoptic summits of a quadrilateral, 335.
Orthosphere, 280.
Osculating circle, the term, 222. See Curvature.
Pappus, li—liv.
Parabola, 44—61; why so called, xliii, 82, 195; touches the line at infinity, 144, 341; properties of triads of tangents to the, xlv, 55—7, 72, 272, 360—1; conjugate triads with respect to the, 274, 281, 294, 358; point-reciprocal of the, 343, 351—2. See Focus.
Parallel conics, 305.
Parallel projection, 236.
Parallels meet at infinity, lix, lxii.
Pascal, lxiv, lxxix, 286—8; applications of his hexagram, 58, 175, 290, 352.
Peaucellier, lxxxvii, 358.
Pedal curves, 370. See Maclaurin.
Pendlebury, 361, 378.
Perspective, 307, 320, 336; homographic plane figures may be placed in, 328; Serenus on, lv; Desargues on, lxi; Bosse on, lxiv.
Peyrard, xxxv.
Picquet, 280, 335, 360, 379.
Pierce Morton, 196, 205.

Plato, xix, xxx, xl.
Plücker, on tangential coordinates, 156; his definition of foci, lxxxi, 311. See Orthocycle.
Pole and polar, the terms, lxxviii.
Polar equations, 34.
Polars, 30, 90; with respect to the circle, liv; Apollonius on, xlv, liv; Desargues' theory of, lxii, 329; reciprocal, 268—271, 346: metric relation of any point and its polar to two fixed points and their, 339. See Reciprocation.
Polygon, inscribed or circumscribed to a conic so that its sides pass through given points, 295, 302, 349, 359; circumscribed to confocal conics, 139, 140. See Curves.
Polyhedra. See Solids.
Poncelet, lxxiii, lxxxi, lxxxiv, 277, 295, 344, 346; on homology, lxxxv; on the four foci of a conic, lxxxi, 311. See Brianchon.
Porisms, Euclid's three books of, li, liv.
Potts, lvi.
Poudra, xl, lv, lxi.
Proclus Diadochus, his list of early geometers, xviii.
Projection, orthogonal, 229—242; parallel, 236; central or conical, 307—320; of cubics, 321; of solids, lxxxv; Brianchon on, lxxix; Möbius on, 330. See Perspective.
Ptolemy, 357; theorem of the six segments ascribed to, li. See Almagest.
Pythagoras, xviii, xxii—xxvii.
Quadrature, of the circle, xxvi, xxix, xxx, xxxix; of the parabola, xxvi, xxxvii, 59; of the hyperbola, 166, 190, 221.
Quadrics, lxxxi, 280, 305, 333; ruled, lxxxiv, 288; of revolution, xl, 213; polar properties of, lxii, 291, 329; in homology, lxxxv; reversion of, 349.
Quadrilateral, properties of the complete, lii, 254—6; in relation to conics, lxii, lxvii, 274—285, 304, 333, 338; projected into a parallelogram, lxxix; or other quadrilateral, 316.
Quetelet, lxxxiii, 204, 295, 346, 357.
Range, the term, lxii, 249.
Reciprocation, 337—354. See Duality. Polars.
Rectangular hyperbola, xxxii, xlviii. See Equilateral.
Rectification, of the circle, xxxix, xl; of the semi-cubical parabola and the cycloid, 221; quadrature of the hyperbola reduced to the rectification of the parabola, 190, 221.
Renouf, xx.
Renshaw, lxxii, 212.
Reversion, lxxxvi, 321—8; properties of minor foci and directrices proved by, 346—8, 353; of quadrics, 349.
Rhind papyrus, xxii, xxvi.
Robertson, lxxi, 206.
Roberval, xl.

Rouché and de Comberousse, 294—5, 319, 321.
Routh, vi.
Salmon, lxxxvi, 133, 140, 164, 210, 277, 286—7, 290, 295, 302, 312, 319, 333, 343.
Sections, the conic, how discovered, xxxi; why named parabola, ellipse, hyperbola, xliii, 82, 195; of any cone by an arbitrary plane, lv, lxiii. See Cone.
Self-conjugate. See Self-polar.
Self-polar triangle, conics having four common points or tangents have a common, lxii, 274—6, 332; inscribed or circumscribed to a second conic, 272—4, 331, 379; the axes of a conic and the line at infinity determine a, 365, 377. See Orthocycle.
Serenus, liv.
Serret, proof of Gaskin's theorem, 274.
Simplicius, xxix, xxx.
Simson, lii, 256.
Smith, H. J. S., lxxxvi, 379.
Solid loci and problems, xxviii, xxxiii, 189.
Solids, the five regular, xx, xxiv, xxxiii; semi-regular, xxxvi.
Steiner, lxxxii, 377; theorem that the directrix of a parabola inscribed in a triangle passes through the orthocentre, 57, 281, 290, 326, 352; on cross ratio, lxxxv, 257, 262; on triads of concurrent osculating circles, 228, 236, 357.
Stubbs, 357.
Sturm, 277.
Subcontrary sections, 210—2.

Supplemental chords, xliv, xlix, 95.
Sylvester on linkages, lxxxviii.
Talbot, 194, 264, 321.
Tangential coordinates, lxviii, 156, 346.
Taylor, J. P., proof of Feuerbach's theorem, 191, 355.
Thales, xviii, xx, xxxv.
Theætetus, xix, xxxi.
Torry, 224, 379.
Townsend, lxxxviii, 216, 249, 280, 287, 295—7, 330, 344, 351, 358—9.
Transformation, homographic, lxvi, 329. See Homographic.
Triangle, through the axis, lv, lxiii, 206; inscribed or circumscribed to a conic, 271—4; orthogonal projection of any triangle into an equilateral, 237. See Self-polar.
Trisection of an angle, xxvii, lxxiv, 141, 189, 299, 377.
Ubaldi, 178.
Umbilicus, 5. See Ombilic.
Viviani, xxxiii.
Walker, G., lxxii; J., 212.
Wallis, 221—2, 296.
Walton, geometrical problems, 161, 189, 195.
Webb, 345.
Whitworth, 308.
Wolstenholme, problems by, 121, 163, 184; proof that the diameter of curvature at any point of a rectangular hyperbola is equal to the normal chord, 377.
Wren, property of the parabola, lxxxv, 296; rectification of the cycloid, 221.

THE END.

www.ingramcontent.com/pod-product-compliance
Lightning Source LLC
Chambersburg PA
CBHW030322020526
44117CB00030B/551